Quantum Mechanics: Concepts, Symmetries, and Recent Developments

Quantum Mechanics: Concepts, Symmetries, and Recent Developments

Guest Editor
Tuong Trong Truong

Basel • Beijing • Wuhan • Barcelona • Belgrade • Novi Sad • Cluj • Manchester

Guest Editor
Tuong Trong Truong
Physics Department
CY Cergy Paris Université
Cergy-Pontoise
France

Editorial Office
MDPI AG
Grosspeteranlage 5
4052 Basel, Switzerland

This is a reprint of the Special Issue, published open access by the journal *Symmetry* (ISSN 2073-8994), freely accessible at: www.mdpi.com/journal/symmetry/special_issues/Quantum_Mechanics_Concepts_Symmetries_Recent_Developments.

For citation purposes, cite each article independently as indicated on the article page online and using the guide below:

Lastname, A.A.; Lastname, B.B. Article Title. *Journal Name* **Year**, *Volume Number*, Page Range.

ISBN 978-3-7258-3240-8 (Hbk)
ISBN 978-3-7258-3239-2 (PDF)
https://doi.org/10.3390/books978-3-7258-3239-2

© 2025 by the authors. Articles in this book are Open Access and distributed under the Creative Commons Attribution (CC BY) license. The book as a whole is distributed by MDPI under the terms and conditions of the Creative Commons Attribution-NonCommercial-NoDerivs (CC BY-NC-ND) license (https://creativecommons.org/licenses/by-nc-nd/4.0/).

Contents

About the Editor vii

Preface ix

Miloslav Znojil
Features, Paradoxes and Amendments of Perturbative Non-Hermitian Quantum Mechanics
Reprinted from: *Symmetry* **2024**, *16*, 629, https://doi.org/10.3390/sym16050629 1

Kenichi Konishi and Hans-Thomas Elze
The Quantum Ratio
Reprinted from: *Symmetry* **2024**, *16*, 427, https://doi.org/10.3390/sym16040427 23

Orion Ciftja
A Charged Particle with Anisotropic Mass in a Perpendicular Magnetic Field–Landau Gauge
Reprinted from: *Symmetry* **2024**, *16*, 414, https://doi.org/10.3390/sym16040414 47

David H. Oaknin
The GHZ Theorem Revisited within the Framework of Gauge Theory
Reprinted from: *Symmetry* **2023**, *15*, 1327, https://doi.org/10.3390/sym15071327 61

Lulin Xiong, Xin Tan, Shikun Zhong, Wei Cheng and Guang Luo
A New Solvable Generalized Trigonometric Tangent Potential Based on SUSYQM
Reprinted from: *Symmetry* **2022**, *14*, 1593, https://doi.org/10.3390/sym14081593 74

Michael B. Heaney
A Time-Symmetric Resolution of the Einstein's Boxes Paradox
Reprinted from: *Symmetry* **2022**, *14*, 1217, https://doi.org/10.3390/sym14061217 92

Collins Okon Edet, Emmanuel Benjamin Ettah, Syed Alwee Aljunid, Rosdisham Endut, Norshamsuri Ali and Akpan Ndem Ikot et al.
Global Quantum Information-Theoretic Measures in the Presence of Magnetic and Aharanov-Bohm (AB) Fields
Reprinted from: *Symmetry* **2022**, *14*, 976, https://doi.org/10.3390/sym14050976 100

Tuong Trong Truong
Dynamical Symmetries of the 2D Newtonian Free Fall Problem Revisited
Reprinted from: *Symmetry* **2021**, *14*, 27, https://doi.org/10.3390/sym14010027 114

Fritz Wilhelm Bopp
The Surjective Mapping Conjecture and the Measurement Problem inQuantum Mechanics
Reprinted from: *Symmetry* **2021**, *13*, 2155, https://doi.org/10.3390/sym13112155 135

Maedeh Mollai and Seyed Majid Saberi Fathi
An Application of the Madelung Formalism for Dissipating and Decaying Systems
Reprinted from: *Symmetry* **2021**, *13*, 812, https://doi.org/10.3390/sym13050812 141

Hung T. Diep
Quantum Spin-Wave Theory for Non-Collinear Spin Structures, a Review
Reprinted from: *Symmetry* **2022**, *14*, 1716, https://doi.org/10.3390/sym14081716 153

About the Editor

Tuong Trong Truong

Dr T. T. Truong is an emeritus professor of Physics at CY Cergy Paris Université in the Department of Physics and is a member of UMR 8089 - Laboratoire de Physique Théorique et Modélisation, a research unit of the French National Scientific Research Organization—CNRS. He graduated from Columbia University in 1973 with a Ph.D degree in theoretical physics under John R. Klauder, who was the head of the Theoretical Physics Group of Bell Telephone Laboratories at Murray Hill, New Jersey, USA. His research started with the quantum canonical formulation of spin one field theory in his thesis. He has published works in integrable field theories in low dimensions and two-dimensional statistical physics during his time at the Free University of Berlin, Germany. After a short tenure at the University of Tours in central France, he moved to the University of Cergy-Pontoise near Paris, France. During this period, his interests have covered several topics, such as anyons, the physics of quantum dots, and lately, imaging inverse problems by Compton scattered radiation.

Preface

Quantum mechanics has brought a gigantic breakthrough to physics since its introduction more than a century ago. It has led to many discoveries at the microscopic level, initiated new pathways to unsuspected properties of matter, and produced new and amazing usable applications. This Special Issue of Symmetry is devoted to recent advances in theoretical developments in quantum mechanics. It is part of the global effort to provide a continuous supply of information to the research community. This volume gathers contributions in various directions and their contents are best described chronologically by their abstracts, which are hereafter listed for the convenience of the reader.

We thank the authors for their significant contributions to this Special Issue "Quantum mechanics: Concepts, Symmetries and Recent Developments", the reviewers and editors for their invaluable work and help insuring the quality of the papers, and finally, the professional and responsive staff of MDPI for making this Special Issue a success.

Tuong Trong Truong
Guest Editor

Article

Features, Paradoxes and Amendments of Perturbative Non-Hermitian Quantum Mechanics

Miloslav Znojil [1,2,3]

1 The Czech Academy of Sciences, Nuclear Physics Institute, Hlavní 130, 250 68 Řež, Czech Republic; znojil@ujf.cas.cz
2 Department of Physics, Faculty of Science, University of Hradec Králové, Rokitanského 62, 500 03 Hradec Králové, Czech Republic
3 Institute of System Science, Durban University of Technology, Durban 4001, South Africa

Abstract: Quantum mechanics of unitary systems is considered in quasi-Hermitian representation and in the dynamical regime in which one has to take into account the ubiquitous presence of perturbations, random or specific. In this paper, it is shown that multiple technical obstacles encountered in such a context can be circumvented via just a mild amendment of the so-called Rayleigh–Schrödinger perturbation–expansion approach. In particular, the quasi-Hermitian formalism characterized by an enhancement of flexibility is shown to remain mathematically tractable while, on the phenomenological side, opening several new model-building horizons. It is emphasized that they include, i.a., the study of generic random perturbations and/or of multiple specific non-Hermitian toy models. In parallel, several paradoxes and open questions are shown to survive.

Keywords: unitary quantum mechanics; non-Hermitian Schrödinger picture; generalized perturbation theory; ambiguity of physical Hilbert space

Citation: Znojil, M. Features, Paradoxes and Amendments of Perturbative Non-Hermitian Quantum Mechanics. *Symmetry* **2024**, *16*, 629. https://doi.org/10.3390/sym16050629

Academic Editor: Tuong Trong Truong

Received: 14 April 2024
Revised: 3 May 2024
Accepted: 6 May 2024
Published: 19 May 2024

Copyright: © 2024 by the author. Licensee MDPI, Basel, Switzerland. This article is an open access article distributed under the terms and conditions of the Creative Commons Attribution (CC BY) license (https://creativecommons.org/licenses/by/4.0/).

1. Introduction

The exact or approximate solutions of the time-independent Schrödinger equation

$$H|\psi_n\rangle = E_n|\psi_n\rangle, \quad |\psi_n\rangle \in \mathcal{H}, \quad n = 0, 1, \ldots \quad (1)$$

play a key role in our understanding of the structure of quantum bound states or resonances. Often, it is believed that up to some truly exotic exceptions the division line which separates the case of bound states from the case of resonances also separates Equation (1) in which H is Hermitian from Equation (1) and in which H is non-Hermitian. Incidentally, the latter belief has been shattered after 1998 when Bender with Boettcher [1] revealed that the class of the "anomalous" non-Hermitian Equation (1) yielding bound states can be larger than expected, also incorporating certain models in which the Hamiltonians have the form of superposition of the most common kinetic energy $\sim p^2$ with an equally standard (but complex) local interaction potential.

In the latter models, widely known as "$\mathcal{PT}$–symmetric" [2–6], the manifest non-Hermiticity of the Hamiltonian

$$H \neq H^\dagger$$

has been found to coexist with the reality of the spectrum. Thus, it was immediate to conclude that the unitarity of the evolution can be guaranteed not only in the conventional textbook spirit (i.e., via the self-adjointness of the Hamiltonian [7]), but also under certain additional technical conditions [8], via the Dieudonné's [9] quasi-Hermiticity requirement

$$H^\dagger \Theta = \Theta H. \quad (2)$$

One can then speak about quantum mechanics of unitary systems which are slightly modified and reformulated in the so-called quasi-Hermitian representation.

In this framework, one of the most important features of the modification may be seen in its innovative approach to the concept of perturbation, which is found to be, in this setting, counterintuitive. This is for three reasons. The first one is that in this formalism (cf., e.g., its reviews in [10] or [11]), we are allowed to change the physical Hilbert-space norm. Thus, in a preselected "perturbed" Hamiltonian $H(\lambda) = H_0 + \lambda H_1$ the size (and, hence, influence) of the perturbation cannot always be kept under reliable control. Often, an enhanced sensitivity to perturbations is observed. For this reason, in open quantum systems, a few more remarks on this subtlety will be added below.

The second reason and paradox emerges when we consider just a closed quantum system in which the influence of $H_1 \neq H_1^\dagger$ is guaranteed to be small. Still, the correct probabilistic interpretation of the system remains ambiguous, mainly again due to the non-uniqueness of the physical Hilbert-space inner-product metric Θ (again, a more detailed support of this observation will be given below).

Thirdly, even if we decide to ignore the latter ambiguity and even if we pick up just any one of the eligible metrics (which would reduce the scope of the theory in a not quite predictable manner of course), such a choice of the geometry of the physical Hilbert space would still vary with λ. This is, probably, the most challenging problem that is also to be addressed in our present paper.

Preliminarily we may notice and emphasize that in the language of mathematics, the problem may be formulated easily because the underlying auxiliary, unitarity-of-evolution-guaranteeing operator Θ (if it exists [8]) can be perceived as representing just an invertible and positive definite *ad hoc* physical-Hilbert-space inner-product metric, $\Theta = \Theta^\dagger > 0$. In the related reformulation of quantum theory, all of the measurable predictions only require, therefore, the evaluation of the following metric-dependent matrix elements:

$$a_n = \langle \psi_n | \Theta A | \psi_n \rangle. \tag{3}$$

The knowledge of the wave function and of the operator A representing an observable of interest must be complemented here by the guarantee of observability $A^\dagger \Theta = \Theta A$ of course [8,10].

One of the most influential sources of interest in certain special classes of non-Hermitian Hamiltonians with real spectra lied in quantum field theory [12] and, in particular, in the role played there by perturbation theory [13–15]. One of the most important subsets of the underlying phenomenological Hamiltonians H is formed, therefore, by the one-parametric families

$$H = H(\lambda) = H(0) + \lambda V \neq H^\dagger \tag{4}$$

where λ is a complex number and where the component V representing the perturbation should not be, in some sense, too large [15].

Under this assumption, a powerful tool of the construction of the solutions of Equation (1) lies in the use of power-series ansatzs

$$E_n = E_n(\lambda) = E_n(0) + \lambda E_n^{(1)} + \lambda^2 E_n^{(2)} + \ldots \tag{5}$$

and

$$|\psi_n\rangle = |\psi_n(\lambda)\rangle = |\psi_n(0)\rangle + \lambda |\psi_n^{(1)}\rangle + \lambda^2 |\psi_n^{(2)}\rangle + \ldots. \tag{6}$$

A serious obstacle emerges when we turn our attention to the unconventional quasi-Hermitian models. In light of Equation (2), the metric will become manifestly λ- dependent in general, $\Theta = \Theta(\lambda)$. In contrast to the conventional perturbation-expansion constructions, it becomes necessary to complement the standard pair (5) and (6) of the Rayleigh–Schrödinger perturbation–expanison ansatzs by their new, operator–expansion partner, say, of the power-series form

$$\Theta(\lambda) = \Theta(0) + \lambda \Theta^{(1)} + \lambda^2 \Theta^{(2)} + \ldots. \tag{7}$$

This means that the non-Hermiticity (4) of Hamiltonian makes a consequent implementation of the Rayleigh–Schrödinger perturbation-expansion approach to the quasi-Hermitian bound state quantum physics complicated.

The consequent theory requires an explicit or implicit reference to as many as five separate but mutually interrelated Hilbert spaces in general (cf. [16]). The main theoretical benefit of such a five-Hilbert-space reformulation of quantum mechanics lies in an exhaustive classification of admissible perturbations. In this sense, our present paper can be read as a more pragmatically oriented follow up of [16].

As an introduction, a few basic features of the theory may be found summarized in Appendices A–D. With this background in mind, Section 2 will add two illustrative examples explaining not only an overall motivation of using non-Hermitian Hamiltonians but also the existence of the deep mathematical differences between the use of perturbation expansions in the closed and open quantum systems.

In Section 3, we will turn our attention to the physical consequences of these differences. We will point out that in the related literature the necessity of an unambiguous separation of the closed-system quantum physics from the open-system quantum physics is not always sufficiently carefully observed. This note will be complemented by an outline of the role of random perturbations in realistic models. A critique of a few recent results will be given in which the depth of the difference between the closed and open systems has been underestimated. This will be followed by a clarification of one of the related paradoxes connected with the usage of the concept of the so-called pseudospectra [17]. For the description of the influence of the random perturbations, the usefulness of the concept of pseudospectra remains strictly restricted to the studies of the open quantum systems. In the quasi-Hermitian models, the transition from spectra to pseudospectra cannot be recommended because it does not lead to any enrichment of the information about the dynamics of the underlying closed quantum systems.

In Section 4, we will finally return to the quasi-Hermitian perturbation theory. We will recall the mathematical challenge represented by the necessity of the construction of an additional operator expansion (7). In the climax of our paper, we will offer a new, alternative, upgraded formulation of the quasi-Hermitian version of the Rayleigh–Schrödinger perturbation series in which the latter necessity will be circumvented.

An extensive discussion and summary of our results will be presented in the last two sections, Sections 5 and 6. The essence of the innovation (and, first of all, of a significant simplification of the formalism) will be shown to lie in an implementation of the biorthogonal-basis ideas [18] as taken from their application in a different, non-stationary quantum dynamics context [19].

2. Merits of Non-Hermitian Hamiltonians

From a purely pragmatic point of view, Schrödinger Equation (1) can be perceived as a linear eigenvalue problem in which, in the majority of applications, the possible non-Hermiticity of the Hamiltonian would make the construction of solutions less stable and technically more difficult. This is a generic statement which is well known [11,17]. People often seem surprised when they encounter a quantum system for which the technically friendliest representation of Hamiltonian happens to be non-Hermitian.

2.1. Dyson-Inspired Simplifications of Schrödinger Equations

A compact account of history of the recent quick enhancement of interest in the closed and stable quantum systems controlled by an "anomalous" Hamiltonian $H \neq H^\dagger$ can be found in [11]. The emergence of such a class of models can look, at first sight, surprising. Nevertheless, one of the oldest demonstrations of the technical advantages of using a non-Hermitian H emerged many years ago, viz., during Dyson's entirely pragmatic, well-motivated, and purely numerical study of a specific real-world many-body problem [20].

An impact of the latter quantum-many-body result remained, for a couple of years, restricted just to nuclear physics [21]. The idea only acquired a new life and broader response after Bender with Milton [12] revealed that the study of non-Hermitian models may be also of immediate relevance in quantum field theory.

In such a broadened methodical context, a particularly elementary and fully non-numerical sample $H^{(JM)} = p^2 + V^{(JM)}(x)$ of such a Hamiltonian has been proposed, in 2006, by Jones with Mateo [22]. Via an exactly solvable toy model, these authors demonstrated that, in some cases, given a conventional self-adjoint Hamiltonian $\mathfrak{h}^{(JM)} = p^2 + \mathfrak{v}^{(JM)}(x)$, one can profit from its replacement by an isospectral alternative $H^{(JM)} = p^2 + V^{(JM)}(x)$, which is non-Hermitian. Indeed, the former operator where the potential contained two components

$$\mathfrak{v}^{(JM)}(x) = -2x + 4x^4 \tag{8}$$

could be perceived as more complicated than their avatar $H^{(JM)}$ with

$$V^{(JM)}(x) = -x^4 \tag{9}$$

containing just the single interaction term. Moreover, the single-term potential (9) is symmetric with respect to the product of parity and time-reversal, i.e., in comparison, it is less complicated than its left–right-asymmetric two-term partner (8). One can really speak about a simplification $\mathfrak{h}^{(JM)} \to H^{(JM)}$, in principle at least.

Both of the latter Hamiltonians predict the same real (i.e., measurable and stable) bound-state spectrum which is discrete and bounded from below. The conventional one, viz., operator $\mathfrak{h}^{(JM)}$ is self-adjoint while its non-Hermitian avatar $H^{(JM)}$ is merely quasi-Hermitian (cf. definition (2)). From an experimentalist's point of view the isospectrality of the two alternative Hamiltonians makes the two representations of the same closed quantum system indistinguishable. For mathematicians, the differences are also not too deep because the main source of difference, viz., the inner-product metric needed in Equation (2), has been found, in [22], in an exact, closed and really elementary operator form

$$\Theta^{(JM)} = \exp\left[p^3/48 - 2p\right]. \tag{10}$$

This makes the non-Hermiticity of $H^{(JM)}$ just a minor, easily surmounted complication.

2.2. Analytic Continuations and Non-Unitary Open Systems

From the point of view of experimental physics, the truly exceptional exact solvability of the Jones' and Mateo's interaction (9) is not so impressive because the coordinate x is complex (see its definition in [22]). This makes the standard probabilistic interpretation of the "simplified" system unclear because the value of x (tentatively treated as the position of a particle) ceases to be a measurable quantity.

A new physics has to be then sought in a return to differential Schrödinger equations in which the coordinate x remains real. In the light of the paradox (or rather of the danger) of the non-observability of coordinates, one is forced to consider the asymptotically less anomalous potentials in which the line of coordinate x could still be kept real. One of such illustrative examples can be found in our older paper [23] where we studied the perturbation expansions (5) of the energies generated by the two-parametric imaginary cubic oscillator Hamiltonian

$$H^{(IC)}(f,g) = -\frac{d^2}{dx^2} + \frac{f^2}{4}x^2 + \mathrm{i}gx^3. \tag{11}$$

Indeed, such a differential–operator model is still non-Hermitian and $\mathcal{PT}$– symmetric, i.e., it is formally closely analogous to Equation (9). Moreover, knowing that after identification $\lambda = g$, i.e., in the weak-coupling regime, the conventional small-anharmonicity expansions would diverge, we were able to transfer the role of a small parameter to the other coupling and choose $\lambda = f$. As a consequence, we achieved a very good convergence of our resulting perturbative strong-coupling series (5) for the energies.

Later, we found a complementary inspiration in [24] in which Bender and Dunne set $f = 1$ and $\lambda = g$ and studied the alternative, divergent but resummable weak-coupling expansion. They were interested in just the ground state energy, and they managed to construct the Rayleigh–Schrödinger perturbation series

$$E^{(BD)}(\lambda) \sim \frac{1}{2} + \sum_{n=1}^{\infty} b_n \lambda^{2n} \qquad (12)$$

up to very large orders by having evaluated the necessary integer-valued coefficients non-numerically,

$$b_1 = 11,\ b_2 = -930,\ b_3 = 158836,\ b_4 = -38501610,\ \ldots. \qquad (13)$$

At $n \gg 1$, they managed to fit these coefficients using an amazing asymptotic formula

$$b_n \sim (-1)^{n+1} \frac{60^{n+1/2}}{(2\pi)^{3/2}} \Gamma\left(n + \frac{1}{2}\right) \left[1 + \mathcal{O}\left(\frac{1}{n}\right)\right]. \qquad (14)$$

Via an appropriate resummation of the divergent series (12), this enabled them to obtain, at any not too large real coupling λ, a very good (they even wrote "excellent") agreement with the known and real numerical value of the ground-state energy $E^{(BD)}(\lambda)$.

As a climax of the story, Bender and Dunne also proposed a phenomenologically meaningful physical output of their considerations. For this purpose, they re-interpreted their asymptotic estimate (14) as a support of the possibility and consistency of an analytic continuation of the function $E^{(BD)}(\lambda)$ to the (cut) complex plane of λ. On these grounds, they were able to evaluate the imaginary part of $E^{(BD)}(i\epsilon)$ and to interpret the result as a prediction of a measurable decay width of another quantum system described by an analytically continued Hamiltonian

$$H = p^2 + x^2/4 - \epsilon x^3 \qquad (15)$$

(cf. Eq. Nr. 5 in [24]). In other words, the initial non-Hermitian operator (11) has been reinterpreted, via analytic continuation, as a more or less standard physical quantum Hamiltonian supporting an unstable (but still observable) ground state.

2.3. Dyson Maps and the Modified Concept of Locality

In the overall framework of quasi-Hermitian quantum mechanics (QHQM) of closed systems as formulated, in the Schrödinger picture, by Scholtz et al. [8], we paid attention, in our recent paper [16], to the consistent applicability of the theory in the presence of perturbations. We pointed out that even in the non-perturbative version of the theory it made sense to realize the description using three separate Hilbert spaces (cf. diagram Nr. (10) in [16] or Equation (A1) in Appendix A).

One of these spaces is denoted here by symbol $\mathcal{L}$. By assumption, it is just a hypothetical and, for practical purposes, inaccessible space. Only the other two are relevant, both hosting operator H and differing just by the respective forms of inner products. The first space (viz., $\mathcal{K}$) is just auxiliary and unphysical. The second one (denoted here as $\mathcal{H}$) is physical and, for this reason, a unitary equivalent to $\mathcal{L}$, with the equivalence mediated by a mapping Ω.

The latter (often called Dyson [25]) mapping is related to the metric by formula

$$\Theta = \Omega^\dagger \Omega \neq I. \qquad (16)$$

In a way dating back to the Dyson's paper [20], the key message as delivered by our paper [16] is that after one makes the Hamiltonian $\lambda-$ dependent and after one implements the perturbation-expansion philosophy, one has to distinguish between the "physics" (represented by the perturbed $H(\lambda)$ at any $\lambda \neq 0$) and "mathematics" (represented by the exactly solvable $H(0)$). In other words, both of the Hamiltonian-supporting Hilbert spaces $\mathcal{K}$ and $\mathcal{H}$ become $\lambda-$ dependent.

Incidentally, at both $\lambda = 0$ and $\lambda \neq 0$, the knowledge of factorization (16) would enable us to return also to the above-mentioned toy-model interaction (9) in which the "false coordinate" appeared to be complex, $x \notin \mathbb{R}$. Due to the exact solvability of the model and due to the extreme simplicity of the related metric (10), one could also introduce a closed-form Dyson-map operator

$$\Omega^{(JM)} = \exp\left(p^3/96 - p\right) \qquad (17)$$

and *define* a correct (i.e., by construction, quasi-Hermitian) coordinate-representing operator $Q^{(JM)}$ acting in $\mathcal{K}$ and $\mathcal{H}$ using formula

$$\mathfrak{q}^{(JM)} = \Omega^{(JM)} Q^{(JM)} \left(\Omega^{(JM)}\right)^{-1} = \mathfrak{q}^\dagger. \qquad (18)$$

This is the definition of a suitable (albeit a bit artificial) observable tractable as a coordinate. From the point of view of consistency of the theory, the choice of the self-adjoint $\mathfrak{q}^{(JM)}$ (or, directly, of the quasi-Hermitian $Q^{(JM)}$) is more or less arbitrary.

The relation (18) can itself be re-read as the closest analogue of connection between the more common energy-operators *alias* Hamiltonians (cf. relation (A2) in Appendix B). Such a constraint can be complemented by some additional phenomenological requirements. For example, it is possible to start from the "inaccessible" textbook Hilbert space $\mathcal{L}$ and choose the left-hand side "input information" $\mathfrak{q}^{(JM)}$ as a diagonal operator with the equidistant spectrum simulating the position on a one-dimensional discrete lattice or on its suitable continuous-spectrum limit [26,27].

3. Norm-Ambiguity Paradox and Its Consequences

A concise outline of the non-Hermitian but unitary theory of closed systems is relocated to Appendices A–D. Using the notation of diagram (A1) in Appendix A, let us now emphasize that in most applications the information about dynamics is carried just by the Hamiltonian H acting in an auxiliary Hilbert space $\mathcal{K}$ in which $H \neq H^\dagger$. As a consequence, the choice of metric Θ, compatible with the quasi-Hermiticity condition (2), remains non-unique [8]. The relevant (i.e., physical, $\Theta-$ dependent) size of the perturbations V in (4) is, therefore, indeterminate.

This is a paradox, the relevance of which becomes particularly serious in the realistic models of quantum systems in which one cannot ignore the possible occurrence of random, uncontrolled, statistically distributed perturbations.

3.1. Random Perturbations and Pseudospectra

In the most common textbook version of quantum mechanics of the perturbed unitary systems living in $\mathcal{L}$, the evolution is generated by the perturbed Hamiltonians, which are self-adjoint.

$$\mathfrak{h}(\lambda) = \mathfrak{h}(0) + \lambda \mathfrak{h}_1 = \mathfrak{h}^\dagger(\lambda)$$

(see [28] or Equation (A2) in Appendix B). The stability of the system may then be tested using all perturbations, the norm of which is bounded, $\|\mathfrak{h}_1\| \leq \epsilon$. For this purpose, the spectra of the perturbed Hamiltonians could be calculated using the $\lambda-$ dependent Schrödinger equation in $\mathcal{L}$,

$$\mathfrak{h}(\lambda) |\psi_n(\lambda)\succ = E_n(\lambda) |\psi_n(\lambda)\succ, \quad n = 0, 1, \ldots \qquad (19)$$

plus, say, the Rayleigh–Schrödinger perturbation-series ansatz (5).

As a result, one would obtain, in principle at least, a union of all of the possible perturbed spectra, i.e., the set

$$\bigcup_{\lambda \|\mathfrak{h}_1\|<\epsilon} \sigma(\mathfrak{h}(0) + \lambda\,\mathfrak{h}_1) \qquad (20)$$

which should lie, for stable systems, just inside a small vicinity of $\sigma(\mathfrak{h}(0))$, i.e., of the unperturbed spectrum. In such a setting, it is recommended to recall the Roch's and Silberman's observation [29] that the set (20) coincides with the so-called pseudospectrum $\sigma_\epsilon(\mathfrak{h}(0))$ of $\mathfrak{h}(0)$, i.e., with the set that is defined as the following union of the spectrum and of the domain in which the resolvent of $\mathfrak{h}(0)$ remains large [17],

$$\sigma_\epsilon(\mathfrak{h}(0)) = \sigma(\mathfrak{h}(0)) \cup \left\{ z \in \mathbb{C} \mid \|(\mathfrak{h}(0) - z)^{-1}\| > \epsilon^{-1} \right\}. \qquad (21)$$

One can cite [30] and conclude that "if $\mathfrak{h}$ is self-adjoint ...", the pseudospectra "give no additional information".

3.2. Norms in Non-Hermitian Models

Let us repeat that as long as the Hamiltonians in question are kept self-adjoint, the Roch's and Silberman's observation simplifies the analysis of the influence of random perturbations because it just shows that the smallness of perturbations immediately implies that at the sufficiently small ϵ the difference between the sets $\sigma_\epsilon(\mathfrak{h}(0))$ (pseudospectrum) and $\sigma(\mathfrak{h}(0))$ (spectrum) becomes negligible.

The situation becomes thoroughly different when a quantum Hamiltonian H is chosen "highly non-self-adjoint" because then, "the pseudospectrum $\sigma_\epsilon(H)$ is typically much larger than the ϵ−neighborhood of the spectrum". There is a subtlety in such a proposition (cited from [30]) because in the context of the general non-Hermitian Schrödinger Equation (1), one has to distinguish, in a way already emphasized in the Introduction, between its open-system and closed-system interpretations.

In the former, "resonances-describing" subcase, we would have to complement Equation (1) by the specification of the conventional Hilbert space endowed with the usual, metric-independent norm. In diagram (A1), such a space is denoted by the dedicated symbol $\mathcal{K}$, with the norm of V denoted as $\|V\|$ as usual. Hence, in such a case (not, by the way, of our present immediate interest), we may formally set $\Theta = I$ and $\mathcal{H} = \mathcal{K}$ in Appendix A.

In the other, "bound-states-describing" subcase (which *is* of our present interest) we may still follow the same conventions as introduced in Appendix A. Thus, with $\Theta \neq I$ and with $\mathcal{H} \neq \mathcal{K}$ we have to treat Schrödinger Equation (1) as living in an amended, physical Hilbert space $\mathcal{H}$.

Unless one asks questions about norms, only the dual versions of the vector spaces $\mathcal{K}$ and $\mathcal{H}$ are different. Still, precisely the difference between the operator norm of V in $\mathcal{K}$ (denoted as usual, $\|V\|$) and in $\mathcal{H}$ (to be denoted differently, say, as $\sharp V \sharp$) becomes one of the most essential aspects of the respective alternative definitions of the Hilbert-space-dependent pseudospectra.

3.3. Pseudospectra in Quasi-Hermitian Models

As long as we are not going to study resonances, we may just restrict our attention to the random perturbations in quasi-Hermitian (i.e., by definition, in the hiddenly unitary) closed quantum systems. In principle, their description in the alternative physical Hilbert spaces $\mathcal{L}$ and $\mathcal{H}$ is then equivalent. In practice, nevertheless, one may observe that the predictions of the measurements as constructed in the textbook Hilbert-space representation are impractical and less user-friendly. Then, we are forced to treat the knowledge of the union (20) of the perturbed spectra in $\mathcal{L}$ as "technically inaccessible".

After we decide to move to $\mathcal{H}$, we must also remember that the corresponding physical norm $\sharp V \sharp$ of perturbations becomes different and, first of all, Θ dependent. The key and meaningful question to ask is then the question about the structure of the union

$$\bigcup_{\lambda \sharp V \sharp < \epsilon} \sigma(H(0) + \lambda V) \tag{22}$$

of the spectra of all of the slightly but randomly perturbed systems living in $\mathcal{H}$.

The above-cited theorem can be recalled again. After one defines the pseudospectrum

$$\sigma_\epsilon(H) := \sigma(H) \cup \{z \in \mathbb{C} \mid \sharp (H-z)^{-1} \sharp > \epsilon^{-1}\} \tag{23}$$

in $\mathcal{H}$, one immediately obtains the Roch–Silberman relationship

$$\bigcup_{\lambda \sharp V \sharp < \epsilon} \sigma(H(0) + \lambda V) = \sigma_\epsilon(H(0)). \tag{24}$$

This is our desired ultimate formula. In the correct and physical Hilbert space $\mathcal{H}$ in which the Hamiltonian is made self-adjoint, this formula defines the sensitivity to perturbations in terms of the correct physical pseudospectrum (23). Its explicit numerical construction is facilitated and made useful. Obviously, once we require our random perturbations to be small in $\mathcal{H}$, we may again recall Proposition Nr. 3 in [30] and conclude that in full parallel with the Hermitian models also in the quasi-Hermitian picture of dynamics the spectrum and pseudospectrum carry equivalent information about the sensitivity of bound states to perturbations,

$$\sigma_\epsilon(H(0)) \subseteq \{z \in \mathbb{C} \mid \text{dist}(z, \sigma(H(0))) < \text{const} \times \epsilon\}. \tag{25}$$

At the small values of ϵ, the pseudospectrum is formed just by a small vicinity of the spectrum. In the terminology of [30], such a pseudospectrum is "trivial" because small random perturbations cannot destroy the stability of the underlying closed quantum system.

4. Amended Rayleigh–Schrödinger Construction

Let us temporarily return to the open-system theory where one does not need to define any nontrivial inner-product metric because the evolution is non-unitary (cf., e.g., monographs [31] or [32]). In Section 2, we recalled, for illustration, the work in [24] as a typical sample of a more traditional approach. Bender and Dunne used there a Hamiltonian (11) for the purposes of the description of a complicated physical phenomenon. The physical Hamiltonian itself, as sampled by Equation (15), has only been deduced after an analytic-continuation redefinition of the model.

In our present paper, our strategy is different, with our attention restricted to the unitary, closed and stable quantum systems in which the unitarity of evolution coexists with the non-Hermiticity of H. In this setting, we intend to describe an amendment of the QHQM perturbation-expansion recipe in which the metric-related technical obstacles will be circumvented using a reformulation of the theory as recently proposed, in a different context, in [19].

4.1. The Choice-of-Space Problem Revisited

The requirement of unitarity of the evolution may make the QHQM perturbation theory discouragingly complicated, mainly due to the operator–expansion nature of the newly emerging series (7) representing the metric. In a way outlined in Appendix C, the theory has to be formulated in as many as five Hilbert spaces (cf. our present diagram (A13) or analogous diagram Nr. (20) in [16]). The standard, reference-providing space $\mathcal{L}$ of textbooks has to be accompanied by the doublet of the preferred representation spaces, viz., by $\mathcal{K}(\lambda)$ pertaining to the ultimate dynamical scenario and by $\mathcal{K}(0)$ representing the solvable unperturbed system. The remaining pair of their amended physical partners consists of the predictions-offering $\mathcal{H}(\lambda)$ (carrying the ultimate picture of physics) and $\mathcal{H}(0)$ (i.e., its unperturbed $\lambda = 0$ partner).

The five-Hilbert-space pattern looks complicated. Concerning its applicability, one has to be a bit skeptical. In what follows, we intend to show that a fairly efficient remedy of the skepticism can be based on a more or less straightforward reformulation of the theory in which the specification of the metric will be re-interpreted as an upgraded form of transition from $\mathcal{K}$ to the correct physical Hilbert space $\mathcal{H}$. A motivation of our present modification of the theory lies in an undeniable appeal of the Rayleigh–Schrödinger perturbation-approximation philosophy, which may be characterized, in the conventional textbook setting, by its enormous technical simplicity. In this sense, we intend to show that this simplicity need not get lost after one moves to the innovative QHQM framework.

Our attention will be concentrated upon the mathematical consistency aspects of the theory. We will emphasize that it is possible to overcome the most unpleasant conceptual complications emerging when one deals with a realistic quasi-Hermitian Hamiltonian of a unitary quantum system which is allowed to vary with a parameter. The theory will be re-analyzed in a way inspired by several publications, a sample of which is recalled in Appendix C.

Attention will be paid to the models in which the parameter-dependence remains weak and tractable by the techniques of perturbation theory [15]. In the first step of amendment of the conventional approaches, we will modify the very concept of a state, keeping in mind that in conventional textbooks, the state is usually characterized by a ket–vector element of a physical Hilbert space (i.e., by $|\psi\rangle \in \mathcal{H}$). The most immediate inspiration of a change in such a definition may be deduced from Equation (3), in which it is sufficient to abbreviate

$$\langle \psi(\lambda)|\Theta(\lambda) := \langle\!\langle \psi(\lambda)| \in \mathcal{K}' \tag{26}$$

or, after the Hermitian conjugation in our mathematical representation space,

$$\Theta(\lambda)|\psi(\lambda)\rangle := |\psi(\lambda)\rangle\!\rangle \in \mathcal{K}. \tag{27}$$

These abbreviations enable us to rewrite Equation (3) in a more compact form,

$$a(\lambda) = \langle\!\langle \psi_n(\lambda)|A|\psi_n(\lambda)\rangle \tag{28}$$

out of which the metric $\Theta(\lambda)$ seems to have "disappeared".

An easy resolution of such an apparent paradox is that we moved back from auxiliary $\mathcal{K}$ to physical $\mathcal{H}$. After some elementary algebra, we also reveal that the parallels between the "old" ket vectors $|\psi_n(\lambda)\rangle \in \mathcal{K}$ and their "new" partners of Equation (27) (which could be called "ketkets") can even be extended, yielding an identically satisfied "parallel" eigenvalue problem

$$H^\dagger(\lambda)|\psi_n(\lambda)\rangle\!\rangle = E_n(\lambda)|\psi_n(\lambda)\rangle\!\rangle, \quad |\psi_n(\lambda)\rangle\!\rangle \in \mathcal{K}, \quad n = 0, 1, \ldots \tag{29}$$

(with the same real spectrum of course) or, after the mere Hermitian conjugation in $\mathcal{K}$, equivalently,

$$\langle\!\langle \psi_n(\lambda)|H(\lambda) = \langle\!\langle \psi_n(\lambda)|E_n(\lambda), \quad \langle\!\langle \psi_n(\lambda)| \in \mathcal{K}', \quad n = 0, 1, \ldots. \tag{30}$$

Now, we are prepared to realize that for vectors, the "physical" Hermitian conjugation as defined, hypothetically, in the "hidden" Hilbert space $\mathcal{H}$ just replaces the ket $|\psi_n(\lambda)\rangle \in \mathcal{H}$ by the "brabra" $\langle\!\langle \psi_n(\lambda)| \in \mathcal{H}'$.

To summarize, we come to the conclusion that in the correct physical Hilbert space $\mathcal{H}$, the most natural representation of an n-th bound state of the quantum system in question will not be provided by any ket but rather by the elementary projector

$$\varrho_n(\lambda) = |\psi_n(\lambda)\rangle\!\rangle \frac{1}{\langle\!\langle \psi_n(\lambda)|\psi_n(\lambda)\rangle\!\rangle} \langle\!\langle \psi_n(\lambda)|. \tag{31}$$

The main advantage of such an upgrade of conventions is two-fold. First, formula (31) remains the same in both of the Hilbert-space representations in $\mathcal{K}$ and in $\mathcal{H}$, and second, using the standard definition

$$a_n(\lambda) = \text{Tr}[A\varrho_n(\lambda)] \tag{32}$$

of the probability density, one immediately rediscovers the above-mentioned equivalent measurement-predicting prescription (3). Moreover, the use of formula also opens the way from pure states to mixed states and quantum statistical physics [33–35].

4.2. Rayleigh–Schrödinger Construction Revisited

In the light of our preceding considerations, the essence of our present innovation of the QHQM Rayleigh–Schrödinger construction of the series (5)–(7) [with an implicit reference to the "measurement-prediction" Formula (3) *alias* (28) *alias* (32)] can be seen to lie simply in the replacement of the almost prohibitively complicated operator–expansion formula (7) by the alternative and formally sufficient new ketket-expansion ansatz

$$|\psi(\lambda)\rangle\rangle = |\psi(0)\rangle\rangle + \lambda\,|\psi^{(1)}\rangle\rangle + \lambda^2\,|\psi^{(2)}\rangle\rangle + \ldots. \tag{33}$$

In other words, we will still have to start from the entirely conventional decomposition (4) of the Hamiltonian and from the related order-by-order re-arrangement

$$\left[H - E(0) + \lambda\,(V - E^{(1)}) - \lambda^2\,E^{(2)} - \ldots\right]\left[|0\rangle + \lambda\,|\psi^{(1)}\rangle + \lambda^2\,|\psi^{(2)}\rangle + \ldots\right] = 0 \tag{34}$$

of our initial perturbed form of Schrödinger Equation (1). The innovation only comes when we reject the recipe of our previous proposal [16] (based on the reconstruction of $\Theta(\lambda)$ via the clumsy power-series ansatz (7)) as unnecessarily (and, what is worse, more or less prohibitively) complicated.

In our present upgraded recipe, one simply complements Equation (34) by its associated partner for ketkets,

$$\left[H^\dagger - E(0) + \lambda\,(V^\dagger - E^{(1)}) - \lambda^2\,E^{(2)} - \ldots\right]\left[|0\rangle\rangle + \lambda\,|\psi^{(1)}\rangle\rangle + \lambda^2\,|\psi^{(2)}\rangle\rangle + \ldots\right] = 0. \tag{35}$$

Obviously, an enormous simplification of the construction of the measurable predictions (32) is achieved. Indeed, in comparison with the complicated formulae in [16], the construction of the necessary recurrences for the sequence of corrections becomes immediate, making use only of the slightly upgraded projector

$$\Pi = I - |0\rangle\langle\langle 0| = \sum_{j>0} |j\rangle\langle\langle j| \tag{36}$$

and leading to the easily deduced formulae for the energies, say,

$$E^{(1)} = \langle\langle 0|V|0\rangle, \quad E^{(2)} = \langle\langle 0|V\Pi|\psi^{(1)}\rangle, \quad \ldots \tag{37}$$

as well as to the kets

$$|\psi^{(1)}\rangle = \Pi\,\frac{1}{E(0) - \Pi H \Pi}\,\Pi V|0\rangle, \tag{38}$$

$$|\psi^{(2)}\rangle = \Pi\,\frac{1}{E(0) - \Pi H \Pi}\,\Pi[V - E^{(1)}]\Pi|\psi^{(1)}\rangle, \tag{39}$$

(etc.) and, analogously, for the ketkets,

$$|\psi^{(1)}\rangle\rangle = \Pi^\dagger\,\frac{1}{E(0) - \Pi^\dagger H^\dagger \Pi^\dagger}\,\Pi^\dagger V^\dagger|0\rangle\rangle, \tag{40}$$

$$|\psi^{(2)}\rangle\rangle = \Pi^\dagger\,\frac{1}{E(0) - \Pi^\dagger H^\dagger \Pi^\dagger}\,\Pi^\dagger[V^\dagger - E^{(1)}]\Pi^\dagger|\psi^{(1)}\rangle\rangle, \tag{41}$$

etc. Summarizing, one only has to remind the readers that the full-fledged version of the present amended QHQM perturbation theory is only needed when we really have to predict the results of measurements of the observable represented by a preselected operator A. In applications, we are often interested in just the evaluation of only one of the values of the energy (which is, moreover, defined as one of the eigenvalues of the Hamiltonian). In practice, such a value is often known to be real. In such a case, naturally, what is needed is just the more or less standard construction of the single power series (5). We may conclude that precisely such simplified calculations were performed in papers [23,24], with the details recalled in Section 2 and in Section 2.2 above.

5. Discussion

5.1. Key Role Played by the Proof of Reality of Spectrum

In the early studies of non-Hermitian Hamiltonians with real spectra [1,2,12,36], the authors admitted that the non-Hermiticity of $H(\lambda)$ could make the standard probabilistic closed-system interpretation of the states questionable. For example, Bender and Dunne [24] circumvented the problem by claiming that their expansion (12) offers only "strong evidence" that the quantity $E(\lambda)$ is an analytic function, which can be continued to the cut complex plane of couplings $g = \lambda^2$.

Later, emphasis has been shifted to the requirement of the reality *alias* potential observability of the would-be bound-state energy-level spectrum of H representing a necessary condition of existence of an amended inner product. A direct and truly innovative closed-system physical interpretation of models started to be sought in the reconstruction of metric $\Theta = \Theta(H)$ [3,10].

In the context of QHQM perturbation theory, for several reasons (some of which have been discussed above), the necessity of the proof of the reality of the energy spectra also acquired a new urgency. In its analysis, as performed in our preceding paper [16], we emphasized that the scope of the QHQM perturbation theory is in fact "too broad". In comparison with the constructive strategy of conventional textbooks (where the trivial physical inner-product metric is chosen in advance), the more flexible QHQM theoretical framework forced us to admit that our Θ must be treated as perturbation dependent. The two conventional Rayleigh–Schrödinger power series (5) and (6) had to be complemented by the third item (7) representing the metric and making the construction of the model (i.e., of its correct physical Hilbert space) almost prohibitively difficult.

In this context, one of our present main results is that we managed to simplify the construction by replacing the difficult operator expansion (7) by its mere ketket–vector alternative (33). Nevertheless, even after such an upgrade of the recipe the (rarely easy!) proof of the reality of the spectrum will still keep playing the most important role of a necessary preparatory step in applications.

5.2. The Requirement of Completeness of the Set of Observables

We achieved a simplification of the non-Hermitian version of the Rayleigh–Schrödinger formalism by making the operator–expansion (7) of the metric "invisible". The price to pay was the loss of insight into the correspondence between the reality of spectrum and the choice of the class of admissible perturbations. In fact, the study of this correspondence is nontrivial, requiring, probably, a return to the study of explicit expansions (7).

The question remains to be kept in mind as a truly interesting and challenging future project, nevertheless. One of the reasons is that it is closely related to the paradox of the ambiguity of the metric. Indeed, it is well known that the operator Θ endowing a given Hamiltonian H with a self-adjoint status in $\mathcal{H}$ need not be unique. As a consequence, even the norm of perturbation V in ansatz (4) can vary so that also the conventional condition of its "sufficient smallness" could be difficult, if not impossible, to prove.

The ambiguity of $\Theta = \Theta(H)$ has been identified in [8], resulting from an incompleteness of our information about the system's dynamics. Indeed, the emergence of any independent candidate Λ for an observable (which would have to be quasi-Hermitian with respect to the same metric, $\Lambda^\dagger \Theta = \Theta \Lambda$) would suppress the ambiguity of Θ whenever such a candidate appears not to be reducible to a function of H, $\Lambda \neq \Lambda(H)$. This means that a unique Θ will be obtained only after one specifies a complete set of irreducible observables $H (= \Lambda_0)$ and $\Lambda (= \Lambda_1)$ and, perhaps, Λ_2, etc. [8].

In a way discussed in [8], one is usually forced to work with only an incomplete irreducible set of preselected observables Λ_j. This means that the ambiguity of the metric can only rarely be fully suppressed. One may try to circumvent the problem by making a more or less arbitrary choice of one of the eligible metrics. The same strategy is, after all, widely accepted in the conventional textbooks using trivial $\Theta = I$.

In the framework of unconventional QHQM, an exhaustive explanation of the problem of the ambiguity of the norm can already be found in [8], where one reads that the variability of our choice of the metric just reflects an incompleteness of the input information about dynamics. This means that such an ambiguity disappears when our knowledge of H becomes complemented by the knowledge of a sufficiently large (i.e., in mathematical language, "irreducible") set of some further operator candidates for the observables.

In this sense, we arrive at a new paradox. Either we postulate such a knowledge or not. Naturally, the abstract theory would only be fully satisfactory in the former case. In such a case, nevertheless, the λ dependence of the Hamiltonian and metric would be inherited by an induced and strongly counterintuitive λ dependence of all of the further (i.e., necessarily quasi-Hermitian) observables Λ_j with $j > 0$.

5.3. The Coordinate-Non-Observability Paradox

Among all of the differential–operator candidates for a closed-system quantum Hamiltonian possessing a real energy-like spectrum as sampled by Equation (11) above, one of the most interesting alternative models was studied by Buslaev and Grecchi [2]. One of the truly striking features of their model (which made it qualitatively different from Equation (11)) was that for the purposes of its mathematical consistency it was necessary to keep the "coordinate" complex (i.e., $x \notin \mathbb{R}$, in the asymptotic domain at least). This is a contradictory situation because such a variable cannot in fact be interpreted as an observable quantity.

The puzzle has been clarified by an explicit reference to perturbation theory in combination with the techniques of analytic continuation. In a way discussed also in section 2 above, Buslaev and Grecchi revealed a hidden, perturbation-series-mediated connection between their manifestly non-Hermitian "complex-coordinate" oscillator and the safely self-adjoint Hamiltonian

$$\mathfrak{h}^{(AHO)} = -\Delta + |\vec{r}|^2 + \lambda |\vec{r}|^4 \qquad (42)$$

describing an entirely conventional quartic anharmonic oscillator [14,37]. They were aware of the divergence of the related Rayleigh–Schrödinger perturbation series (5), but their analysis revealed the existence of an intimate relationship between operator (42) (defined as self-adjoint in the most common physical Hilbert space $L^2(\mathbb{R}^d)$) and its specific non-Hermitian isospectral descendant.

In [2], the same idea has been shown to work also in application to another, multiparametric multiplet of ordinary differential Hamiltonian-like operators $H_n^{(BG)}$ with $n = 1, 2, \ldots, K$ (with, incidentally, $K = 7$). A special status has been again enjoyed by the element $H_1^{(BG)} = \mathfrak{h}^{(BG)}$ which was required, in the most conventional Hilbert space $L^2(\mathbb{R})$, to be self-adjoint. The last element $H_K^{(BG)}$ of the sequence appeared to be non-Hermitian but parity-time-symmetric *alias* $\mathcal{PT}$– symmetric. For our present purposes, we may abbreviate $H_K^{(BG)} = H^{(BG)}$ (i.e., drop the last subscript) and notice that the above-mentioned Jones and Mateo isospectrality relationship finds a direct analogue in formula

$$\mathfrak{h}^{(BG)} \sim H^{(BG)}.$$

This is not too surprising because the Jones and Mateo Hamiltonian $\mathfrak{h}^{(JM)}$ is just a parameter-free special case of the Buslaev and Grechi multiparametric operator $\mathfrak{h}^{(BG)}$. Thus, after a multiparametric generalization of Jones' and Mateo's Dyson operator (17), a new light could be thrown upon the concept of locality in non-Hermitian physics (cf. [38] and also formula (18) in Section 2.3 above).

Incidentally, Jones' and Mateo's "direction of simplification" becomes inverted since the evolution controlled by $H^{(BG)}$ has to be reclassified as a more complicated picture of dynamics. Still, the message which survives is that the physical interpretation is directly provided by the Rayleigh–Schrödinger perturbation series of Equation (5).

5.4. A Detour to Meaningful Complex Spectra

In the conventional applications of perturbation theory, one starts from the knowledge of a preselected family of Hamiltonians

$$H(\lambda) = H_0 + \lambda H_1 \tag{43}$$

in which the unperturbed operator H_0 is assumed maximally user-friendly or even, often, diagonal. The specification of the admissible perturbations λH_1 is then rather routine, made in accordance with both the phenomenological and mathematical model-building needs [15]. Family (43) is chosen, in most textbooks, as a mere sum of two self-adjoint operators.

We have already emphasized that once one admits a manifest non-Hermiticity of one or both of the operator components of the Hamiltonian in an auxiliary Hilbert space $\mathcal{K}$,

$$H_0 \neq H_0^\dagger, \quad H_1 \neq H_1^\dagger \tag{44}$$

the technical costs of such a weakening of the conventional assumptions may be high (cf. [3,8,10,11,39,40]). Even when one decides to keep the working Hilbert space perturbation-independent, $\mathcal{K}(\lambda) = \mathcal{K}(0) = \mathcal{K}$, a number of challenging questions survive. One of the most important ones follows from the possible loss of the reality of eigenvalues,

$$E(\lambda) = E(0) + \lambda E^{(1)} + \lambda^2 E^{(2)} + \ldots \in \mathbb{C}. \tag{45}$$

Then, one has to accept the open-system philosophy and to treat the Rayleigh–Schrödinger expansions just as an ansatz which could work even when $E(\lambda) \notin \mathbb{R}$ and even when the series is divergent.

The feasibility of such an alternative model-building strategy has been confirmed, e.g., by Caliceti et al. [36] (cf. also a more recent review of the field in [41]). In essence, the latter authors revealed that in a number of specific toy models the conventional ansatz (45) may still serve as a productive constructive tool, yielding, at the small and real coupling constants, the real (i.e., energy-like) as well as complex (i.e., resonance-representing) low-lying spectra after standard resummation.

5.5. Real Spectra and the Paradox of Emergent Instabilities

The reality of spectra of the Hamiltonians has independently been noticed in the context of quantum field theory [12]. This attracted the attention of the physics community to the applicability of expansions (45) in the non-Hermitian setting of Equation (44). The authors of the innovated studies of imaginary cubic anharmonic-oscillator Hamiltonians

$$H^{(CAHO)}(\mu,\nu) = -\frac{d^2}{dx^2} + \frac{\mu^2}{4}x^2 + i\nu x^3 \tag{46}$$

identified either $\lambda = \mu$ (say, in the "strong-coupling expansions" of Ref. [23]) or $\lambda = \nu$ (say, in the "weak-coupling expansions" in [24]).

Even when having the strictly real bound-state-like spectra, the latter model-building efforts were criticized by mathematicians [17,30]. They recommended a replacement of the mere search for eigenvalues (characterized as "fragile") by a more ambitious construction of pseudospectra. We have to point out that the mathematically well-founded latter recommendation has been based on a conceptual misunderstanding. Fortunately, a disentanglement of the misunderstanding was straightforward. It proved sufficient to distinguish between the closed and open systems and to show that the construction of pseudospectra only makes sense (and offers new information) in the latter case (for more details, see also Section 3 above).

From a purely mathematical point of view, one should not be too surprised by the latter conclusion and, in particular, by the "wild" [30] behavior of open systems exhibiting emergent instabilities because the theory behind the closed systems is different. For them, the constructions and predictions obtained in the two alternative "physical" Hilbert spaces $\mathcal{L}$ and $\mathcal{H}$ are, by definition, equivalent. Thus, no paradox can be seen in the existence of the mechanism due to which the pseudospectra of closed systems remain well behaved even when the representation of dynamics itself is non-Hermitian.

Another, even more straightforward explanation of the existence of the emergent open-system instabilities becomes best visible when the system in question happens to lie close to a Kato's exceptional-point singularity [42–44]. In such a vicinity, indeed, the operator of metric Θ becomes singular and dominated by a projector [45]. Perturbations H_1, which are small with respect to the correct physical Hilbert-space norm (in our present notation this means that $\sharp H_1 \sharp \ll 1$) may still be, simultaneously, very large with respect to the conventional open-system norm as defined in the standard regular limit of $\Theta \to I$ (i.e., $\|H_1\| \gg 1$). As a consequence, perturbations may be expected to lead to the "wild" forms of pseudospectrs (20) as sampled, via a number of elementary examples, in [30].

5.6. Ultimate Challenge: Models Where the Metric Does Not Exist

To a compact introduction in the overall QHQM theory, as provided in Appendix A, it makes sense to add that a truly enormous increase in the popularity of the formalism has been inspired by Bender's and Boettcher's claim [1] that the reality of spectra is a phenomenon which can be observed in an unexpectedly broad class of models which are not only phenomenologically attractive but also mathematically user-friendly. These results set the scene for an intensive subsequent study. It is of no surprise that whenever an operator H proves non-Hermitian (in $\mathcal{K}$) while its spectrum $\{E_n\}$ is "bound-state-like" (i.e., real, discrete, and bounded from below), one feels tempted to consider the possibility of its quantum quasi-Hermitian Hamiltonian-operator interpretation.

In [1], the temptation has been further supported by the detailed analysis of the specific ordinary differential Hamiltonian-like operators

$$H^{(BB)}(n) = -\frac{d^2}{dx^2} - (ix)^{n+2}, \quad n \geq 0. \qquad (47)$$

These operators are, in general, complex and manifestly non-Hermitian but still possessing the strictly real and discrete bound-state-like spectra. On these grounds, Bender and Boettcher conjectured that such operators could be treated as Hamiltonians in certain unconventional, "analytically continued" quantum theories.

In 2012, Siegl and Krejčiřík [46] opposed the claim. Using the rigorous methods of functional analysis, they proved that for at least some of the toy models where H had the elementary differential–operator form (47), an acceptable metric Θ which would satisfy relation (2) does not exist at all. This weakened the enthusiasm because at least some of the local interaction benchmark models cannot be endowed with *any* admissible physical-Hilbert space $\mathcal{H}$.

One of the ways of circumventing such a mathematical disproof of quasi-Hermiticity has been found in a transition to the open system reinterpretation of the models [5]. As a benefit, such a change of strategy simplified the mathematics because one could simply set $\Theta = I$. A return to the old open-system philosophy behind models (47) appeared even productive in mathematics. In a way outlined in Section 5.5 above, it led to the discovery of certain unexpected spurious approximate solutions of Equation (1) emerging at the energies which were far from the spectrum [30]. Thus, it was immediate to conclude that in place of the spectrum, the much more useful descriptive tool can be sought in the pseudospectra.

In a way which we described in Section 3 above, the pseudospectra directly characterize the influence of random perturbations upon dynamics of the systems. Incidentally, their analysis has been shown to make sense only in the open-system cases in which the spectra of H are not real. As long as $\text{Im } E_n \neq 0$ at some n, the Hamiltonian cannot be Hermitian, $H \neq H^\dagger$. Thus, we may set $\Theta = I$ and identify $\mathcal{H} = \mathcal{K}$. In contrast, the knowledge of pseudospecta is not needed in the other, stable-bound-state scenario, partially because the implication $[\text{Im } E_n \neq 0] \implies [H \neq H^\dagger]$ cannot be inverted.

It is possible to conclude that for the closed systems with the Hamiltonians sampled by Equation (47), the question of their acceptability is still open. The reason is that the physical Hilbert space defined in terms of the correct inner-product metric $\Theta = \Theta(H)$ need not exist. Thus, whenever we decide to stay inside the QHQM theory and require that

$$[H \neq H^\dagger] \,\&\, [\text{Im } E_n = 0], \quad \forall n$$

we must keep in mind that the status of many popular illustrative examples has to be reconsidered as inconclusive, with an acceptable physical interpretation being still sought in several new directions [47–55].

In fact, Scholtz et al. [8], were probably already aware of the similar mathematical subtlety because they complemented the quasi-Hermiticity requirement (2) by a few further consistency-supporting sufficient conditions. Among them, the most prominent amendment of the theory seems to be their mysterious requirement of the boundedness of H. Unmotivated and counterintuitive as it might have looked in the past, it was probably one of the lucky parts of the formulation of QHQM by Scholtz et al. [8] because, in retrospect, it excludes the contradictory differential unbounded-operator models (47) as unacceptable.

6. Summary

In the textbooks on quantum theory, the authors have to distinguish between the models supporting, and not supporting, the presence of resonances. The notion of perturbation plays a fundamental role in both of these implementations of the theory. This is for two reasons. The first one is realistic: Whenever one tries to prepare and study a quantum system, stable or unstable, it is hardly possible to achieve its absolute isolation from an uncontrolled environment. One has to guarantee the negligibility of influence of such an environment using, typically, non-Hermitian Hamiltonians and open-system models with complex spectra and random perturbations.

The second reason is mathematical: Even if we manage to guarantee that the system in question is, up to negligible errors, isolated, perturbation theory re-appears as a powerful tool suitable for calculations and for an efficient evaluation of predictions. Naturally, a consistency of perturbation-related constructions needs a guarantee of a "sufficient smallness" of the perturbation. Such a guarantee is a task, an explicit formulation of which depends on the model-building details. Our present attention has mainly been devoted, therefore, to the physics of stable bound states (and just marginally to unstable resonances) in a way motivated by the recently increasing popularity of the so-called non-Hermitian Schrödinger representations of the stable and unitary quantum systems.

In the literature, the presentation of this subject may be found accompanied by the emergence of multiple new and unanswered questions. In our paper, we picked up a few of such questions which we were able to answer. Basically, our answers may be separated in several groups. In the first one, we felt inspired by the authors who studied the pseudospectra. We imagined that in such an area of research the application of innovative mathematics is not always accompanied by a clear explanation of physics. In this setting, we conjectured that a key to the resolution of certain emerging apparent paradoxes can be found in distinguishing, more consequently, between the traditional non-Hermitian quantum models with $\Theta = I$ (for the study of which the pseudospectra have been found truly indispensable [17]) and the more recent and sophisticated closed-system theories in which the specification of the correct inner-product metric proves nontrivial and Hamiltonian-dependent, $\Theta = \Theta(H) \neq I$.

In this way, we managed to explain that in certain applications (sampled by the random-perturbation studies) there is practically no difference between the use of the QHQM and standard quantum mechanics. A slightly different conclusion has been achieved when we turned attention to a more explicit study of the Rayleigh–Schrödinger perturbation expansions. The differences, not too well visible in the mere calculations of energies [23,24], appeared immediately deeply relevant when one becomes interested in practically any other observable quality/quantity of the system.

The latter feature of the theory has already been observed (and not found to be too welcome) in our preceding paper [16]. Our detailed analysis of the structure of the QHQM-version of perturbation theory led us there to a few rather skeptical conclusions concerning the applicability of the formalism in its full generality. In our present paper, we managed to show that the strength of the latter discouraging results can perceivably be weakened when one reconsiders the theory and after one reduces its scope to just the description of its experimentally verifiable predictions.

In this spirit, we proposed replacing the next-to-prohibitively difficult operator-valued solution of Equation (2) (specifying the perturbation-dependent metric $\Theta(\lambda)$ needed in Equation (3)) by the vector-valued solution of Equation (29) entering the modified form (28)) of the same prediction which is, even by itself, much easier to evaluate.

In conclusion, it is probably worth adding that along the same methodical lines one could also get beyond the framework of the Schrödinger picture in which the opereators of observables are mostly assumed time independent, $A \neq A(t)$. In the future, perhaps, the same methodical ideas might prove applicable also in the non-stationary context and models and in the interaction-picture extension of the hiddenly unitary-evolution formalism as proposed a couple of years ago in [56].

Funding: This research received no external funding.

Data Availability Statement: No new data were created in this study.

Acknowledgments: The author was supported by Faculty of Science of University of Hradec Kralove.

Conflicts of Interest: The author declares no conflicts of interest.

Appendix A. Quantum Mechanics in Quasi-Hermitian Representation

A comprehensive outline of the formulation of unitary quantum mechanics in which the conventional requirement of Hermiticity of the Hamiltonian is replaced by an apparently weaker, metric-dependent quasi-Hermiticity constraint (2) can be found not only in the older review by Scholtz et al. [8] but also in a few newer papers (e.g., [3,10,57]) and books (e.g., [4,11]). In the interpretation of review [39], the formalism is based on the simultaneous use of a triplet of Hilbert spaces (say, $[\mathcal{L}, \mathcal{K}, \mathcal{H}]$) connected by the Dyson-inspired [20] mutual correspondences, as displayed in the following diagram:

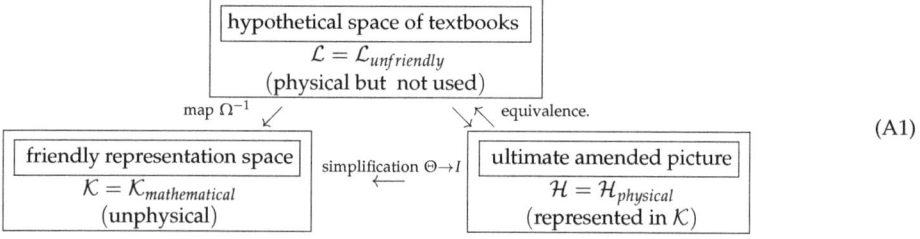

(A1)

In such an arrangement, by assumption, the two lower-line Hilbert spaces $\mathcal{K}$ and $\mathcal{H}$ coincide as linear modules or vector-space sets of the ket–vector elements $|\psi\rangle$. Thus, we can write $|\psi\rangle \in \mathcal{K}$ and/or $|\psi\rangle \in \mathcal{H}$ and treat the Hamiltonian H (carrying the input information about dynamics) as an operator defined and acting in both of these two spaces.

The difference between $\mathcal{K}$ and $\mathcal{H}$ lies in two conventions. First, the former, auxiliary, manifestly unphysical Hilbert space $\mathcal{K}$ is definitely preferred as the more user-friendly one. The standard Dirac's notation is applied to the bra–vector elements of its dual marked by a prime, $\langle\psi| \in \mathcal{K}'$. Second, the correct physical Hilbert space $\mathcal{H}$ is only treated as represented in $\mathcal{K}$ using the mere change of the inner product,

$$\langle\psi_a|\psi_b\rangle = \text{the inner product in } \mathcal{K}, \quad \langle\psi_a|\Theta|\psi_b\rangle = \text{the (mimicked) product in } \mathcal{H}.$$

The bra–vector elements of the dual physical vector space are, in the notation of Ref. [58], denoted as "brabras", $\langle\!\langle\psi| \in \mathcal{H}'$. They have the metric-dependent representation

$$\langle\!\langle\psi| = \langle\psi|\Theta$$

in $\mathcal{K}'$. Thus, we can treat these brabras as the Hermitian conjugates of the kets in the physical Hilbert space $\mathcal{H}$. In parallel, we can also introduce the "ketkets" $|\psi\rangle\!\rangle = \Theta|\psi\rangle$ as the Hermitian conjugates of the brabras with respect to the simpler, conventional inner-products in the unphysical but preferred representation-Hilbert-space $\mathcal{K}$.

Appendix B. Rayleigh–Schrödinger Construction in $\mathcal{L}$

A factorization $\Theta = \Omega^\dagger \Omega$ of the metric enables us to define the textbook $\mathcal{L}$–space self-adjoint avatar of our Hamiltonian

$$\mathfrak{h} = \Omega H \Omega^{-1} = \mathfrak{h}^\dagger. \tag{A2}$$

It acts in the upper component $\mathcal{L}$ of diagram (A1), which is just the conventional physical Hilbert space of textbooks. The latter Hilbert space can be perceived as the set of the "spiked-ket" elements $|\psi\succ = \Omega|\psi\rangle \in \mathcal{L}$ and of their Hermitian-conjugate "spiked-bra" duals $\prec\psi| = \langle\psi|\Omega^\dagger \in \mathcal{L}'$. By definition, the hypothetical and practically inaccessible operator $\mathfrak{h}$ is an $\mathcal{L}$-space image of our preselected Hamiltonian H. Hence, the the abovementioned links of $\mathcal{L}$ to the other two spaces imply that the Hermiticity of $\mathfrak{h}$ in $\mathcal{L}$ is equivalent to the (hidden) Hermiticity of our H in $\mathcal{H}$. In contrast, the same operator H is non-Hermitian in the mathematical manipulation space $\mathcal{K}$.

In the context of perturbation theory with $\mathfrak{h} = \mathfrak{h}(\lambda) = \mathfrak{h}_0 + \lambda\mathfrak{v}$ in Schrödinger equation

$$\mathfrak{h}(\lambda)|\psi_n(\lambda)\succ = E_n(\lambda)|\psi_n(\lambda)\succ, \quad n = 0, 1, \ldots \tag{A3}$$

the standard power-series ansatz for energies (5) is complemented by its wave-function analogue

$$|\psi_n(\lambda)\succ = |\psi_n(0)\succ + \lambda|\psi_n^{(1)}\succ + \lambda^2|\psi_n^{(2)}\succ + \ldots. \tag{A4}$$

The Hermiticity (A2) is then an important mathematical advantage. In particular, this property enables us to treat the unperturbed Schrödinger equation

$$(\mathfrak{h}_0 - E_n(0))|\psi_n(0)\succ = 0 \tag{A5}$$

as a standard eigenvalue problem, preferably solvable in closed form. Next, we may recall any textbook and write down the first-order-approximation extension of Equation (A5),

$$(\mathfrak{h}_0 - E_n(0))|\psi_n^{(1)}\succ + (\mathfrak{v} - E_n^{(1)})|\psi_n(0)\succ = 0 \tag{A6}$$

as well as its second-order extension

$$(\mathfrak{h}_0 - E_n(0))|\psi_n^{(2)}\succ + (\mathfrak{v} - E_n^{(1)})|\psi_n^{(1)}\succ + (-E_n^{(2)})|\psi_n(0)\succ = 0 \tag{A7}$$

etc. In this manner, we may reconstruct the sequence of the corrections to the energy,

$$E_n^{(1)} = \prec \psi_n(0)|\mathfrak{v}|\psi_n(0) \succ , \tag{A8}$$

$$E_n^{(2)} = \prec \psi_n(0)|\mathfrak{v}|\psi_n^{(1)} \succ \tag{A9}$$

(etc.) as well as the analogous sequence of the corrections to the wave-function ket–vectors

$$|\psi_n^{(1)} \succ = Q \frac{1}{E_n(0) - Q\mathfrak{h}_0 Q} Q\mathfrak{v}|\psi_n(0) \succ , \tag{A10}$$

$$|\psi_n^{(2)} \succ = Q \frac{1}{E_n(0) - Q\mathfrak{h}_0 Q} Q(\mathfrak{v} - E_n^{(1)})|\psi_n^{(1)} \succ \tag{A11}$$

(etc.) where the symbol

$$Q = I - |\psi_n(0) \succ \prec \psi_n(0)| \tag{A12}$$

denotes an elementary projector "out of model space".

Appendix C. Open Questions behind Quasi-Hermitian Perturbations

In the ultimate physical Hilbert space $\mathcal{H}$ in which H is self-adjoint, it would be possible to introduce a dedicated superscript marking the space-characterizing conjugation and to rewrite Equation (2) as follows,

$$H = H^{\ddagger} := \Theta^{-1} H^{\dagger} \Theta.$$

Nevertheless, once we move to the preferred representation space $\mathcal{K}$, the latter notation becomes redundant because the relation $H = H^{\ddagger}$ finds its rephrasing in the quasi-Hermiticity constraint (2) in $\mathcal{K}$.

In applications, we have to re-read Equation (2) as restricting an assignment of metric Θ to a preselected non-Hermitian operator H. Such a metric will necessarily vary with the Hamiltonian in general, $\Theta = \Theta(H)$. The same observation applies to its Dyson-map factor, $\Omega = \Omega(H)$. Both of these comments have already been formulated in [16]. We pointed out there that whenever one decides to consider any one-parametric family of Hamiltonians $H = H(\lambda)$ (including also the perturbed Hamiltonians of Equation (4) as a special case), the physical meaning of the quantum system can only be deduced from its textbook probabilistic interpretation in $\mathcal{L}$ at every λ.

This means that the change in the parameter will imply the change of diagram (A1). The independence of the unperturbed and perturbed versions of Schrödinger Equation (1) lead to the necessity of working, at every non-vanishing parameter λ, with as many as six separate Hilbert spaces. Even though we can merge $\mathcal{L}(\lambda) = \mathcal{L}(0) = \mathcal{L}$ and use the single textbook space for reference, the union of the two respective diagrams (A1) with $\lambda = 0$ and $\lambda \neq 0$ still has to be replaced by their five-Hilbert-space concatenation.

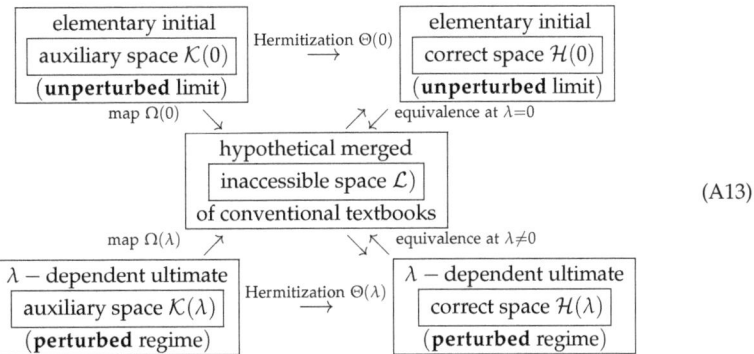

(A13)

In [16], we emphasized that the general QHQM formalism remains consistent and applicable even when the λ−dependence of the Hilbert space metric $\Theta(\lambda)$ is not smooth. Nevertheless, we proposed that for the perturbed models of Equation (4) characterized by a smooth λ−dependence of the Hamiltonian, it makes sense to also postulate the analyticity of $\Theta(\lambda)$. Still, our concluding comments concerning the practical feasibility of the calculations were skeptical. In contrast, one of the key messages, as delivered by our present paper, can be seen in a significant suppression of the latter skepticism.

Appendix D. Biorthonormalized Unperturbed Bases

In Hilbert space $\mathcal{K}$, our Hamiltonians are assumed non-Hermitian, $H \neq H^\dagger$. In Section 4, we emphasized that we therefore have to complement the conventional Schrödinger equation (i.e., Equation (1) for the ket vectors) by its conjugate partner specifying their $\mathcal{H}$−space duals. This goal is achieved either via Equation (30) for the "brabra" vectors or, equivalently, via Equation (29) for the "ketket" vectors.

Temporarily let us simplify the mathematics and assume that $\dim \mathcal{K} < \infty$ [59]. Then, for the reasons explained in diagram (A13) of Appendix C, we must distinguish between the equations at $\lambda = 0$ (the unperturbed limit) and at $\lambda \neq 0$ (the perturbed regime). In the former case, let us now rewrite both of the $\lambda = 0$ Schrödinger equations in a more compact notation,

$$H |n\rangle = E_n |n\rangle, \quad n = 0, 1, \ldots, \dim \mathcal{K} - 1, \tag{A14}$$

$$H^\dagger |n\rangle\rangle = E_n |n\rangle\rangle, \quad n = 0, 1, \ldots, \dim \mathcal{K} - 1. \tag{A15}$$

In the framework of perturbation theory in its most elementary form, the solutions of such an advanced, "doubled" quasi-Hermitian bound-state problem are usually assumed available in closed form. We will also require that all of the unperturbed eigenvectors form a biorthonormalized set (i.e., one has $\langle\langle \psi_m | \psi_n \rangle = \delta_{mn}$), which is complete. Thus, we will have, at our disposal, the spectral decomposition of the identity operator,

$$\sum_{n=0}^{\dim \mathcal{K}-1} |n\rangle \langle\langle n| = I. \tag{A16}$$

Formally, one can even postulate the validity of a spectral representation of the unperturbed Hamiltonian,

$$H(0) = \sum_{n=0}^{\dim \mathcal{K}-1} |n\rangle E_n(0) \langle\langle n|. \tag{A17}$$

Finally, recalling [58], one can write down also the multiparametric definition

$$\Theta(0) = \sum_{n=0}^{\dim \mathcal{K}-1} |n\rangle\rangle \, |\kappa_n(0)|^2 \, \langle\langle n| \qquad (A18)$$

of all of the metrics which would be formally compatible with the Dieudonné's quasi-Hermiticity constraint (2) at $\lambda = 0$. In parallel, the related Dyson-map factor $\Omega = \Omega(0)$ appearing in Equation (16) and in diagrams (A1) and/or (A13), as well as in the explicit definition $|\psi \succ = \Omega \, |\psi\rangle \in \mathcal{L}$ of the elements of the hypothetical space of textbooks can be formally represented by the sum

$$\Omega(0) = \sum_{n=0}^{\dim \mathcal{K}-1} |n \succ \kappa_n(0) \, \langle\langle n|. \qquad (A19)$$

Depending on the representation, one can insert here $\langle\langle \psi| \in \mathcal{H}'$ or $\langle\langle \psi| = \langle \psi | \Theta \in \mathcal{K}'$.

References

1. Bender, C.M.; Boettcher, S. Real Spectra in Non-Hermitian Hamiltonians Having PT Symmetry. *Phys. Rev. Lett.* **1998**, *80*, 5243. [CrossRef]
2. Buslaev, V.; Grecchi, V. Equivalence of unstable anharmonic oscillators and double wells. *J. Phys. A Math. Gen.* **1993**, *26*, 5541–5549. [CrossRef]
3. Bender, C.M. Making sense of non-Hermitian Hamiltonians. *Rep. Prog. Phys.* **2007**, *70*, 947–1118. [CrossRef]
4. Bender, C.M. *PT Symmetry in Quantum and Classical Physics*; World Scientific: Singapore, 2018; (with contributions from Dorey, P.E.; Dunning, C.; Fring, A.; Hook, D.W.; Jones, H.F.; Kuzhel, S.; Levai, G.; Tateo, R.).
5. Christodoulides, D.; Yang, J.-K. (Eds.) *Parity-Time Symmetry and Its Applications*; Springer: Singapore, 2018.
6. Bagchi, B.; Ghosh, R.; Sen, S. Analogue Hawking Radiation as a Tunneling in a Two-Level PT-Symmetric System. *Entropy* **2023**, *25*, 1202. [CrossRef]
7. Stone, M.H. On one-parameter unitary groups in Hilbert Space. *Ann. Math.* **1932**, *33*, 643–648. [CrossRef]
8. Scholtz, F.G.; Geyer, H.B.; Hahne, F.J.W. Quasi-Hermitian Operators in Quantum Mechanics and the Variational Principle. *Ann. Phys.* **1992**, *213*, 74–101. [CrossRef]
9. Dieudonne, J. Quasi-Hermitian Operators. In Proceedings of the International Symposium on Linear Spaces, Jerusalem, Israel, 5–12 July 1960; Pergamon: Oxford, UK, 1961; pp. 115–122.
10. Mostafazadeh, A. Pseudo-Hermitian Representation of Quantum Mechanics. *Int. J. Geom. Meth. Mod. Phys.* **2010**, *7*, 1191–1306. [CrossRef]
11. Bagarello, F.; Gazeau, J.-P.; Szafraniec, F.; Znojil, M. (Eds.) *Non-Selfadjoint Operators in Quantum Physics: Mathematical Aspects*; Wiley: Hoboken, NJ, USA, 2015.
12. Bender, C.M.; Milton, K.A. Nonperturbative Calculation of Symmetry Breaking in Quantum Field Theory. *Phys. Rev. D* **1997**, *55*, R3255. [CrossRef]
13. Bender, C.M.; Wu, T.T. Anharmonic Oscillator. *Phys. Rev.* **1969**, *184*, 1231–1260. [CrossRef]
14. Turbiner, A.; del Valle, D.C. Anharmonic oscillator: A solution. *J. Phys. A Math. Theor.* **2021**, *54*, 295404. [CrossRef]
15. Kato, T. *Perturbation Theory for Linear Operators*; Springer: Berlin/Heidelberg, Germany, 1966.
16. Znojil, M. Theory of response to perturbations in non-hermitian systems using five-Hilbert-space reformulation of unitary quantum mechanics. *Entropy* **2020**, *22*, 80. [CrossRef]
17. Trefethen, L.N.; Embree, M. *Spectra and Pseudospectra: The Behavior of Nonnormal Matrices and Operators*; Princeton University Press: Princeton, NJ, USA, 2005.
18. Brody, D.C. Biorthogonal quantum mechanics. *J. Phys. A Math. Theor.* **2013**, *47*, 035305. [CrossRef]
19. Znojil, M. Quasi-Hermitian formulation of quantum mechanics using two conjugate Schroedinger equations. *Axioms* **2023**, *12*, 644. [CrossRef]
20. Dyson, F.J. General theory of spin-wave interactions. *Phys. Rev.* **1956**, *102*, 1217. [CrossRef]
21. Janssen, D.; Dönau, F.; Frauendorf, S.; Jolos, R.V. Boson description of collective states. *Nucl. Phys. A* **1971**, *172*, 145–165. [CrossRef]
22. Jones, H.F.; Mateo, J. An Equivalent Hermitian Hamiltonian for the non-Hermitian $-x^4$ Potential. *Phys. Rev. D* **2006**, *73*, 085002. [CrossRef]
23. Fernández, F.; Guardiola, R.; Ros, J.; Znojil, M. Strong-coupling expansions for the PT-symmetric oscillators $V(r) = aix + b(ix)^2 + c(ix)^3$. *J. Phys. A Math. Gen.* **1998**, *31*, 10105–10112. [CrossRef]
24. Bender, C.M.; Dunne, G.V. Large-order perturbation theory for a non-Hermiitan PT-symmetric Hamiltonian. *J. Math. Phys.* **1999**, *40*, 4616–4621. [CrossRef]
25. Koukoutsis, E.; Hizanidis, K.; Ram, A.K.; Vahala, G. Dyson maps and unitary evolution for Maxwell equations in tensor dielectric media. *Phys. Rev. A* **2023**, *107*, 042215. [CrossRef]

26. Jones, H.F. Interface between Hermitian and non-Hermitian Hamiltonians in a model calculation. *Phys. Rev. D* **2008**, *78*, 065032. [CrossRef]
27. Znojil, M. Scattering theory using smeared non-Hermitian potentials. *Phys. Rev. D* **2009**, *80*, 045009. [CrossRef]
28. Messiah, A. *Quantum Mechanics*; North Holland: Amsterdam, The Netherlands, 1961.
29. Roch, S.; Silberman, B. C^*-algebra techniques in numerical analysis. *J. Oper. Theory* **1996**, *35*, 241–280.
30. Krejčiřík, D.; Siegl, P.; Tater, M.; Viola, J. Pseudospectra in non-Hermitian quantum mechanics. *J. Math. Phys.* **2015**, *56*, 103513. [CrossRef]
31. Rotter, I. A non-Hermitian Hamilton operator and the physics of open quantum systems. *J. Phys. A Math. Theor.* **2009**, *42*, 153001. [CrossRef]
32. Moiseyev, N. *Non-Hermitian Quantum Mechanics*; Cambridge Univ. Press: Cambridge, UK, 2011.
33. Jakubský, V. Thermodynamics of Pseudo-Hermitian Systems in Equilibrium. *Mod. Phys. Lett. A* **2007**, *22*, 1075–1084. [CrossRef]
34. Znojil, M. Non-Hermitian interaction representation and its use in relativistic quantum mechanics. *Ann. Phys.* **2017**, *385*, 162–179. [CrossRef]
35. Moise, A.A.A.; Cox, G.; Merkli, M. Entropy and entanglement in a bipartite quasi-Hermitian system and its Hermitian counterparts. *Phys. Rev. A* **2023**, *108*, 012223. [CrossRef]
36. Caliceti, E.; Graffi, S.; Maioli, M. Perturbation theory of odd anharmonic oscillators. *Commun. Math. Phys.* **1980**, *75*, 51–66. [CrossRef]
37. Eremenko, A.; Gabrielov, A. Analytic continuation of eigenvalues of a quartic oscillator. *Comm. Math. Phys.* **2009**, *287*, 431. [CrossRef]
38. Liu, Y.X.; Jiang, X.P.; Cao, J.P.; Chen, S. Non-Hermitian mobility edges in one-dimensional quasicrystals with parity-time symmetry. *Phys. Rev. B* **2020**, *101*, 174205. [CrossRef]
39. Znojil, M. Three-Hilbert-space formulation of Quantum Mechanics. *Symmetry Integr. Geom. Methods Appl. SIGMA* **2009**, *5*, 001.
40. Ju, C.Y.; Miranowicz, A.; Chen, Y.N.; Chen, G.Y.; Nori, F. Emergent parallel transport and curvature in Hermitian and non-Hermitian quantum mechanics. *Quantum* **2024**, *8*, 1277. [CrossRef]
41. Caliceti, E.; Graffi, S. Criteria for the Reality of the Spectrum of PT-Symmetric Schrödudinger Operators. In *Non-Selfadjoint Operators in Quantum Physics: Mathematical Aspects*; Bagarello, F., Gazeau, J.-P., Szafraniec, F., Znojil, M., Eds.; Wiley: Hoboken, NJ, USA, 2015; Chapter 4, pp. 189–240.
42. Zezyulin, D.A.; Kartashov, Y.V.; Konotop, V.V. Metastable two-component solitons near an exceptional point. *Phys. Rev. A* **2021**, *104*, 023504. [CrossRef]
43. Bagchi, B.; Ghosh, R.; Sen, S. Exceptional point in a coupled Swanson system. *Europhys. Lett.* **2022**, *137*, 50004. [CrossRef]
44. Guria, C.; Zhong, Q.; Ozdemir, S.K.; Patil, Y.S.S.; El-Ganainy, R.; Harris, J.G.E. Resolving the topology of encircling multiple exceptional points. *Nat. Commun.* **2024**, *15*, 1369. [CrossRef] [PubMed]
45. Henry, R.A.; Batchelor, M.T. Exceptional points in the Baxter-Fendley free parafermion model. *Scipost Phys.* **2023**, *15*, 016. [CrossRef]
46. Siegl, P.; Krejčiřík, D. On the metric operator for the imaginary cubic oscillator. *Phys. Rev. D* **2012**, *86*, 121702. [CrossRef]
47. Berry, M.V. Physics of Nonhermitian Degeneracies. *Czech. J. Phys.* **2004**, *54*, 1039–1047. [CrossRef]
48. Bagarello, F.; Fring, A. A non selfadjoint model on a two dimensional noncommutative space with unbound metric. *Phys. Rev. A* **2013**, *88*, 042119. [CrossRef]
49. Krejčiřík, D.; Siegl, P. Elements of spectral theory without the spectral theorem. In *Non-Selfadjoint Operators in Quantum Physics: Mathematical Aspects*; Bagarello, F., Gazeau, J.-P., Szafraniec, F., Znojil, M., Eds.; Wiley: Hoboken, NJ, USA, 2015; Chapter 5, pp. 241–292.
50. Günther, U.; Stefani, F. IR-truncated PT-symmetric ix^3 model and its asymptotic spectral scaling graph. *arXiv* **2019**, arXiv:1901.08526.
51. Ramirez, R.; Reboiro, M.; Tielas, D. Exceptional Points from the Hamiltonian of a hybrid physical system: Squeezing and anti-Squeezing. *Eur. Phys. J. D* **2020**, *74*, 193. [CrossRef]
52. Brody, D.C.; Hughston, L.P. Quantum measurement of space-time events. *J. Phys. A Math. Theor.* **2021**, *54*, 235304. [CrossRef]
53. Alase, A.; Karuvade, S.; Scandolo, C.M. The operational foundations of PT-symmetric and quasi-Hermitian quantum theory. *J. Phys. A Math. Theor.* **2022**, *55*, 244003. [CrossRef]
54. Feinberg, J.; Riser, B. Pseudo-Hermitian random-matrix models: General formalism. *Nucl. Phys.* **2022**, *B 975*, 115678. [CrossRef]
55. Semorádová, I.; Siegl, P. Diverging eigenvalues in domain truncations of Schroedinger operators with complex potentials. *SIAM J. Math. Anal.* **2022**, *54*, 5064–5101. [CrossRef]
56. Znojil, M. Time-dependent version of cryptohermitian quantum theory. *Phys. Rev. D* **2008**, *78*, 085003. [CrossRef]
57. Wang, W.H.; Chen, Z.L.; Li, W. The metric operators for pseudo-Hermitian Hamiltonian. *ANZIAM J.* **2023**, *65*, 215–228. [CrossRef]
58. Znojil, M. On the role of the normalization factors κ_n and of the pseudo-metric P in crypto-Hermitian quantum models. *Symmetry Integr. Geom. Methods Appl. SIGMA* **2008**, *4*, 001. [CrossRef]
59. Ballesteros, A.; Ramírez, R.; Reboiro, M. Non-standard quantum algebras and finite dimensional PT-symmetric systems. *J. Phys. A Math. Theor.* **2024**, *57*, 035202. [CrossRef]

Disclaimer/Publisher's Note: The statements, opinions and data contained in all publications are solely those of the individual author(s) and contributor(s) and not of MDPI and/or the editor(s). MDPI and/or the editor(s) disclaim responsibility for any injury to people or property resulting from any ideas, methods, instructions or products referred to in the content.

Article

The Quantum Ratio

Kenichi Konishi [1,2,*] and Hans-Thomas Elze [1]

[1] Department of Physics "E. Fermi", University of Pisa, Largo Pontecorvo 3, Ed. C, 56127 Pisa, Italy; elze@df.unipi.it
[2] INFN, Pisa Section, Largo Pontecorvo 3, Ed. C, 56127 Pisa, Italy
[*] Correspondence: kenichi.konishi@unipi.it

Abstract: The concept of quantum ratio has emerged from recent efforts to understand how Newton's equations appear for the center of mass (CM) of an isolated macroscopic body at finite body temperatures as a first approximation of quantum mechanical equations. It is defined as $Q \equiv R_q / L_0$, where the quantum fluctuation range R_q is the spatial extension of the pure-state CM wave function, whereas L_0 stands for the body's linear size (the space support of the internal bound-state wave function). The two cases $R_q/L_0 \lesssim 1$ and $R_q/L_0 \gg 1$ roughly correspond to the body's CM behaving classically or quantum mechanically, respectively. In the present note, we elaborate on this concept and illustrate it through several examples. An important notion following from introduction of the quantum ratio is that the elementary particles (thus, the electron and the photon) are quantum mechanical even when environment-induced decoherence places them into a mixed state. Thus, decoherence and classical state should not be identified. This simple observation, further illustrated by consideration of a few atomic and molecular processes, may have significant implications for the way that quantum mechanics works in biological systems.

Keywords: quantum mechanics; classical mechanics; macroscopic bodies; quantum fluctuations

Citation: Konishi, K.; Elze, H.-T. The Quantum Ratio. *Symmetry* **2024**, *16*, 427. https://doi.org/10.3390/sym16040427

Academic Editors: Tuong Trong Truong and Sergei D. Odintsov

Received: 21 February 2024
Revised: 29 March 2024
Accepted: 1 April 2024
Published: 4 April 2024

Copyright: © 2024 by the authors. Licensee MDPI, Basel, Switzerland. This article is an open access article distributed under the terms and conditions of the Creative Commons Attribution (CC BY) license (https://creativecommons.org/licenses/by/4.0/).

1. Introduction: The Quantum Ratio

The concept of quantum ratio emerged during the efforts to understand the conditions under which the center of mass (CM) of an isolated macroscopic body possesses a unique classical trajectory. It is defined as

$$Q \equiv \frac{R_q}{L_0}, \tag{1}$$

where R_q is the quantum fluctuation range of the CM of the body under consideration and L_0 is the body's (linear) size. The criterion proposed to tell whether the body behaves quantum mechanically or classically is [1]

$$Q \gg 1, \quad \text{(quantum)}, \tag{2}$$

or

$$Q \lesssim 1, \quad \text{(classical)}, \tag{3}$$

respectively.

Let us assume that the total wave function of the body has a factorized form

$$\Psi(\mathbf{r}_1, \mathbf{r}_2, \ldots \mathbf{r}_N) = \Psi_{CM}(\mathbf{R}) \, \psi_{int}(\hat{\mathbf{r}}_1, \hat{\mathbf{r}}_2, \ldots \hat{\mathbf{r}}_{N-1}), \tag{4}$$

where Ψ_{CM} is the CM wave function, the N-body bound state is described by the internal wave function ψ_{int}, $\{\hat{\mathbf{r}}_1, \hat{\mathbf{r}}_2, \ldots \hat{\mathbf{r}}_{N-1}\}$ are the internal positions of the component atoms or molecules, $\mathbf{R}$ is the CM position, and $\mathbf{r}_i = \mathbf{R} + \hat{\mathbf{r}}_i$ ($i = 1, 2, \ldots, N$). In the case of a macroscopic body, N can be as large as $N \sim 10^{25}, 10^{50}$, etc.

1.1. The Size of the Body

L_0 is determined by ψ_{int}. A possible definition of L_0 is

$$L_0 = \text{Max}_i \, \bar{r}_i, \qquad \bar{r}_i \equiv (\langle \psi_{int} | (\hat{\mathbf{r}}_i)^2 | \psi_{int} \rangle)^{1/2}, \tag{5}$$

though the detailed definition is not important here. L_0 is the spatial support (extension) of the internal wave function describing the bound state a macroscopic body, and from some scales upwards, might well be described as a classical bound state due to gravitational or electromagnetic forces; however, their size is always well defined). It is the (linear) size of the body. Even though L_0 might somewhat depend on the body temperature T (the average internal excitation energy), it is well defined even in the $T \to 0$ limit. It represents the extension of the ground-state wave function of the bound state describing the body.

For an atom, the definition (5) correctly provides the outermost orbit in the electronic configuration. L_0 varies from 0.5 Å to hundreds of Å for atoms and molecules. The atomic nuclei (composed of protons and neutrons), which are bound more strongly by short-range nuclear forces, have smaller size L_0 on the order of Fermi $\sim 10^{-13}$ cm. For mesoscopic to macroscopic bodies, L_0 varies vastly depending on the composition, the types of the forces which bind them, and their particular molecular or crystalline structures. For the earth (the radius) $L_0 \sim 6400$ km.

An exception is the case of the elementary particles, which have $L_0 = 0$. This has a simple implication according to (2): having $Q = \infty$, the elementary particles are quantum-mechanical.

It might be argued that length scales (i.e., small or large) are relative concepts in physics; at distances much larger than L_0, any body looks pointlike. More generally, when changing the scales of distances or energies, the physics might look similar. A more rigorous formulation of this idea (scale invariance) is that of the renormalization group in relativistic quantum field theories in four dimensions, e.g., theories of the fundamental interactions [2–5], or in lower-dimensional models of critical phenomena [6].

Scale invariance holds if the system possesses no fixed length scale (nonrelativistic quantum mechanics, having only $\hbar$ with the dimension of an action as the fundamental constant in its formulation, shares this property [7]). The so-called quantum nonlocality is one of its consequences. However, in specific problems the masses and the potential explicitly break scale invariance in general. For a class of potentials, such as the delta-function or $1/r^2$ potentials in $D = 2$ space dimensions, the system possesses exact scale invariance [8]. From the point of view of the theory of fundamental interactions, the absence of a fixed length scale means that physics at low energies does not depend on the ultraviolet cutoff Λ. Thus, we need to introduce the regularization (and renormalization) of the theory because of the presence of ultraviolet divergences. In other words, the theory is of renormalizable type, that is, a quantum field theory without any a priori mass (or length) parameter.

For the questions of interest in the present work, however, it is important, and we do know, that the world we live in has definite length scales, such as Bohr's radius and the size of atomic nuclei. In other words, terms such as microscopic (from elementary particles and nuclei to atoms and molecules) and macroscopic (much larger than these) have well-defined concrete meanings.

These fixed sizes (or length scales) characterizing our world are set by the fundamental constants of nature ($\hbar$, e, c) and by the parameters in the theory of the fundamental (strong and electroweak) interactions [2–5], namely, the quark and lepton masses, W and Z masses. See Section 2.1 for more on this topic.

1.2. Quantum Range R_q

The quantum fluctuation range R_q is determined by $\Psi_{CM}(\mathbf{R})$. In principle, it is just the (spatial) extension of the pure-state wave function $\Psi_{CM}(\mathbf{R})$ describing the CM of the body.

However, it is a much more complex quantity than L_0, and depends on many factors. In quantum mechanics (QM) there is no a priori upper limit to R_q. Take for instance the wave function of a free particle, ψ; while it might be thought that the normalization condition $||\psi|| = 1$ necessarily sets a finite quantum fluctuation range, this is not the case. As is well known (Weyl's criterion), a particle can be in a state arbitrarily close to a plane-wave state,

$$\psi \propto e^{i\mathbf{p}\cdot\mathbf{r}/\hbar}, \qquad (6)$$

i.e., in a momentum eigenstate, which has $R_q = \infty$. This fact, the absence of a priori upper limit for R_q, is another consequence of the fact that QM laws contain no fundamental constant with the dimension of a length.

Given a body, R_q will in general depend on the internal structures, excitation modes, and body temperature. These cause self-induced (or thermal) decoherence due to the emission of photons which carry away information and seriously reduce R_q. If the body is not isolated, its R_q is severely affected by the action of environment-induced decoherence [9–14] upon the surrounding temperature, flux, etc. Moreover, R_q depends on the external electromagnetic fields, which may split the wave packets, as in the Stern–Gerlach setup, as well as on possible quantum-mechanical correlations (entanglement) among distant particles. R_q may depend also on time.

An important question concerns the width of the wave packet of the CM of an isolated (microscopic or macroscopic) particle, Δ_{CM}. This should not be confused with L_0. Being the spread of a single-particle wave function, Δ_{CM} is a measure of the quantum fluctuation range

$$R_q \gtrsim \Delta_{CM}, \qquad (7)$$

though R_q can be much larger than Δ_{CM} in general.

As for the relation between Δ_{CM} and L_0, Δ_{CM} corresponds to the uncertainty of the CM position of the body. For a macroscopic body, an experimentalist who is capable of measuring and determining its size L_0 with some precision will certainly be able to measure the CM position R with

$$\Delta_{CM} \lesssim L_0, \quad \text{or even with} \quad \Delta_{CM} \ll L_0. \qquad (8)$$

Nevertheless, such a relation neither holds necessarily nor is required in general.

A macroscopic body, especially at exceedingly low temperatures near $T = 0$, may well be in a state of position uncertainty (the width of the wave packet)

$$\Delta_{CM} \gg L_0. \qquad (9)$$

Using (7), such a system is seen to have $Q \gg 1$, and as such is quantum mechanical. Many attempts to realize macroscopic quantum states experimentally by bringing the system temperatures close to $T = 0$ have been made recently [15–26].

Vice versa, a well-defined CM position (8) set up at time $t = 0$ does not in itself tell whether the system will behave quantum mechanically or classically.

A free wave packet of an atom or molecule with initial position uncertainty Δ_{CM} will quickly diffuse (the diffusion rate depends on the mass) and acquire $R_q \sim \Delta_{CM} \gg L_0$ (see Table 1, taken from [1]). In the Stern–Gerlach setup, with an inhomogeneous magnetic field, the (transverse) wave packet of an atom or a molecule with spin will be split in two or more wave packets, which can become separated even by a macroscopic distance (R_q) such that $R_q \gg L_0$, $Q \gg 1$ (see Section 2.3 for more about this).

On the other hand, a macroscopic body does not diffuse (see Table 1). Its CM wave packet does not split under an inhomogeneous magnetic field either [1]. Therefore, if the CM position of a macroscopic body is measured with precision (8) at time $t = 0$, the relation

$R_q \lesssim L_0$ ($Q \lesssim 1$) is maintained in time. Such a body evolves classically with a well-defined trajectory obeying Newton's equations [1].

Table 1. Diffusion time of the free wave packet for different particles. Conventionally, we take the initial wave packet size of 1 μ = 10^{-6} m and define the diffusion time as the Δt needed for its size to double. For a macroscopic particle of 1 g, the doubling time is 10^{19} s $\sim 10^{11}$ yrs, which exceeds the age of the universe.

Particle	Mass (in g)	Diffusion Time (in s)
electron	9×10^{-28}	10^{-8}
hydrogen atom	1.6×10^{-24}	1.6×10^{-5}
C_{70} fullerene	8×10^{-22}	8×10^{-3}
a stone of 1 g	1	10^{19}

1.3. The Microscopic Degrees of Freedom Inside of a Macroscopic Body Are Quantum Mechanical

The present discussion on the quantum ratio is concerned with the question of how classical behavior for the CM of a macroscopic body emerges from QM. An important fact to be kept in mind is the following: even if a macroscopic (or a mesoscopic) body might behave classically as a whole due to environment-induced or self-induced (or thermal) decoherence and to its large mass [1], the internal microscopic degrees of freedom, electrons, atomic nuclei, and photons remain quantum mechanical (see Sections 2.1, 2.2 and 3). All sorts of quantum-mechanical processes (e.g., tunneling) continue to be active inside the body, even if various decoherence effects may be significant. These quantum phenomena constitute the essence of the physics of polymers and general macromolecules, and consequently of biology. They hold the key to the answers to many questions in biology, genetics, and neuroscience that are unanswered today (see for instance [27,28]). The consideration of the present note has nothing to add directly to these questions; however, see a few related general comments in Section 3 below.

2. The Quantum Ratio Illustrated

In this section, the quantum ratio is illustrated via several examples.

2.1. Elementary Particles

The elementary particles known today (as of the year 2024, see [29]) are the quarks, leptons (electron, muon, τ lepton), the three types of neutrinos, and the gauge bosons (the gluons, W, Z bosons, and photon), with the masses listed below (Tables 2–4).

Table 2. The quark masses in MeV/c^2; the errors are not indicated. 1 MeV/$c^2 \simeq 1.782661 \times 10^{-27}$ g.

| 2.16 (u) | 4.67 (d) | 93.4 (s) | 1.27×10^3 (c) | 4.18×10^3 (b) | 172.7×10^3 (t) |

Table 3. The lepton masses, with the e, μ, and τ masses in MeV/c^2.

| 0.51099895 (e) | 105.658 (μ) | 1776.86 (τ) | $m_\nu \neq 0$; $m_\nu < 0.8$ eV/c^2 |

Table 4. Gauge bosons and their masses.

photon	gluons	$W^{\pm}$ (GeV/c^2)	Z (GeV/c^2)
0	0	80.377 ± 0.012	91.1876 ± 0.0021

The fact that the processes involving these particles are very accurately described by the local quantum field theory $SU(3)_{QCD} \times \{SU(2)_L \times U(1)\}_{GWS}$ up to the energy range of $O(10)$ TeV means that

$$L_0 \lesssim O(10^{-18}) \text{ cm} . \tag{10}$$

In future, these elementary particles might well turn out to be made of some constituents unknown today, bound by some new forces yet to be discovered. For present-day physics, their size can be taken to be

$$L_0 = 0 \quad \therefore \quad Q = \infty \,. \tag{11}$$

Thus, the elementary particles are quantum mechanical. The virtual emission and absorption of a particle of mass m provides the physical "size" h/mc, known as the Compton length, of any quantum-mechanical particle; however, this should be distinguished from the size L_0 defined as the extension of its internal wave function.

This notion is generally taken for granted by physicists, even if no justification is (was) known as such. Here, as we are inquiring as to whether a certain "particle", be it an atom, molecule, macromolecule, a piece of crystal, etc., behaves quantum mechanically or classically, and under which conditions, it perhaps makes sense to ask whether or why an elementary particle is quantum mechanical. Introduction of the quantum-ratio criterion offers an immediate (affirmative) answer to the first question and explains the second. By definition, the elementary particles have no internal structures, and consequently no internal excitations. Thus, there is no sense in talking about their body temperature or thermal decoherence.

Note that any quantum particle, such as an electron, behaves "classically" under certain conditions (the Ehrenfest theorem), e.g., when it is well-localized, free or under homogeneous electromagnetic field, and within the diffusion time; however, this is not what we mean by a classical particle.

The observation that the elementary particles (Tables 2–4) are quantum mechanical in the light of the quantum ratio (10) and (11) might sound new; however, it is really not. Actually, it reflects the common understanding that matured around 1970 in the high-energy physics community (e.g., 't Hooft, Cargese lecture [30]) that the laws of nature at the microscopic level are expressed by a renormalizable and relativistic local gauge theory (a quantum field theory) of the elementary particles. Such a theory describes pointlike particles ($L_0 = 0$). Quantum gravity or string theory effects, possibly relevant near the Planckian energies $M_{Pl} \sim 10^{19}\, GeV$, do contain a length scale $\sim 10^{-32}$ cm, but it is beyond the scope of the present work to consider whether and how these affect the discussion of quantum or classical physics at the larger distances ($\geq 10^{-18}$ cm) that we are concerned with here.

The scale or dilatation-invariance of this type of theories is broken by the necessity of introducing an ultraviolet cutoff, a mass scale Λ_{UV}, to regularize, renormalize, and define a finite theory (quantum anomaly). Remarkably, the scale invariance is restored by the introduction of the renormalized coupling constants (defined conveniently at some reference mass scale μ) and by giving them an appropriate μ dependence (the renormalization-group equations); see, e.g., Coleman's 1971 Erice lecture [31].

The fixed length or mass scales of our world, mentioned in Section 1.1, concern the infrared fate of such dilatation invariance. These fixed scales can ultimately be traced to (i) the vacuum expectation value $\langle \phi^0 \rangle \simeq 246\, GeV$ of the Higgs scalar field in the $SU(2) \times U(1)$ Glashow–Weinberg–Salam electroweak theory, and (ii) the mass scale $\Lambda_{QCD} \simeq 150\, MeV$, dynamically generated by the strong interactions (quantum chromodynamics). These break scale-invariance. All fixed length scales, from the microscopic to the macroscopic world we live in, follow from these and from some dimensionless coupling constants in the $SU(3)_{QCD} \times \{SU(2) \times U(1)\}_{GWS}$ theory; for instance, the nuclear size is typically on the order of the pion's Compton length, $h/m_\pi c$, and $m_\pi^2 \sim m_{u,d} \Lambda_{QCD}$. The proton and neutron masses ($\sim 940\, MeV/c^2$) are mainly provided by the strong interaction effects $\sim \Lambda_{QCD}$. The Bohr radius is $\hbar^2/m_e e^2$.

2.2. Hadrons and Atomic Nuclei

Atomic nuclei, together with various hadrons (the mesons and baryons), are the smallest composite particles known today. Until around 1960, the mesons (π, K, ...) and baryons (p, n, ...) used to be part of the list of "elementary particles", together with leptons. As the theory of strong interactions (the quark model, and subsequently quantum chromodynamics, a non-Abelian $SU(3)$ gauge theory of quarks and gluons) was established around 1974–1980, they were replaced by the quarks and gluons as more fundamental constituents of nature.

The atomic nuclei are bound states of the nucleons, i.e., protons (p) and neutrons (n). They are bound by the strong interactions, and their size is on the order of

$$L_0 \sim A^{1/3} \text{ fm}, \qquad 1\,\text{fm} \equiv 10^{-13}\,\text{cm} = 10^{-5}\,, \tag{12}$$

where A is the mass number. The Coulombic wave functions in the atoms and molecules have extension (R_q) of the order of Å; thus,

$$Q = \frac{R_q}{L_0} \gtrsim O(10^5)\,. \tag{13}$$

The atomic nuclei are quantum mechanical.

To say that the atomic nuclei are quantum mechanical because of the atomic extension (13), however, is certainly too reductive. The atomic nuclei indeed may appear without being bound in atoms. For instance, the α particle is the nucleus of the helium atom, but may come out of a metastable nucleus through α decay and propagate as a free particle. It possesses a size on the order of (12), much larger than the typical size of an elementary particles (10), but for processes typically involving the distance scales much larger than 1 fm, it behaves as a pointlike particle, i.e., quantum mechanically, just as any elementary particle does. Similarly for the proton, the nucleus of the hydrogen atom, with $L_0 \sim 0.84$ fm.

2.3. Stern–Gerlach Experiment

The next smallest composite particles known in nature are the atoms. They are Coulombic composite states made of electrons moving around a positively charged atomic nuclei, almost pointlike (at the atomic scales) and $O(10^3 \sim 10^5)$ times heavier than the electron.

Let us consider the well known Stern–Gerlach process of atoms with a magnetic moment in an inhomogeneous magnetic field. To be concrete, we take as an example the very original Stern–Gerlach experiment with the silver atom [32] Ag, with mass and size

$$m_{Ag} \simeq 1.79 \times 10^{-22}\,\text{g}\,, \qquad L_0 \simeq 1.44 \times 10^{-8}\,\text{cm}\,. \tag{14}$$

With the electronic configuration and the global quantum number

$$[Kr]\,4d^{10}\,5s^1\,, \qquad {}^2S_{1/2}\,; \tag{15}$$

the magnetic moment of the atom is dominated by the spin of the outmost electron,

$$\mu = \frac{e\hbar}{2m_e c}\,g\,\mathbf{s}\,, \tag{16}$$

where g is the gyromagnetic ratio $g \simeq 2$ of the electron. [Kr] above indicates the zero angular momentum-spin ($L = S = 0$) closed shell of the Kripton electronic configuration describing the first 36 electrons.

The question is whether the silver atom, which is certainly a quantum-mechanical bound state of 47 electrons, 47 protons, and 51 neutrons and has a mass $\sim$100 times that of the hydrogen atom, behaves as a whole (i.e., its CM) as a QM particle with spin 1/2 or as a classical particle of magnetic moment μ.

The beam of Ag is sent into the region, 3.5 cm long, of an inhomogeneous magnetic field $\mathbf{B} = (0, 0, B(z))$, $dB(z)/dz \neq 0$, as it proceeds in, e.g., the $\hat{x}$ direction. The beam width, which reflects the apertures of the two slits used to prepare the well-collimated beam, is about $\sim$0.02–0.03 mm wide (the transverse wave packet size of the atom can be taken to be of this order; the silver atom, having a mass roughly 100 times that of the hydrogen atom, has a diffusion time on the order of 10^{-2} s (see Table 1), meaning that the diffusion during the travel of 3.5 cm is entirely negligible). After passing the region of the magnetic field, the image of the atoms on the glass screen shows two bands clearly separated by about 0.2 mm in the direction of $\hat{z}$. In other words, at the end of the region with the magnetic field, the atom is described by a split wave packet of the form

$$\psi = \psi_1(\mathbf{r})|\uparrow\rangle + \psi_2(\mathbf{r})|\downarrow\rangle \tag{17}$$

with the centers of the two subpackets ψ_1 and ψ_2 at $\mathbf{r} = \mathbf{r}_1$ and $\mathbf{r} = \mathbf{r}_2$, respectively, where $|z_1 - z_2| \simeq 0.2$ mm. The spatial support of the wave function ψ can be taken as about that size. It follows that

$$Q = \frac{R_q}{L_0} \gtrsim \frac{0.2\,\text{mm}}{1.4 \cdot 10^{-8}\,\text{cm}} \simeq 10^6 \gg 1, \tag{18}$$

with the Ag atom as a whole behaving as a perfectly quantum-mechanical particle.

Actually, the fact that the wave packets are divided in two by an inhomogeneous magnetic field does not necessarily mean that the system is in a pure state of the form (17); the wave function of the form (17) corresponds to a 100% polarized beam, where all incident atoms are in the same spin state

$$c_1|\uparrow\rangle + c_2|\downarrow\rangle = \begin{pmatrix} c_1 \\ c_2 \end{pmatrix}, \qquad |c_1|^2 + |c_2|^2 = 1. \tag{19}$$

If the beam is partially polarized or unpolarized, the spin state is described by a density matrix ρ. The pure state (19) corresponds to the density matrix

$$\rho^{(pure)} = \begin{pmatrix} |c_1|^2 & c_1 c_2^* \\ c_2 c_1^* & |c_2|^2 \end{pmatrix}, \tag{20}$$

whereas a general mixed state is described by a generic Hermitian 2×2 matrix ρ with

$$\text{Tr}\,\rho = 1, \qquad \rho_{ii} \geq 0, \qquad i = 1, 2. \tag{21}$$

In an unpolarized beam, $\rho = \frac{1}{2}\mathbf{1}$.

What the Stern–Gerlach experiment measures is the relative frequency with which the atom arriving at the screen happens to have spin $s_z = \frac{1}{2}$ or spin $s_z = -\frac{1}{2}$. Let

$$\Pi_\uparrow = |\uparrow\rangle\langle\uparrow| = \begin{pmatrix} 1 & 0 \\ 0 & 0 \end{pmatrix}, \qquad \Pi_\downarrow = |\downarrow\rangle\langle\downarrow| = \begin{pmatrix} 0 & 0 \\ 0 & 1 \end{pmatrix} \tag{22}$$

be the projection operators on the spin-up (down) states; according to QM, the relative intensities of the upper and lower blots on the screen are

$$\bar{\Pi}_\uparrow = \text{Tr}\,\Pi_\uparrow \rho = \rho_{11}\,; \qquad \bar{\Pi}_\downarrow = \text{Tr}\,\Pi_\downarrow \rho = \rho_{22}, \qquad \rho_{11} + \rho_{22} = 1. \tag{23}$$

The prediction about the relative intensities of the two narrow atomic image bands from the wave function (17) and from the density matrix (23) is in general indistinguishable, as is well known. In other words, what the Stern–Gerlach experiment shows is not whether the atom is in a pure state of the form (17) or in a generic (spin-) mixed state (23), but rather that the silver atom is a quantum mechanical particle.

Indeed, the prediction for a classical particle is qualitatively different. If classical, each atom would move depending on the orientation of its magnetic moment, tracing a well-defined trajectory

$$m\dot{\mathbf{r}} = \mathbf{p}, \quad \frac{dp_x}{dt} = \frac{dp_y}{dt} = 0, \quad \frac{dp_z}{dt} = F_z = -\frac{\partial}{\partial z}\boldsymbol{\mu}\cdot\mathbf{B}. \tag{24}$$

It would arrive at some generic point on the screen. If the initial orientation of the magnetic moment were random, after many classical atoms had arrived they would leave a continuous band of atomic images, rather than two narrow well-separated bands, as has been experimentally observed and as QM predicts.

Vice versa, if the orientation of the magnetic moment/spin is fixed (and the same) for all incident atoms, then classical atoms will produce only one narrow band, whereas QM atoms will leave two separate image bands. These considerations clearly tell that the concepts of mixed state and classical particle should be distinguished. We extend these discussions further in Section 3, taking into account the effects of environment-induced decoherence as well as the large spin limit and their respective relations to the classical limit (24).

2.4. Atomic and Molecular Interferometry

Many beautiful experiments exhibiting the quantum mechanical feature (wave character) of atoms and molecules have been performed (or proposed) using various types of interferometers [33–42]. One of the most powerful approaches uses Talbot–Lau interferometry [33,36–39].

The essential part of all these experiments makes use of the so-called Talbot effect [43]. In a typical setting, an atomic or molecular beam passing through the first slit is sent to the diffraction grating (G_2 in Figure 1), which consists of many slit apertures set with period d. After passing through the diffraction grating, the wave function of the atom or molecule, which originated from a point source, has the form

$$\psi(x_2) \simeq \sum_i \psi_i(x_2), \tag{25}$$

where $\psi_i(x_2)$ is the (transverse) wave packet of the atom (molecule) which has passed through the i-th slit. Just behind the diffraction grating, the distribution of the atom (molecule) has a modulation such as in Figure 2, with each peak corresponding to the position of a slit opening.

The coherent components of the wave function (25) corresponding to the different paths shown in Figure 1 interfere constructively or destructively depending on the vertical position x_3 on the imaging screen, which is set at a distance L_2 from the diffraction grating. In the so-called near-field diffraction–interference effects, the intensity modulation behind the diffraction grating (Figure 2) is reproduced (self-imaging) on the screen G_3 [33,36–39,43], where L_2 takes definite values related to the Talbot length (here, d is the period of the slits in the diffraction grating and $\lambda_{dB} = \frac{h}{p}$, is the de Broglie wavelength (with p the longitudinal momentum) of the atom or molecule),

$$L_T = \frac{d^2}{\lambda_{dB}}, \tag{26}$$

as shown in Figure 1. The details of the calculation for the sum over paths can be found in [36]. By introducing the geometrical magnification factors $M_1 \equiv (L_1 + L_2)/L_2$, $M_2 \equiv (L_1 + L_2)/L_1$, the sum over the different paths of Figure 1 is shown to provide (for instance, at $L_2 = 2M_2L_T$) the intensity pattern $|\psi(x)|^2$ behind the diffraction grating reproduced at G_3, with an enlarged period M_2d. For $L_2 = M_2L_T$, the same intensity pattern appears except shifted by a half-period; thus, $x_3 \to x_3 + M_2d/2$.

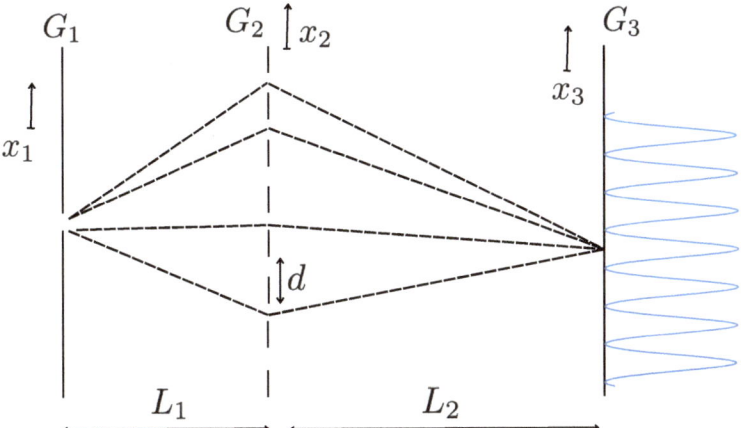

Figure 1. The Talbot effect. The intensity modulation of the molecules immediately after the passage of the diffraction grating G_2 (Figure 2) is reproduced due to the sum over paths at an imaging plane G_3 placed at definite distances L_2 related to the Talbot length (26) from G_2.

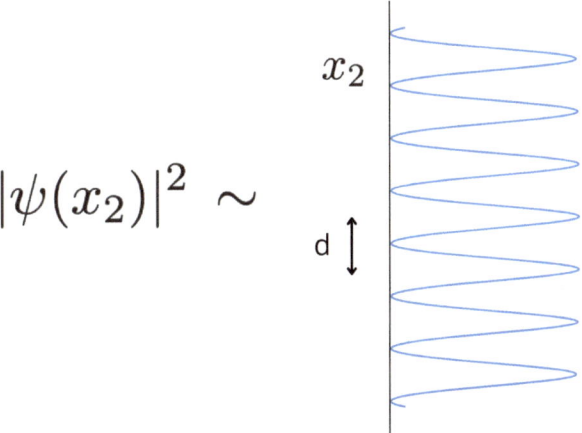

Figure 2. The intensity pattern of the atom (or molecule) behind the diffraction grating G_2. Each peak corresponds to a slit opening.

In a Talbot–Lau interferometer, which is a variation of the above, the imaging screen is replaced by a vertically movable (i.e., in the x_3 direction) transmission-scanning grating with an appropriate period d' (see Figure 3). In this way, the occurrence of the interference fringes—the Talbot self-imaging—is converted to the total transmission rate of the molecules (atoms), which varies periodically as a function of the vertical (x_3) position of the scanning grating G_3 as a whole. Another advantage of the Talbot–Lau interferometer is the possibility of introducing the incoherent source beam hitting the first grating. Even though the coherent sum over paths is relevant only for the atoms (or the molecules) which have originated from a definite source slit, the use of an incoherent source can increase the total counts after the third transmission grating, significantly improving the statistics of the experiments.

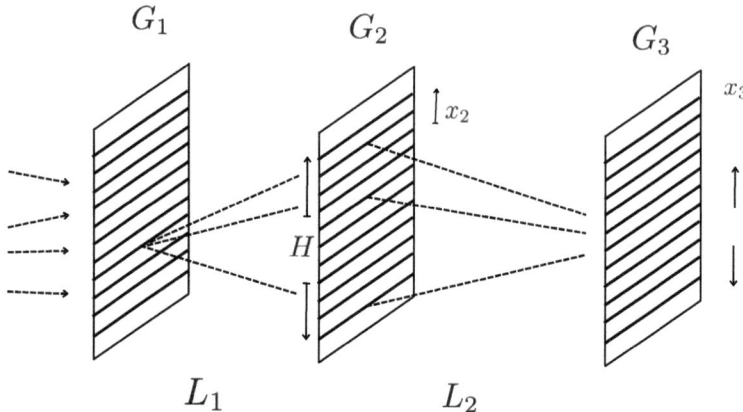

Figure 3. Talbot–Lau interferometer; G_2 is the diffraction grating, the thick lines are the slit openings, G_3 is a vertically movable transmission-scanning grating, and G_1 are the source slits.

For the purpose of discussing the quantum fluctuation range and the quantum ratio, these details of the setup are not really fundamental; we need simply to know the transverse spatial extension of the wave function (25). This in turn can be taken as the height of the diffraction grating G_2, H in Figure 3. The detection of the Talbot effect or Talbot–Lau fringe visibility is a proof that the transverse wave packets in (25) are indeed in coherent superposition, i.e., that it is a pure state. Thus, we take the quantum fluctuation range R_q in Table 5 from the experimental total height H of the diffraction grating G_2 (Figure 3).

A large quantum ratio implied by such a quantum range (along with the size) is certainly an indication that these atoms and molecules are quantum mechanical, even though such an observation does not supersede the direct evidence of quantum coherence and interference effects presented in [33,36–39].

Table 5. The size (L_0), quantum fluctuation range (R_q), and quantum ratio ($Q \equiv R_q/L_0$) of atoms and molecules in various experiments. The mass is in atomic units (au), L_0 is in Angströms (Å), and R_q is in mm. In all cases, the momentum of the atom (molecule), its masses, the size of the whole experimental apparatus, and consequently the time interval involved are all such that the quantum diffusion of the (transverse) wave packets are negligible.

Particle	Mass	L_0	R_q	Q	Exp	Miscl
Ag	108	1.44	0.2	$\sim 10^6$	[32]	Stern-Gerlach
Na	23	2.27	0.5/0.75	$\sim 10^6$	[39]	
C_{70}	840	9.4	16	$\sim 10^7$	[36–38]	$T \ll 2000$ K
C_{70}	840	9.4	~ 0.001	$\sim 10^3$	[38]	$T \geq 3000$ K

On the "Matter Wave"

A familiar concept used in articles on atomic and molecular interferometry [34–42] is the "matter wave". This might appear to summarize well the characteristic feature of quantum-mechanics, namely, "wave-particle duality". Actually, such an expression is more likely to obscure the essential quantum mechanical features of these processes rather than clarifying them. It appears to imply that beams of atoms or molecules somehow behave as a sort of wave; however, this is not quite an accurate description of the processes studied in [34–42].

The wave–particle duality of de Broglie, the core concept of quantum mechanics, refers to the property of each single quantum-mechanical particle, and not to any unspecified collective motion of particles in the beam (the "wave nature" of atoms or molecules observed in the interferometry [34–42] must be distinguished from many-body collective quantum phenomena, such as Bose–Einstein condensed ultra-cold atoms described by a macroscopic wave function). This point was demonstrated experimentally by Tonomura et al. [44] in a double-slit electron interferometry experiment à la Young, with exemplary clarity.

Exactly the same phenomena occur in the case of any atomic or molecular interferometry. For each single incident atom or molecule, it amounts to the position measurement at the third imaging screen. For each incident particle, the result for the exact final vertical position at G_3 is not known; in accordance with QM, it cannot be predicted. Only after the data with many incident particles are collected do we observe the interference effect reflecting the coherence among the components of the extended wave function (25), in accordance with QM laws.

From the data in [34–42], it is not difficult to verify that the average distance between the successive atoms (or the molecules) as compared to the size of their longitudinal wave-packet (which can be deduced from the momentum uncertainty Δp) is many orders of magnitude larger. For instance, in the case of the sodium atom experiment [39], the ratio is about 6 cm/47 Å $\sim 10^7$. In the case of C_{70} [38] this ratio seems to be even greater. The atoms or molecules do arrive one by one.

As the correlation among the atoms or molecules in the beam is negligible (as it should be) and the position of each final atom/molecule is apparently random, the resulting interference fringes, such as are manifested in Talbot (or Talbot–Lau) interferometers, is all the more surprising and interesting. What these experiments show goes much deeper into the heart of QM than the words "matter wave" or "wave–particle duality" might suggest.

3. Decoherence Versus Classicality

The atomic and molecular experiments discussed in Sections 2.3 and 2.4 are all performed in a high-quality vacuum [32–42]. This is necessary lest the scattering of the atom or molecule under study with environmental particles, e.g., air molecules, destroy their pure quantum-state nature and destroy their ability to exhibit typical quantum phenomena such as diffraction, coherent superposition, and interference. These processes are known as environment-induced decoherence [9–14]. Under environment-induced decoherence, the object under consideration becomes a mixture, and the diffraction, coherent superposition, and interference phenomena typical of pure quantum states are lost.

However, this does not necessarily mean that the system becomes classical. Being in a mixed state is necessary in order for the system to behave classically, but is in general not sufficient [1]. Unfortunately, there seems to be a widespread and inappropriate identification in the literature between the two concepts of mixed (decohered) states and classical states.

Consider the free electron; its decoherence rate/time has been studied under various types of environments [9–14]. For instance, in the 300 K atmosphere at 1 atm pressure, a free electron decoheres in 10^{-13} s [11]. When it interacts subsequently with other systems, however, it does so quantum mechanically, not as a classical particle. When it leaves the region with "environment", it emerges as a free particle, in a pure quantum state. The same can be said of the photon, as well as of any other elementary particle.

A related remark may be made about cosmic rays. The gamma rays (photons), neutrinos, protons, etc., which are produced in the hot and dense interiors of stars, once outside of stars, travel through cosmic space (a good approximation of vacuum) as free quantum particles in pure states.

In the experiment of [38], C_{70} molecules are excited by a laser beam before they enter the interferometer. When the average temperature of the molecules exceeds 3000 K, the Talbot–Lau interference fringe signals are found to disappear, showing that the molecules became mixed states, in agreement with the decoherence theory (here, decoherence is caused by the excitation of the molecules and the ensuing photon emission, so it is more appropriate to talk about thermal (or self-) decoherence [38,45] rather than environment-induced decoherence). The quantum fluctuation range R_q takes the order of the diffraction grating period d; this value is provided in Table 5. However, this does not mean that the C_{70} molecules become classical; rather, what we can conclude from [38] is that thermal decoherence places the molecule in an incoherent mixed state.

Below, we consider two more test cases in which the difference among the pure state, decohered mixed state, and classical state can be seen very clearly. These considerations can have far-reaching consequences; for instance, they may indicate a way out of the "no-go" verdict for the relevance of quantum mechanics in brain dynamics [12]. They may even be fundamental in all microscopic processes underlying biological systems (see Section 1.3) [27,28].

3.1. Stern–Gerlach Setup, Decoherence, and Classical Limit

The original Stern–Gerlach (SG) process has been discussed already (Section 2.3). What this experimental result shows is that the silver atom behaves as a quantum mechanical particle, either in a pure or (spin-) mixed state.

Here, we discuss the SG process again in more detail in three different regimes: (i) for a pure QM process; (ii) under environmental decoherence (i.e., for an incoherent mixed state); and (iii) for a classical particle. The main aim of this discussion is to highlight the differences between these different physical situations as sharply as possible.

3.1.1. Pure QM State

For definiteness, let us take an incident atom with spin $s = \frac{1}{2}$ directed in a definite but generic direction $\mathbf{n} = (\sin\theta\cos\phi, \sin\theta\sin\phi, \cos\theta)$, i.e.,

$$\Psi = \psi(\mathbf{r},t)|\mathbf{n}\rangle, \tag{27}$$

where

$$|\mathbf{n}\rangle = c_1|\uparrow\rangle + c_2|\downarrow\rangle, \quad c_1 = e^{-i\phi/2}\cos\tfrac{\theta}{2}, \quad c_2 = e^{i\phi/2}\sin\tfrac{\theta}{2}, \tag{28}$$

and $\psi(\mathbf{r},t)$ describes the wave packet of the atom moving towards the $\hat{x}$ direction before entering the region with an inhomogeneous magnetic field $\mathbf{B}$. The Hamiltonian is provided by

$$H = \frac{\mathbf{p}^2}{2m} + V, \quad V = \boldsymbol{\mu}\cdot\mathbf{B}, \tag{29}$$

$$\boldsymbol{\mu} = \frac{e\hbar}{2m_e c}g\,\mathbf{s}, \quad \mathbf{B} = (0,0,B(z)), \quad dB(z)/dz \neq 0. \tag{30}$$

and the time evolution of the system is described by the Schrödinger equation

$$i\hbar\,\partial_t \Psi = H\,\Psi, \tag{31}$$

where the total energy is conserved.

After the atom enters the inhomogeneous field $\mathbf{B}$, the upper and lower spin components of the wave function split; thus, we write

$$\Psi_\uparrow = \psi_1(\mathbf{r},t)|\uparrow\rangle + \psi_2(\mathbf{r},t)|\downarrow\rangle. \tag{32}$$

The up- and down-spin components $\psi_{1,2}$ satisfy the Schrödinger equation ($\mu_B = \frac{e\hbar}{2m_e c}$ is the Bohr magneton, and recall the well known fact that the gyromagnetic ratio $g \simeq 2$ of the electron and the spin magnitude $1/2$ approximately cancel)

$$i\hbar\frac{\partial}{\partial t}\psi_{1,2} = \left(\frac{\mathbf{p}^2}{2m} \pm \mu_B B(z)\right)\psi_{1,2}. \tag{33}$$

We assume that, prior to entering the region with the magnetic field, the wave function $\psi(\mathbf{r},t)$ is a compact wave packet (e.g., a Gaussian with width a) moving towards the $\hat{x}$ direction.

From (33) and their complex conjugates, the Ehrenfest theorems for the spin-up and spin-down components follow separately:

$$\frac{d}{dt}\langle \mathbf{r}\rangle_1 = \langle \mathbf{p}/m\rangle_1\,,\quad \frac{d}{dt}\langle \mathbf{p}\rangle_1 = -\langle \nabla(\mu_B B(z))\rangle_1\,, \tag{34}$$

$$\frac{d}{dt}\langle \mathbf{r}\rangle_2 = \langle \mathbf{p}/m\rangle_2\,,\quad \frac{d}{dt}\langle \mathbf{p}\rangle_2 = +\langle \nabla(\mu_B B(z))\rangle_2\,, \tag{35}$$

where $\langle \mathbf{r}\rangle_1 \equiv \langle \psi_1|\mathbf{r}|\psi_1\rangle$, etc. That is, for a sufficiently compact initial wave packet $\psi(\mathbf{r},t)$, the expectation values of $\mathbf{r}$ and $\mathbf{p}$ in the up and down components $\psi(\mathbf{r},t)_{1,2}$ respectively follow the classical trajectories of a spin-up or spin-down particle. At the end of the region with the magnetic field $\mathbf{B}$ (we assume that the transit time of the whole process and the mass of the atom are such that the free quantum diffusion of the wave packets is negligible; see also Appendix A) it is described as a split wave packet of form (17). Even though the two subpackets might be well-separated by a macroscopic distance $|\mathbf{r}_1-\mathbf{r}_2|$, they are still in coherent superposition. Its pure-state nature can be verified by reconverging them using a second magnetic field with the opposite gradients and studying their interferences (i.e., the quantum eraser).

A variational solution of (33) in the case of a linear potential $B(z) = b_0 z$ is provided in Appendix A. The splitting of the wavepacket is illustrated in Figure 4a.

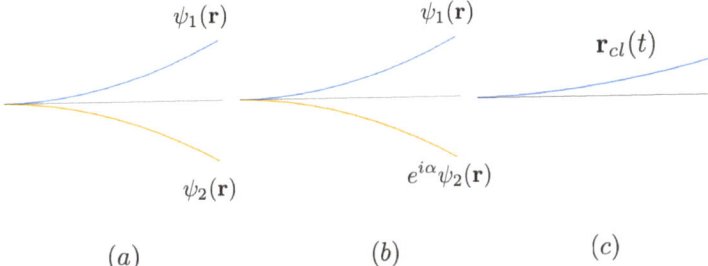

Figure 4. The spin-up and spin-down sub-wavepackets of the silver atom evolve independently under the Schrödinger equation both in vacuum ((**a**), pure state) and in a weak environment (37)–(39) (**b**), where the decoherence is represented by an unknown and to-be-averaged-over relative phase α between $\psi_1(\mathbf{r})$ and $\psi_2(\mathbf{r})$. Here, (**c**) represents a unique classical trajectory.

3.1.2. Environment-Induced Decoherence

Let us now consider the SG process (27)–(30) again, this time in a poor vacuum, e.g., in the presence of non-negligible background. The decoherence of a well-separated split wave packet such as (17) due to interactions with environmental particles has been the subject of intense study [9–14]. The upshot of the results of these investigations is that environment-induced decoherence causes the pure state (17) to become a mixed state at $t \gg 1/\Lambda$, described by a diagonal density matrix

$$\psi(\mathbf{r})\psi(\mathbf{r}')^* \to \psi_1(\mathbf{r})\psi_1(\mathbf{r}')^*|\uparrow\rangle\langle\uparrow| + \psi_2(\mathbf{r})\psi_2(\mathbf{r}')^*|\downarrow\rangle\langle\downarrow|\,,\quad |\mathbf{r}_1-\mathbf{r}_2| \gg \lambda\,, \tag{36}$$

where Λ is the decoherence rate [9–14] and λ is the de Broglie wavelength of the environmental particles. The density matrix (36) means that each atom is now *either near* $\mathbf{r}_1$ *or near* $\mathbf{r}_2$. The prediction for the SG experiment, however, is the same as in the case of the spin-mixed state (partially polarized source atoms) (23); it cannot be distinguished from the prediction $|c_1|^2 : |c_2|^2$ for the relative intensities of the two image bands in the case of the pure state.

Actually, the study of the effects of environmental particles on any given quantum process is a complex and highly nontrivial problem involving many factors, including the density and flux of the particles, the pressure, the average temperature, the kinds of particles present, the type of interactions, and more [9–14]. Thus, a simple statement such as (36) might seem an oversimplification.

Without going into the detailed features of the environment, we may nevertheless attempt to clarify the basic conditions under which a result such as (36) can be considered reliable. Following [11], we introduce the *decoherence time* $\tau_{dec} \sim 1/\Lambda_{dec}$ as a typical timescale over which decoherence takes place. In addition, the *dissipation time* τ_{diss} may be considered as the timescale in which the loss of the (energy, momentum) of the atom under study due to interactions with environmental particles become significant. Unlike [11], however, we shall not consider τ_{dyn}, the typical timescale of the internal motion of the object under study; roughly speaking, the size L_0 that we introduced in defining the quantum ratio (1), that is, the space support of the internal wave function, corresponds to this ($\tau_{dyn} \propto L_0$). The quantum–classical criteria suggested by [11] might appear to have some similarities with (2) and (3); however, the former seems to leave unanswered questions such as "what happens to a quantum particle ($\tau_{dyn} < \tau_{dec}$), at $t > \tau_{dec}$?" This is precisely the sort of question which we are trying to address here.

We also need to consider the typical *quantum diffusion time* τ_{diff}, and finally the *transition time* τ_{trans}, which is the interval of time that the atom spends between the source slit and the image screen (or at least the final reference position in the direction of motion).

First of all, we assume that the velocity of the incident atom, its mass, and the size of the whole apparatus are such that the free quantum diffusion (the spreading of the wave packets) is negligible during the process under study. Furthermore, the environment is assumed to be sufficiently weak that the effects of energy loss, momentum transfer, etc., can be neglected to a good approximation. As shown in [9–14], the loss of phase coherence is a much more rapid process than the dynamical effects affecting the motion of the particle under consideration.

In other words, we consider the time scales

$$\tau_{dec} \ll \tau_{trans} \ll \tau_{diff}, \tau_{diss} . \tag{37}$$

The first inequality tells us that the motion of the wave packets is much slower than the typical decoherence time. Considering the atom at some point, where it is described by a split wave packet of the form (17) with the centers separated by

$$|\mathbf{r}_1 - \mathbf{r}_2| \gg a , \tag{38}$$

where a is the size of the original wavepacket, we may treat such an atom as if it were at rest and first take into account the rapid decoherence processes studied in [9–14] (a sort of Born–Oppenheimer approximation). Furthermore, we can take the environment particles with a typical de Broglie wavelength λ such that

$$a \ll \lambda \ll |\mathbf{r}_1 - \mathbf{r}_2| ; \tag{39}$$

the environment particle can resolve between the split wave packets but not between the interior of each of the subpackets $\psi_1(\mathbf{r})$ or $\psi_2(\mathbf{r})$.

In conclusion, under conditions (37)–(39), each of the split wave packets proceeds just as in the pure case (no environment) reviewed in Section 3.1.1, and their average position and momentum (i.e., the expectation values) obey Newton's Equations (34) and (35). Each of the subpackets describes a quantum particle in a (position) mixed state, that is, near either $\mathbf{r}_1$ or $\mathbf{r}_2$. After leaving the region of the SG magnets, it is just a (pure-state) wave packet $\psi_1(\mathbf{r})$ or $\psi_2(\mathbf{r})$. The two possibilities no longer interfere, in contrast to the pure split wave packet studied in Section 3.1.1 (see Figure 4b).

Needless to say, if any of the conditions (37)–(39) are violated, the motion of the atom would be very different; for instance, $\tau_{diss} \ll \tau_{trans}$ would mean totally random motion for the atom. Even in such a case, however, the effects of the environment-induced decoherence/disturbance are quite distinct from the motion of a classical particle, with a unique smooth trajectory, as discussed below.

3.1.3. Classical (or Quantum?) Particle

A classical particle with the magnetic moment directed towards

$$\mathbf{n} = (\sin\theta\cos\phi, \sin\theta\sin\phi, \cos\theta) \tag{40}$$

is described by Newton's Equation (24). The way in which the unique trajectory for a classical particle emerges from quantum mechanics has been discussed in [1], where the magnetic moment is an expectation value

$$\sum_i \langle \Psi | (\hat{\mu}_i + \frac{e_i \ell_i}{2m_i c}) | \Psi \rangle = \mu \tag{41}$$

taken in the internal bound-state wave function Ψ and where μ_i and $\frac{e_i \ell_i}{2m_i c}$ denote the intrinsic magnetic moment and that due to the orbital motion of the i-th constituent atom (molecule) with $i = 1, 2, \ldots, N$. Clearly, in general, the considerations made in Sections 3.1.1 and 3.1.2 for a spin 1/2 atom with a doubly split wave packet cannot be generalized simply to (or compared with) a classical body (41) with $N \sim O(10^{23})$.

Nevertheless, logically one cannot exclude particular systems (e.g, a magnetized metal piece) with all spins directed in the same direction, for instance. Thus, one might wonder how a quantum mechanical particle of spin S behaves under an inhomogeneous magnetic field in the large spin limit, i.e., $S = \frac{N}{2}$, $N \to \infty$.

The question is *whether the conditions discussed in [1] for the emergence of classical mechanics (with a unique trajectory) for a macroscopic body (and see Section 5 below) are sufficient*, or whether some extra condition or mechanism is needed to suppress possible wide spreading of the wave function into many subpackets (see Figure 5) under an inhomogeneous magnetic field.

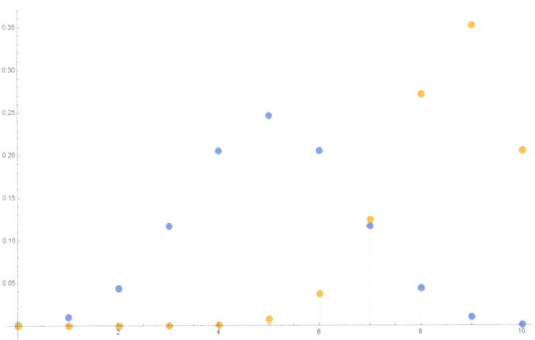

Figure 5. The distribution $|c_k|^2$ in k, i.e., in possible values of S_z, $-S \leq S_z \leq S$, for a spin $S = 5$ particle in the state (42) with $\theta = \pi/2$ (center, bue dots) and $\theta = \pi/4$ (right, orange dots).

The answer is simple but somewhat unexpected. Consider the state of a particle with spin S directed towards a direction $\mathbf{n}$:

$$(\mathbf{S} \cdot \mathbf{n}) |\mathbf{n}\rangle = S |\mathbf{n}\rangle, \qquad \mathbf{n} = (\sin\theta\cos\phi, \sin\theta\sin\phi, \cos\theta). \tag{42}$$

Assuming that the magnetic field is in the z direction (along with its gradients), we need to express $|\mathbf{n}\rangle$ as a superposition of the eigenstates of S_z,

$$|\mathbf{n}\rangle = \sum_{k=0}^{N} c_k |S_z = M\rangle, \quad M = -\frac{N}{2} + k, \quad (k = 0, 1, \ldots, N). \tag{43}$$

The expansion coefficients c_k are known (to obtain (44), consider (42) as a direct product state of N spin $\frac{1}{2}$ particles, all oriented in the same direction (28)). Collecting terms with a fixed k (the number of spin-up particles) provides (44)

$$c_k = \binom{N}{k}^{1/2} e^{-iM\phi/2} \left(\cos\frac{\theta}{2}\right)^k \left(\sin\frac{\theta}{2}\right)^{N-k}, \quad \sum_{k=0}^{N} |c_k|^2 = 1, \tag{44}$$

where $\binom{N}{k}$ are binomial coefficients, $N!/k!(N-k)!$. Using Stirling's formula, we find the distribution in various $S_z = M$ at large N and k with fixed $x = k/N$ to be

$$|c_k|^2 \simeq e^{Nf(x)}, \quad x = k/N, \tag{45}$$

with

$$f(x) = -x\log x - (1-x)\log(1-x) + 2x\log\cos\frac{\theta}{2} + 2(1-x)\log\sin\frac{\theta}{2}. \tag{46}$$

The saddle-point approximation, valid at $N \to \infty$, provides us with

$$f(x) \simeq -\frac{(x-x_0)^2}{x_0(1-x_0)}, \quad x_0 = \cos^2\frac{\theta}{2}; \tag{47}$$

thus,

$$|c_k|^2 \longrightarrow \delta(x - x_0) \tag{48}$$

in the $N \to \infty$ ($x = k/N$ fixed) limit. The narrow peak position $x = x_0$ (see (43)) corresponds to

$$S_z = M = N(x - \tfrac{1}{2}) = S(2\cos^2\tfrac{\theta}{2} - 1) = S\cos\theta. \tag{49}$$

Thus, a large-spin ($S \gg \hbar$) quantum particle with the spin directed towards $\mathbf{n}$ in a Stern–Gerlach setup with an inhomogeneous magnetic field $\mathbf{B} = (0, 0, B(z))$ moves along the single trajectory of a classical particle with $S_z = S\cos\theta$ instead of spreading over a wide range of split subpacket trajectories covering $-S \leq S_z \leq S!$ (see Figures 5 and 6).

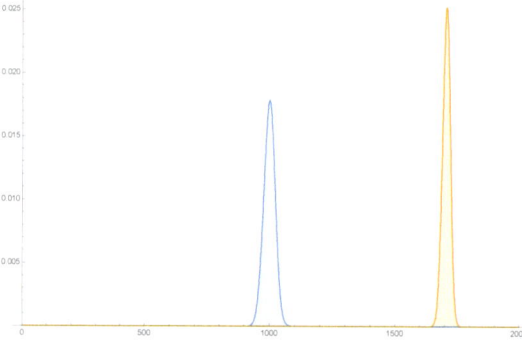

Figure 6. The distribution in possible values of S_z for a spin $S = 10^3$ particle in the state (42), with $\theta = \pi/2$ (center, blue) and $\theta = \pi/4$ (right, orange); the particle starts to appear classical.

This somewhat surprising result means that QM takes care of itself in showing that a large-spin particle ($S/\hbar \to \infty$) follows a classical trajectory which is consistent with the known general behavior of the wave function in the semi-classical limit ($\hbar \to 0$) (of course, this does not mean that the classical limit necessarily requires or implies $S \to \infty$). If the value $S \gg \hbar$ is understood as being due to the large number of spin $1/2$ particles composing it (see the previous footnote), the spikes (47) and (48) can be understood as being due to the accumulation of an enormous number of microstates giving $S_z = M$).

3.2. Tunneling Molecules

As another example, let us consider a toy version of an atom (or a molecule) of mass m moving in the z direction with momentum p_0, now with a split wave packet in the transverse (x, y) plane:

$$\Psi = e^{ip_0 z/\hbar}\psi(x,y), \qquad \psi(x,y) = c_1\psi_1(x,y) + c_2\psi_2(x,y), \tag{50}$$

where ψ_1 and ψ_2 are narrow (free) wave packets centered at $\mathbf{r}_1 = (x_1, y_1)$ and $\mathbf{r}_2 = (x_2, y_2)$, respectively. This is somewhat analogous to the wave function of the silver atom (17) or of the C_{70} molecule (25). Actually, we take a wave packet $\chi_{p_0}(p, z)$ for the longitudinal wave function by considering a linear superposition of the plane waves $e^{ipz/\hbar}$, with the momentum p narrowly distributed around $p = p_0$. For instance, a Gaussian distribution $\sim e^{-(p-p_0)^2/b^2}$ in p yields a Gaussian longitudinal wave packet in z of width $\sim 2\hbar/b$. At times much smaller than the characteristic diffusion time $t \ll \frac{2m\hbar}{b^2}$, the particle is approximately described by the wave function shown below. The exact answer has the Gaussian width in the exponent replaced as $\frac{b^2}{4\hbar^2} \to \frac{b^2}{4\hbar^2(1+ib^2 t/2m\hbar)}$ and the overall wave function multiplied by $(1+ib^2 t/2m\hbar)^{-1/2}$. These are the standard diffusion effects of a free Gaussian wave packet of width $a = 2\hbar/b$. Moreover, if the longitudinal wave packet and the transverse subwave packets are taken to be of a similar size, then the free diffusion of the transverse wave packets (and consequently the t-dependence of $\psi(x,y)$) can be neglected).

$$\Psi_{asymp} \sim e^{ip_0 z/\hbar} e^{-ip_0^2 t/2m\hbar} e^{-\frac{b^2}{4\hbar^2}(z-\frac{p_0 t}{m})^2} \psi(x,y). \tag{51}$$

Assuming that such a particle is incident from $z = -\infty$ ($t = -\infty$), moves towards right (increasing z), and hits a potential barrier (Figure 7)

$$V = \begin{cases} 0, & |z| > a, \\ V(z), & -a < z < a \end{cases} \tag{52}$$

with a height that is above the energy of the particle (approximately provided by the longitudinal kinetic energy $E \simeq \frac{p_0^2}{2m}$), as the longitudinal and transverse motions are factorized, the relative frequencies (probabilities) of finding the particle on both sides of the barrier (barrier penetration and reflection) at large t can be calculated by standard one-dimensional QM. (In [7,46], it was proposed to use the phrase "(normalized) relative frequency" instead of the word "probability", as the traditional probabilistic Born rule places human intervention as the central element of its formulation, and distorts the way quantum mechanical laws (the laws of nature!) look. To the authors' view, this has been the origin of innumerable puzzles, apparent contradictions, and conundrums entertained in the past; see [7,46] for a new perspective and a more natural understanding of the QM laws).

The answer is well known; the tunneling frequency is provided, in the semi-classical approximation, by

$$P_{tunnel} = |c|^2, \qquad c \sim e^{-\int_{-a_0}^{a_0} dz \sqrt{2m(V(z)-E)}/\hbar}, \tag{53}$$

$(V(z) - E > 0, -a_0 < z < a_0)$. The particle on the right of the barrier is described by the wave function

$$\Psi_{penetrated} \simeq c \Psi_{asymp} = c\, e^{i p_0 z/\hbar} e^{-i p_0^2 t/2m\hbar} e^{-\frac{b^2}{4\hbar^2}(z - \frac{p_0 t}{m})^2} \psi(x,y)\,, \tag{54}$$

where c is the transmission coefficient of (53). The transverse coherent superposition of the two components (sub-wavepackets) (50) remains intact (see Figure 7).

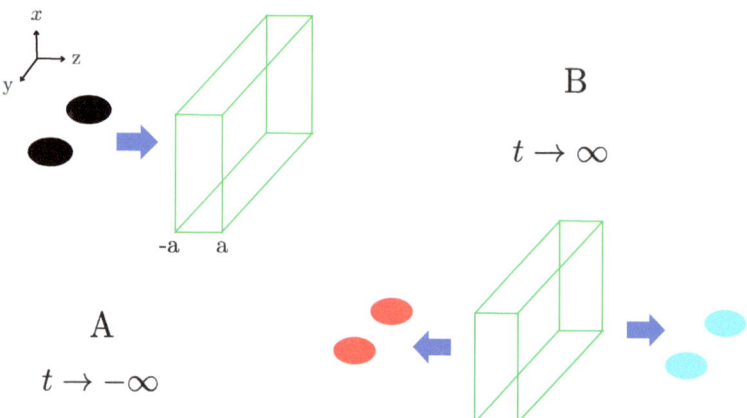

Figure 7. In the left-hand figure (**A**), an atom (molecule) arrives from $z = -\infty$ and moves towards the potential barrier $V(z)$ at $-a < z < a$ (independent of x and y). It is described by a wave packet (split in the transverse direction as in (50)). The wave function of the particle at $t \to \infty$, shown on the right-hand side (**B**), contains both the reflected and transmitted waves. The coherent superposition of the two sub-wavepackets in the (xy) plane remains intact.

Now, we can reconsider the whole process with the region left of the barrier ($z < -a$) immersed in air (or, as in the C_{70} experiments [38], the incident molecules can be bombarded by laser beams, become excited, and emit photons before they reach the potential barrier). While the precise decoherence rate depends on several parameters, in general the incident particles are decohered in a very short time, as in (36) [9–14]. The particle to the left of the barrier (we assume that the environmental particles (air molecules) have energies much less than the barrier height, and remain confined in the region left of the barrier) is now a mixture; each atom (molecule) is either near $\mathbf{r}_1 = (x_1, y_1)$ or $\mathbf{r}_2 = (x_2, y_2)$ in the transverse plane. When the particle hits the potential barrier, it tunnels through it with relative frequencies (53) and emerges on the other side of the barrier as a free particle. It has the wave function (54), with $\psi(x,y)$ replaced by $\psi_1(x,y)$ with relative frequency $|c_1|^2/(|c_1|^2 + |c_2|^2)$ and by $\psi_2(x,y)$ with frequency $|c_2|^2/(|c_1|^2 + |c_2|^2)$. While this is a statistical mixture, each part is a pure quantum mechanical particle. See Figure 8.

Our discussion above assumes that the air molecules (the environmental particles) are just energetic enough (i.e., their de Broglie wave length is small enough) to resolve the transverse split wave packets (see (36)) while at the same time being much less energetic than the longitudinal kinetic energy $\frac{p_0^2}{2m}$ and with sufficiently small flux. In writing (54), we have assumed that the effects of the environmental particles on the longitudinal wave packet are small, even though the tunneling frequency may be somewhat modified, as it is very sensitive to its energy.

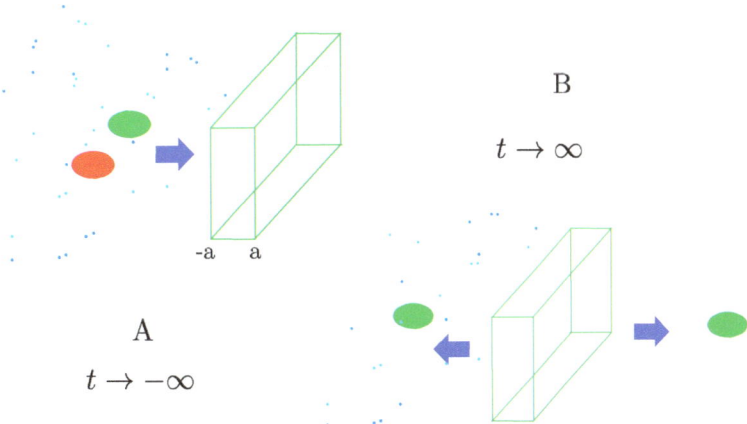

Figure 8. In the left-hand figure (**A**), an atom (molecule) arrives from $z = -\infty$ and moves towards the potential barrier at $-a < z < a$, as in Figure 7. In contrast to the process in vacuum shown in Figure 7, however, this time the half-space on the left of the potential barrier contains air. The molecule is now in a mixed state due to the environment-induced decoherence. Its (transverse) position density matrix becomes diagonal; it is either near (x_1, y_1) or near (x_2, y_2). The wave function of the particle at $t \to \infty$, shown in the right part (**B**), still contains a small transmitted wave as well as the reflected wave, however, without coherent superposition of the two transverse wave packets.

Obviously, in a much warmer and denser environment the effects of scattering on our molecule would be more severe, and the tunneling rate would become considerably smaller. Even then, our atom (or molecule) would remain quantum mechanical (the situation is reminiscent of the α particle track in a Wilson chamber; α is scattered by atoms, ionizing them on the way, but traces a roughly straight trajectory. When it arrives at the end of the chamber, it is the same α particle, and has not become classical).

4. The Abstract Concept of a "Particle of Mass *m*"

It is customary to consider an otherwise unspecified "particle of mass *m*" in order to discuss model systems in both quantum mechanics and classical mechanics. We will see that considerations based on such an abstract concept of a particle cannot be used to discriminate classical objects from quantum mechanical systems or as a way to explain the emergence of classical mechanics from QM.

Let us consider a $1D$ particle of mass m moving in a harmonic-oscillator potential

$$H = \frac{p^2}{2m} + \frac{m\omega^2}{2}x^2 .\tag{55}$$

The coherent state is defined by $a\,|\beta\rangle = \beta\,|\beta\rangle$, where a is the annihilation operator. Its well known solution in the coordinate representation is just the Gaussian wave packet

$$x_0 = \sqrt{\frac{\hbar}{2m\omega}}(\beta + \beta^*) \equiv A\cos\varphi, \quad p_0 = i\sqrt{\frac{\hbar m\omega}{2}}(\beta^* - \beta) = m\omega A\sin\varphi, \tag{56}$$

$$\psi(x) = \langle x|\beta\rangle = \mathcal{N}\exp\left[-\frac{(x-x_0)^2}{4D} + i\frac{p_0 x}{\hbar}\right], \tag{57}$$

with

$$D = \langle(\Delta x)^2\rangle = \frac{\hbar}{2m\omega} . \tag{58}$$

The Schrödinger time evolution can be expressed as the time variation of the center of mass and its mean momentum, $x_0 \to x_0(t)$, $p_0 \to p_0(t)$,

$$x_0(t) = A \cos(\varphi + \omega t) \qquad p_0(t) = m\omega A \sin(\varphi + \omega t), \qquad (59)$$

while the wave packet shape and size (58) remain unchanged in time. This looks exactly like the motion of a classical oscillator of mass m and size D!

It is sometimes thought that such a behavior on the part of the coherent states carries the key to understanding the emergence of classical mechanics from QM. However, there are reasons to believe that this may not be quite the correct way of reasoning.

In order to see that such an identification/analogy cannot be pushed too far, consider quenching, i.e., suddenly turning off the oscillator potential, setting $\omega = 0$ at $t = t_0$. The particle starts moving freely, with the initial condition $(x_0(t_0), p_0(t_0))$.

The problem is that there is no way to tell what happens at $t \geq t_0$. A quantum mechanical particle would diffuse with a rate depending on its mass, as in Table 1. A classical particle does not diffuse. The expression "a particle of mass m" does not tell us whether it is a quantum or a classical particle, or what the true size of the body L_0 (unrelated to D) is.

Note that by describing this body as a "particle" it is tacitly assumed that its physical size is irrelevant (i.e., $L_0 = 0$) to the modeled harmonic oscillator problem. However, the physical size L_0 of the particle does matter. If its (unspecified) size were truly zero it would be quantum mechanical, as $Q = R_q/L_0 = \infty$.

The lesson to be drawn from this discussion is that a model system based on an abstract "particle" concept in which the information about L_0 is lacking cannot be used to study the emergence of classical physics from quantum mechanics. Allowing for decoherence effects and selecting a particular class of mixed states as privileged ones by introducing some criteria may not lead to a satisfactory understanding of how classical physics emerges from QM.

5. Discussion

An immediate implication of the introduction of the quantum ratio concept is that elementary particles are quantum mechanical. This is the case even if under certain conditions, such as environment-induced decoherence, they may be reduced to mixed states. They remain quantum mechanical. The distinction between the concept of mixed (quantum) states and classical states is essential. As the electron and photon are elementary particles, they remain quantum mechanical even in the warm and dense environments of biological systems.

We have studied the quantum ratio of some larger particles (atoms and molecules) via examination of various interferometry experiments, which indeed show that these particles behave quantum mechanically in vacuum.

In Section 3, we have provided an extensive discussion of several real and model examples involving atoms, molecules, and elementary particles in order to highlight the reasons that environment-induced decoherence [9–14] in itself does not make the affected particle classical, as is often stated or tacitly assumed.

Though in a slightly different context, the so-called negative-result experiments or null measurements [47,48] tell a similar story. There, the exclusion of some of the possible experimental outcomes (a non-measurement) by use of an intentionally biased measurement setup, implies the loss of the original superposition of states. However, the predicted state of the system remains a perfectly quantum mechanical one, even though it now exists in a more restricted region of Hilbert space; see [49] for a recent review and careful re-examination of the interpretation of these negative-result experiments.

All of these discussions naturally lead us back to the recurrent theme in quantum mechanics, namely, mixed states versus pure quantum states. As is widely acknowledged, there are no differences of principle. As famously noted by Schrödinger, complete knowledge of the total closed system Σ (its wave function, the pure state vector) does not necessarily mean the same for a part of the system ($A \subset \Sigma$).

Only in exceptional situations in which the interactions *and correlations* between the subsystem of interest ("local", A) and the rest of the world ("rest", Σ/A) can be neglected, and as such where the total wave function has a factorized form

$$\Psi^\Sigma \simeq \psi^A \otimes \Phi^{\Sigma/A}, \qquad (60)$$

can we describe the local system A in terms of a wave function. Whenever the factorization (60) fails, system A is a mixture described by a density matrix.

Quantum measurement is a process in which the factorized state (60), where ψ^A is the quantum state of interest and the measurement device Φ^0 is part of $\Phi^{\Sigma/A}$, is brought into an entangled state, triggered by a spacetime pointlike interaction event [7,46,49].

Even a macroscopic system can be brought to the pure-state form ψ^A, as in (60), at sufficiently low temperatures. At $T = 0$, any system is in its quantum-mechanical ground state; see [15–26] for efforts to realize such macroscopic quantum states experimentally at very low temperatures.

Vice versa, the classical equations of motion describe the CM of a macroscopic body *at finite temperatures*. When the following three conditions are met,

(i) for macroscopic motions (i.e., $\hbar \simeq 0$), the Heisenberg relation does not limit the simultaneous determination (i.e., the initial condition) for the position and momentum

(ii) there is a lack of quantum diffusion due to large mass (i.e., a large number of atoms and molecules composing the body)

(iii) there is a finite body temperature, implying thermal decoherence and the mixed-state nature of the body

then the CM of a body has a unique trajectory [1]. Newton's equations for it then follow from the Ehrenfest theorem. If the quantum fluctuation range R_q is not larger than the size of the body, i.e., if $Q = R_q/L_0 \lesssim 1$, then such a trajectory can be regarded as the classical trajectory of that body.

To summarize, introduction of the quantum ratio is an attempt to move beyond the familiar ideas on the emergence of classical physics from QM, such as large action, the semiclassical limit (or $\hbar \to 0$), and Bohr's correspondence principle (or environment-induced decoherence) [9–14]. Clearly, there is no sharp boundary between where and when QM or classical mechanics respectively describe a given system more appropriately. The quantum ratio is a proposal for an approximate but simple universal criterion for characterizing the two kinds of physical systems: quantum ($Q \gg 1$) and classical ($Q \lesssim 1$).

Author Contributions: Conceptualization, K.K. and H.-T.E.; methodology, K.K.; validation, H.-T.E.; formal analysis, K.K.; original draft preparation, K.K.; review and editing, K.K. and H.-T.E. All authors have read and agreed to the published version of the manuscript.

Funding: The work by K.K. is supported by the INFN special initiative project grant GAST (Gauge and String Theories).

Acknowledgments: We thank Francesco Cappuzzello, Giovanni Casini, Marco Matone, Pietro Menotti, and Arkady Vainshtein for discussions.

Conflicts of Interest: The authors declare no conflicts of interest.

Appendix A. Variational Solution for SG Wavepackets

The Schrödinger Equation (33) can be solved by separation of the variables

$$\Psi(\mathbf{r}, t) = \chi(x, t) \eta(y, t) \psi(z, t). \qquad (A1)$$

As the motions in the x and y directions are free ones, we focus on $\psi(z,t)$.

We recall Dirac's variational principle [50]. Consider the effective action

$$\Gamma[\psi] = \int dt \, \langle \psi(t) | (i\partial_t - \hat{H}) | \psi(t) \rangle . \tag{A2}$$

Then, the variation with respect to $|\psi\rangle$ and $\langle\psi|$

$$\frac{\delta \Gamma[\psi]}{\delta \psi} = 0 \quad \text{for all } \psi \text{ with } \langle \psi | \psi \rangle = 1 , \tag{A3}$$

i.e., requiring that the effective action Γ be stationary against arbitrary variations of the normalized wave function, which vanish at $t \to \pm\infty$, is equivalent to the exact Schrödinger equation (this has been applied in a study of semiquantum chaos in a double-well oscillator in [51], but can be used in quantum field theory as well with suitable wave functionals; see [50,52] for example).

An important property which follows immediately is that orthogonal superpositions of variational trial eigenstates of the Hamiltonian evolve independently without interfering with each other. Let $|\psi\rangle = |\psi_1\rangle|\uparrow\rangle + |\psi_2\rangle|\downarrow\rangle$ be the sum of the two orthogonal spin-up and spin-down eigenstates of

$$\hat{H} = \frac{1}{2m}\hat{p}_z^2 + \mu\, b_0\, z\sigma_z \tag{A4}$$

in the Stern–Gerlach setup (with $\mathbf{B} = (0,0,b_0 z)$). Then, the effective action becomes a sum of two independent terms, $\int \langle \psi_1 | \ldots | \psi_1 \rangle$ and $\int \langle \psi_2 | \ldots | \psi_2 \rangle$, which can be varied separately.

We choose the following normalized Gaussian trial wave functions, which are suitable for the effectively one-dimensional problem of particles with mass m and with the magnetic moment μ moving in a magnetic field $\propto b_0 z$ transverse to the beam direction (i.e., $\hat{x}$):

$$\psi(z,t) = (2\pi G(t))^{-\frac{1}{4}} \exp\left\{ -\left(\tfrac{1}{4G(t)} - i\sigma(t)\right)(z - \bar{z}(t))^2 + i\bar{p}(t)(z - \bar{z}(t)) \right\} \tag{A5}$$

where $G(t), \sigma(t), \bar{p}(t), \bar{z}(t)$ are the variational-parametric functions, $\bar{p}(t), \bar{z}(t)$ describe the momentum and position of the wave packet, and $G(t), \sigma(t)$ describe the quantum diffusion. Substituting this into (A2) yields

$$\Gamma[\psi] = \int dt \left\{ \bar{p}\dot{\bar{z}} - \frac{1}{2m}\bar{p}^2 \mp \mu b_0 \bar{z} + \hbar\left[\sigma\dot{G} - \frac{2}{m}\sigma^2 G - \frac{1}{8m}G^{-1}\right] \right\} . \tag{A6}$$

Independent variations with respect to $G(t), \sigma(t), \bar{p}(t), \bar{z}(t)$ give

$$\dot{\bar{z}} = \frac{1}{m}\bar{p} , \quad \dot{\bar{p}} = \mp\mu b_0 , \tag{A7}$$

$$\dot{G} = \frac{4}{m}\sigma G , \quad \dot{\sigma} = -\frac{2}{m}\sigma^2 + \frac{1}{8m}G^{-2} . \tag{A8}$$

Note that, in the magnetic field of linear inhomogenuity $\mathbf{B} = (0,0,b_0 z)$ under consideration here, the center of the wave packet $\bar{p}(t), \bar{z}(t)$ moves as a classical particle and the diffusion effects $G(t), \sigma(t)$ are the same as for a free wave packet.

The solution of Equations (A7) and (A8) is

$$\bar{z}(t) = \frac{1}{m}(\mp\tfrac{1}{2}\mu b_0 t^2 + \bar{p}_0 t) + \bar{z}_0 ; \quad \bar{p}(t) = \mp\mu b_0 t + \bar{p}_0 ,$$

$$G(t) = \frac{i}{m}t + G_0 , \quad \sigma(t) = \frac{i}{4}(\tfrac{i}{m}t + G_0)^{-1} , \tag{A9}$$

where $\bar{x}_0, \bar{p}_0,$ and G_0 set the initial conditions at $t = 0$. The diffusion of the wave packet $\tfrac{1}{4}G^{-1} - i\sigma = \tfrac{1}{2}(\tfrac{i}{m}t + G_0)^{-1}$ is the same as in the free case, as noted already. We understand this as being due to the fact that, in the linear field $B_z(z) = b_0 z$, the force is constant and the same for each part inside the wave packets $\psi_{1,2}(z)$. The effect of quantum diffusion is negligible for $t \ll mG_0$, where G_0 is the initial wave packet size.

Substituting (A9) into (A5) yields our variational solution of the Schrödinger equation. The wave packets for spin-up and spin-down states remain in coherent superposition but move independently (see Figure 4).

References

1. Konishi, K. Newton's equations from quantum mechanics for a macroscopic body in the vacuum. *Int. Journ. Mod. Phys. A* **2023**, *38*, 2350080. [CrossRef]
2. Weinberg, S. A model of Leptons. *Phys. Rev. Lett.* **1967**, *19*, 1264. [CrossRef]
3. Salam, A. Weak and electromagnetic interactions. In *Elementary Particle Theory*; Svartholm, N., Ed.; Almqvist Forlag AB: Stockholm, Sweden, 1968; p. 367.
4. Glashow, S.L.; Iliopoulos, J.; Maiani, L. Weak Interactions with Lepton-Hadron Symmetry. *Phys. Rev. D* **1970**, *2*, 1285. [CrossRef]
5. Fritzsch, H.; Gell-Mann, M.; Leutwyler, H. Advantages of the color octet gluon picture. *Phys. Lett.* **1973**, *47*, 365. [CrossRef]
6. Wilson, K.G. The Renormalization Group and Critical Phenomena. *Rev. Mod. Phys.* **1983**, *55*, 583. [CrossRef]
7. Konishi, K. Quantum fluctuations, particles and entanglement: A discussion towards the solution of the quantum measurement problems. *Int. J. Mod. Phys. A* **2022**, *37*, 2250113. [CrossRef]
8. Jackiw, R. *Delta-Function Potentials in Two- and Three-Dimensional Quantum Mechanics*; Bég Memorial Volume; Ali, A., Hoodbhoy, P., Eds.; World Scientific: Singapore, 1991.
9. Joos, E.; Zeh, H.D. The emergence of classical properties through interaction with the environment. *Z. Phys. B* **1985**, *59*, 223–243. [CrossRef]
10. Zurek, W.H. Decoherence and the Transition from Quantum to Classical. *Phys. Today* **1991**, *44*, 36. . [CrossRef]
11. Tegmark, M. Apparent wave function collapse caused by scattering. *Found. Phys. Lett.* **1993**, *6*, 571. . [CrossRef]
12. Tegmark, M. Importance of quantum decoherence in brain processes. *Phys. Rev. E* **2000**, *61*, 4194. [CrossRef]
13. Joos, E.; Zeh, H.D.; Kiefer, C.; Giulini, D.; Kupsch, J.; Stamatescu, I.O. *Decoherence and the Appearance of a Classical World in Quantum Theory*; Springer: Berlin/Heidelberg, Germany, 2002.
14. Zurek, W.H. Decoherence, einselection, and the quantum origins of the classical. *Rev. Mod. Phys.* **2003**, *75*, 715–775. . [CrossRef]
15. Leggett, A.J. Macroscopic Quantum Systems and the Quantum Theory of Measurement. *Suppl. Prog. Theor. Phys.* **1980**, *69*, 80. [CrossRef]
16. Courty, J.-M.; Heidmann, A.; Pinard, M. Quantum limits of cold damping with optomechanical coupling. *Eur. Phys. J. D* **2001**, *17*, 399–408. [CrossRef]
17. Armour, A.D.; Blencowe, M.P.; Schwab, K.C. Entanglement and decoherence of a Micromechanical Resonator via Coupling to a Cooper-Pair Box. *Phys. Rev. Lett.* **2002**, *88*, 148301. [CrossRef]
18. Knobel, R.G.; Cleland, A.N. Nanometer-scale displacement sensing using a single electron transistor. *Nature* **2003**, *424*, 17. [CrossRef] [PubMed]
19. LaHaye, M.D.; Buu, O.; Camarota, B.; Schwab, K.C. Approaching the Quantum Limit of a Nanomechanical Resonator. *Science* **2004**, *304*, 74–77. [CrossRef] [PubMed]
20. Cleland, A.N.; Geller, M.R. Superconducting Qubit Storage and Entanglement with Nanomechanical Resonators. *Phys. Rev. Lett.* **2004**, *93*, 070501. [CrossRef]
21. Martin, I.; Shnirman, A.; Tian, L.; Zoller, P. Ground-state cooling of mechanical resonators. *Phys. Rev. B* **2004**, *69*, 125339. [CrossRef]
22. Kleckner, D.; Bouwmeester, D. Sub-kelvin optical cooling of a micromechanical resonator. *Nature* **2006**, *444*, 2. [CrossRef]
23. Regal, C.A.; Teufel, J.D.; Lehnert, K.W. *Measuring Nanomechanical Motion with a Microwave Cavity Interferometer*; Macmillan Publishers Limited: New York, NY, USA, 2008. . [CrossRef]
24. Schliesser, A.; Rivière, R.; Anetsberger, G.; Arcizetandt, O.; Kippenberg, J. *Resolved-Sideband Cooling of a Micromechanical Oscillator*; Macmillan Publishers Limited: New York, NY, USA, 2008. . [CrossRef]
25. Abbott, B.; Abbott, R.; Adhikari, R.; Ajith, P.; Allen, B.; Allen, G.; Amin, R.; Anderson, S.B.; Hanna, C.; LIGO Scientific; et al. Observation of a kilogram-scale oscillator near its quantum ground state. *New J. Phys.* **2009**, *11*, 073032. . [CrossRef]
26. O'Connell, A.D.; Hofheinz, M.; Ansmann, M.; Bialczak, R.C.; Lenander, M.; Lucero, E.; Neeley, M.; Sank, D.; Wang, H.; Cleland, A.N.; et al. Quantum ground state and single-photon control of a mechanical resonator. *Nature* **2010**, *464*, 697. [CrossRef] [PubMed]
27. Kim, Y.; Bertagna, F.; D'souza, E.M.; Heyes, D.J.; Johannissen, L.O.; Nery, E.T.; Pantelias, A.; Sanchez-Pedreno Jimenez, A.; Slocombe, L.; McFadden, J.; et al. Quantum Biology: An Update and Perspective. *Quantum Rep.* **2021**, *3*, 80–126. [CrossRef]
28. Pietra, G.D.; Vedral, V.; Marletto, C. Temporal witnesses of non-classicality in a macroscopic biological system. *arXiv* **2023**, arXiv:2306.12799v1.
29. Particle Data Group; Workman, R.L.; Burkert, V.D.; Crede, V.; Klempt, E.; Thoma, U.; Tiator, L.; Agashe, K.; Aielli, G.; Allanach, B.C. Review of Particle Physics. *Prog. Theor. Exp. Phys.* **2022**, *2022*, 083C01. 2023 update.

30. Hooft, G.'T. Why Do We Need Local Gauge Invariance in Theories With Vector Particles? An Introduction. *NATO Sci. Ser. B* **1980**, *59*, 101–115.
31. Coleman, S. Dilatations (1971). In *Aspect of Symmetry—Selected Erice Lectures*; Cambridge University Press: Cambridge, UK, 1985.
32. Gerlach, W.; Stern, O. Der experimentelle Nachweis der Richtungsquantelung im Magnetfeld. *Z. Phys.* **1922**, *9*, 349. [CrossRef]
33. Keith, D.W.; Ekstrom, C.R.; Turchette, Q.A.; Prichard, D.E. An interferometer for Atoms. *Phys. Rev. Lett.* **1991**, *66*, 2693. [CrossRef]
34. Brand, C.; Troyer, S.; Knobloch, C.; Cheshinovsky, O.; Arndt, M.A. Single, double and triple-slit diffraction of molecular matter waves. *arXiv* **2021**, arXiv:2108.06565v2.
35. Arndt, M.; Nairz, O.; Vos-Andreae, J.; Keller, C.; Zouw, G.v.; Zeilinger, A. Wave-particle duality of C_{60} molecules. *Nature* **1999**, *401*, 680. [CrossRef]
36. Brezger, B.; Arndt, M.; Zeilinger, A. Concepts for near-field interferometers with large molecules. *J. Opt. B Quant. Semiclass. Opt.* **2003**, *5*, S82–S89. [CrossRef]
37. Brezger, B.; Hackermüller, L.; Uttenthaler, S.; Petschinka, J.; Arndt, M.; Zeilinger, A. Matter-Wave Interferometer for Large Molecules. *Phys. Rev. Lett.* **2002**, *88*, 100404. [CrossRef]
38. Hackermüller, L.; Hornberger, K.; Brezger, B.; Zeilinger, A.; Arndt, M. Decoherence of matter waves by thermal emission of radiation. *Nature* **2004**, *427*, 711. [CrossRef]
39. Chapman, M.S.; Ekstrom, C.R.; Hammond, T.D.; Schmiedmayer, J.; Tannian, B.E.; Wehinger, S.; Pritchard, D.E. Near-field imaging of atom diffraction gratings: The atomic Talbot effect. *Phys. Rev. A* **1995**, *51*, R14–R17. [CrossRef]
40. Nowak, S.; Kurtsiefer, C.; Pfau, T.; David, C. High-order Talbot fringes for atomic matter waves. *Opt. Lett.* **1997**, *22*, 1430. [CrossRef]
41. Clauser, J.F.; Li, S. Talbot-vonLau atom interferometry with cold slow potassium. *Phys. Rev. A* **1994**, *49*, R2213. [CrossRef]
42. Bateman, J.; Nimmrichter, S.; Hornberger, K.; Ulbricht, H. Near-field interferometry of a free-falling nanoparticle from a point-like source. *Nat. Commun.* **2014**, *5*, 4788. [CrossRef]
43. Talbot, H.F. LXXVI. Facts relating to optical science. No. IV. *Lond. Edinb. Dublin Philos. Mag. J. Sci.* **1836**, *9*, 401–407. [CrossRef]
44. Tonomura, A.; Endo, J.; Matsuda, T.; Kawasaki, T.; Ezawa, H. Dimonstration of single-electron buildup of interference pattern. *Am. J. Phys.* **1989**, *57*, 117. [CrossRef]
45. Hansen, K.; Campbell, E.E.B. Thermal radiation from small particles. *Phys. Rev. E* **1998**, *58*, 5477. [CrossRef]
46. Konishi, K. Quantum fluctuations, particles and entanglement: Solving the quantum measurement problems. *J. Phys. Conf. Ser.* **2023**, *2533*, 012009. [CrossRef]
47. Renninger, M. Messungen ohne Störung des Meßobjekts. *Z. Phys.* **1960**, *158*, 417. [CrossRef]
48. Elitzur, A.C.; Vaidman, L. Quantum mechanical interaction-free measurements. *Found. Phys.* **1993**, *23*, 987. [CrossRef]
49. Konishi, K. On the negative-result experiments in quantum mechanics. *arXiv* **2023**, arXiv:2310.01955.
50. Jackiw, R.; Kerman, A. Time-dependent variational principle and the effective action. *Phys. Lett.* **1979**, *71A*, 158. [CrossRef]
51. Blum, T.; Elze, H.T. Semiquantum Chaos in the Double-Well. *Phys. Rev.* **1996**, *E 53*, 3123. [CrossRef]
52. Elze, H.T. Quantum Decoherence, Entropy and Thermalization in Strong Interactions at High Energy. *Nucl. Phys.* **1995**, *B 436*, 213. [CrossRef]

Disclaimer/Publisher's Note: The statements, opinions and data contained in all publications are solely those of the individual author(s) and contributor(s) and not of MDPI and/or the editor(s). MDPI and/or the editor(s) disclaim responsibility for any injury to people or property resulting from any ideas, methods, instructions or products referred to in the content.

Article

A Charged Particle with Anisotropic Mass in a Perpendicular Magnetic Field–Landau Gauge

Orion Ciftja

Department of Physics, Prairie View A&M University, Prairie View, TX 77446, USA; ogciftja@pvamu.edu

Abstract: The loss of any symmetry in a system leads to quantum problems that are typically very difficult to solve. Such a situation arises for particles with anisotropic mass, like electrons in various semiconductor host materials, where it is known that they may have an anisotropic effective mass. In this work, we consider the quantum problem of a spinless charged particle with anisotropic mass in two dimensions and study the resulting energy and eigenstate spectrum in a uniform constant perpendicular magnetic field when a Landau gauge is adopted. The exact analytic solution to the problem is obtained for arbitrary values of the anisotropic mass using a mathematical technique that relies on the scaling of the original coordinates. The characteristic features of the energy spectrum and corresponding eigenstate wave functions are analyzed. The results of this study are expected to be of interest to quantum Hall effect theory.

Keywords: charged particle; magnetic field; anisotropic mass; Landau states

Citation: Ciftja, O. A Charged Particle with Anisotropic Mass in a Perpendicular Magnetic Field–Landau Gauge. *Symmetry* **2024**, *16*, 414. https://doi.org/10.3390/sym16040414

Academic Editors: Sergei D. Odintsov and Tuong Trong Truong

Received: 3 January 2024
Revised: 13 March 2024
Accepted: 14 March 2024
Published: 2 April 2024

Copyright: © 2024 by the author. Licensee MDPI, Basel, Switzerland. This article is an open access article distributed under the terms and conditions of the Creative Commons Attribution (CC BY) license (https://creativecommons.org/licenses/by/4.0/).

1. Introduction

Great advances in the fields of nanotechnology and low-dimensional systems have enabled the precise, controlled fabrication of materials at atomic and molecular scales. The electron's quantum mechanical nature is very pronounced in this regime. Therefore, there is a great potential payoff that electronic devices built on nanoscale may manifest many desirable quantum properties. This means that current science and technology may be at the cusp of major developments that can fundamentally change our life for many decades to come. Low-dimensional systems, in particular two-dimensional (2D) systems of electrons are seen as some of the most fascinating systems for meeting the technological challenges of the future [1–5]. The great interest in 2D systems of electrons stems from the fact that the combination of low-dimensionality, confinement, discreteness of the electron's charge, electron's quantum spin and interaction/correlation effects can lead to very intriguing quantum phenomena [6,7]. The application of new and extraordinary experimental tools, in conjunction with the production of novel materials, has created an urgent need for a better understanding of the many novel unexpected physical phenomena that are observed under these conditions [8,9].

The application of a strong uniform constant magnetic field perpendicular to a 2D system of electrons dramatically changes its physics. As a matter of fact, a 2D system of electrons in a perpendicular magnetic field exhibits remarkable quantum phenomena at very low temperatures. Two novel specific physical phenomena, the integer quantum Hall effect (IQHE) [10] and fractional quantum Hall effect (FQHE) [11], stand out as two of the most important discoveries in condensed matter physics over the last decades. The appearance of plateaus in the Hall resistance plot as a function of the magnetic field was an unexpected finding. The Hall resistance on these plateaus is quantized at $R_H = h/(\nu\, e^2)$ values, where h is the Planck constant, e is the magnitude of electron's charge and the quantum number ν is the integer (1, 2, etc.) for the IQHE or fractional (1/3, 1/5, etc.) for the case of the FQHE.

The IQHE has a simpler explanation that originates from the quantum physics of single-particle states in a perpendicular magnetic field. On the other hand, the FQHE represents a particular example of a novel collective quantum liquid state of matter that originates, in a unique way, from strong electronic interactions/correlations. For both cases, the first step in comprehending the phenomena is to consider the solution of the quantum problem of a spinless charged particle in a 2D system subject to a perpendicular magnetic field. This quantum problem was solved by Landau a long time ago [12]. The model typically assumes that charged particles (for instance, electrons) with a fixed given isotropic mass are confined in a 2D system that is subjected to a strong, uniform, constant and perpendicular magnetic field. The main feature of the quantum solution is that the resulting energy spectrum consists of massively degenerated discrete quantum states known as Landau levels which are separated from each other by an energy gap. For a given Landau level, the eigenstate spectrum of the many degenerated wave functions describes states with the same kinetic energy quantized by the application of the magnetic field.

Within the realm of quantum mechanics, the kinetic energy of an electron moving in a perpendicular magnetic field is quantized to values of $\hbar \omega_c/2$, $3\hbar \omega_c/2$, $5\hbar \omega_c/2$, and so on, where $\hbar = h/(2\pi)$ is the reduced Planck's constant and ω_c is the cyclotron angular frequency. The energy gap between two neighboring Landau levels is $\hbar \omega_c$ while $\hbar \omega_c/2$ is known as the lowest Landau level energy. The number of eigenstates in each Landau level represents the degeneracy of that level and is proportional to the value of the magnetic field (B) and to the area (A) of the 2D sample, $N_s = BA/\Phi_0$ where $\Phi_0 = h/e$ is the magnetic flux quantum. Note that the degeneracy, N_s, of each Landau level increases with increasing the magnetic field. A key parameter that controls the properties of the system is the filling factor, ν, which is defined as the ratio of the number of electrons, N, to the degeneracy (number of available states) of each Landau level, $\nu = N/N_s$. In fact, the filling factor represents exactly the quantum number, ν, in the expression for the Hall resistance plateaus, R_H. This means that IQHE occurs when the filling factor, ν, is an integer, while the FQHE happens when ν is fractional.

In the extreme quantum limit of a very high perpendicular magnetic field, the degeneracy of each Landau level becomes so large that all electrons may be accommodated in the lowest Landau level, with a negligible admixture of higher Landau levels. In fact, some of most important FQHE liquid states occur when the lowest Landau level is fractionally filled with electrons. Under these conditions, the kinetic energy of the electrons is essentially quenched to a constant value (that corresponds to the lowest Landau level energy per electron). The electrons also have a quantum spin that couples to the magnetic field. The energy associated with this coupling is known as Zeeman energy and this is smallest when the quantum spin of the electrons is aligned with the magnetic field. For this reason, and to simplify the treatment, one may assume that the quantum spin of electrons is "frozen" by the magnetic field and, therefore, the electrons may be seen as effectively spinless charged particles.

As the magnetic field varies, the stabilization of the novel quantum phases of electrons happens at special filling factors that generally have odd denominators. Among them, the most robust FQHE states correspond to filling factors $\nu = 1/3$ and $1/5$ and are well described by Laughlin's theory in terms of trial wave functions [13]. Differently from odd-denominator-filled states in the lowest Landau level, even-denominator-filled states with filling factors $\nu = 1/2$, $1/4$ and $1/6$ do not show typical FQHE features and behave as isotropic compressible metallic Fermi liquid states [14]. The composite fermion theory [15] for the FQHE sheds light on the Fermi-liquid nature of such even-denominator-filled states. On the other hand, Wigner crystallization occurs when the filling factor becomes around, or less than, $\nu = 1/7$, as seen in various studies [16–20].

As discussed above, the 2D model of a charged particle in a uniform constant perpendicular magnetic field has many applications in quantum mechanics ranging from theories of magnetism a century ago [21,22] to quantum Hall effect phenomena during the last few decades [23–32]. The model in which a charged particle has a given constant isotropic mass was solved exactly by Landau in a work where the so-called Landau gauge was first introduced [12]. The exact solution of the stationary Schrödinger equation in this case is relatively straightforward for such a gauge due to the possibility of separating variables and writing the overall wave function as the product of a plane wave for one position variable and a displaced one-dimensional (1D) harmonic oscillator for the other one.

However, it is well known that any loss of symmetry in a quantum system leads to mathematical problems that generally become much more difficult to solve. A common situation of this nature arises when we deal with particles such as electrons trapped in semiconductor materials. For these conditions, the electrons may possess an anisotropic effective mass. The purpose of this work is to consider the 2D model of a spinless charged particle with anisotropic mass in a uniform constant perpendicular magnetic field and show that this quantum problem has a simple and exact analytic solution, despite the presence of mass anisotropy. The mathematical approach that we use is based on the introduction of "new" scaled distorted coordinates. The method allows one to transform the original problem of a charged particle with anisotropic mass in standard coordinates to that of a charged particle with isotropic effective mass in "new" scaled distorted coordinates.

The paper is organized as follows: In Section 2, we explain the quantum solution for the case of a charged particle with constant isotropic mass in a uniform constant perpendicular magnetic field. In Section 3, we provide the exact solution when the mass is anisotropic and point out the key details of the adopted mathematical method. In Section 4, we discuss the subtle effects that may come from the anisotropic mass of electrons in quantum Hall systems. In Section 5, we provide some concluding remarks.

2. Results—Isotropic Mass

In this section, we focus our attention on the quantum problem of a spinless particle with a constant isotropic mass, $m > 0$, and charge, q, moving in 2D space in the presence of a uniform constant perpendicular magnetic field. We clarify that by a constant quantity we mean one that does not change with time. A particle with a constant isotropic mass is a rather conventional one. For instance, it can be an electron with a bare mass, m_e, and a negative charge in studies of 2D electronic systems under ideal conditions. For more realistic experimental situations, one must take into account the fact that 2D systems of electrons are often created at the interface of a semiconductor heterojunction and/or heterostructure, such as $GaAs/AlGaAs$. Since electrons are typically confined in the conduction band of a given host semiconductor, the mass, m, for this case would represent the electron's effective band mass for those structures that are known to have a constant isotropic effective mass (for example, it is known that the effective band mass of electrons is isotropic and has the value, $m = 0.067\,m_e$, in a $GaAs$ host semiconductor).

Providing certain details to the solution of this known problem is beneficial for understanding how the emerging new problem of a particle with anisotropic mass can be mapped back to the known results. To begin with, the magnetic field perpendicular to the 2D plane is written as follows:

$$\vec{B} = (0, 0, B_z). \tag{1}$$

Any magnetic field is given in terms of a vector potential, so that

$$\vec{B} = \vec{\nabla} \times \vec{A}(x,y), \tag{2}$$

where $\vec{\nabla} = \left(\frac{\partial}{\partial x}, \frac{\partial}{\partial y}, \frac{\partial}{\partial z}\right)$ is the nabla or del operator, $\frac{\partial}{\partial x}, \frac{\partial}{\partial y}, \frac{\partial}{\partial z}$ are partial derivatives and $\vec{A}(x,y)$ is the vector potential for the given magnetic field. The choice of $\vec{A}(x,y)$ is not unique. The simplest choice is the so-called Landau gauge, which may take the following two flavours:

$$\vec{A}(x,y) = B_z\,(0, x, 0), \tag{3}$$

or

$$\vec{A}(x,y) = B_z\,(-y, 0, 0). \tag{4}$$

In this work, we choose the Landau gauge in Equation (3).

The general quantum Hamiltonian is

$$\hat{H} = \frac{1}{2m}\left[\hat{\vec{p}} - q\,\vec{A}(x,y)\right]^2, \qquad (5)$$

where $\hat{\vec{p}} = (\hat{p}_x, \hat{p}_y)$ is the 2D linear momentum operator. The x and y components of the 2D linear momentum operator may be explicitly written as follows:

$$\hat{p}_x = -i\hbar\frac{\partial}{\partial x} \;;\; \hat{p}_y = -i\hbar\frac{\partial}{\partial y}, \qquad (6)$$

where $i = \sqrt{-1}$ is the imaginary unit and $\hbar$ is the reduced Planck's constant. One must notice that the interaction of the particle's quantum spin with the magnetic field (the Zeeman effect) is not included in the Hamiltonian of Equation (5), since, for simplicity, we are assuming a spinless charged particle.

One can write the quantum Hamiltonian as follows:

$$\hat{H} = \frac{1}{2m}\left[\hat{p}_x - q\,A_x(x,y)\right]^2 + \frac{1}{2m}\left[\hat{p}_y - q\,A_y(x,y)\right]^2. \qquad (7)$$

For the Landau gauge in Equation (3), the quantum Hamiltonian in Equation (7) becomes

$$\hat{H} = \frac{\hat{p}_x^2}{2m} + \frac{(\hat{p}_y - q\,B_z\,x)^2}{2m} = \frac{1}{2m}\left[\hat{p}_x^2 + (\hat{p}_y - q\,B_z\,x)^2\right]. \qquad (8)$$

The stationary Schrödinger equation to solve is

$$\hat{H}\,\Psi(x,y) = E\,\Psi(x,y), \qquad (9)$$

where E is the energy and $\Psi(x,y)$ is the wave function. In order to solve this equation, one envisions the particle as being constrained in a 2D area, $L_x\,L_y$, where $-L_x/2 \leq x \leq +L_x/2$, $0 \leq y \leq L_y$ and $L_x \to \infty$ and $L_y \to \infty$. Hence, in the x-direction, one has $-\infty < x < +\infty$. On the other hand, periodic boundary conditions (PBC) for the wave function are imposed in the y-direction:

$$\Psi(x,y) = \Psi(x, y + L_y). \qquad (10)$$

Given the form of the Hamiltonian in Equation (8) and the PBC choice in Equation (10), one searches for a wave function that solves Equation (9) as a product of a plane wave state in the y-direction and a function that depends on coordinate x in the other direction:

$$\Psi(x,y) = \frac{e^{i k_y y}}{\sqrt{L_y}}\,\Phi(x). \qquad (11)$$

The overall normalization of the wave function must be such that

$$\int_{-\infty}^{+\infty} dx \int_0^{L_y} dy\,|\Psi(x,y)|^2 = 1. \qquad (12)$$

A substitution of the expression from Equation (11) into Equation (9) gives

$$\left[\frac{\hat{p}_x^2}{2m} + \frac{(\hbar k_y - q\,B_z\,x)^2}{2m}\right]\Phi(x) = E\,\Phi(x). \qquad (13)$$

One can rewrite Equation (13) as

$$\left[\frac{\hat{p}_x^2}{2m} + \frac{m}{2}\left(\frac{q\,B_z}{m}\right)^2\left(x - \frac{\hbar k_y}{q\,B_z}\right)^2\right]\Phi(x) = E\,\Phi(x). \qquad (14)$$

At this juncture, we introduce the explicit definition of the cyclotron angular frequency:

$$\omega_c = \frac{|q|\,|B_z|}{m} > 0, \tag{15}$$

where $|q| > 0$ is the magnitude of the charge of the particle, $|B_z| > 0$ is the magnitude of the magnetic field perpendicular to the 2D plane and $m > 0$ is the constant isotropic mass of the charged particle. Classically speaking, a charged particle, such as the one described above, moving perpendicular to the direction of a uniform constant magnetic field, will undergo a uniform circular motion with a given rotational cyclotron frequency, $f_c = \omega_c/(2\pi)$. The cyclotron frequency is independent of the radius of the circle of rotation and velocity. All charged particles with the same charge-to-mass ratio will undergo a circular motion with the same frequency.

With help from the definition in Equation (15), one can write Equation (14) as follows:

$$\left[\frac{\hat{p}_x^2}{2m} + \frac{m}{2}\omega_c^2\left(x - \frac{\hbar k_y}{q B_z}\right)^2\right]\Phi(x) = E\,\Phi(x). \tag{16}$$

The PBC in the y-direction, as specified by Equation (10), fixes the allowed values of the k_y wave vector:

$$k_y = \frac{2\pi}{L_y} j\;;\quad j = 0, \pm 1, \pm 2\ldots \tag{17}$$

Note that Equation (16) represents a displaced 1D quantum oscillator centered at $\hbar k_y/(q B_z)$, which has a known solution. The resulting discrete energy eigenvalues are

$$E_n = \hbar \omega_c \left(n + \frac{1}{2}\right)\;;\quad n = 0, 1, \ldots. \tag{18}$$

These are the Landau levels. Note that the discrete Landau level energies are highly degenerate since the quantum number, k_y, does not enter the energy expression. The normalized eigenfunctions corresponding to the above energy levels may be written as follows:

$$\Psi_{n k_y}(X, y) = \frac{e^{i k_y y}}{\sqrt{L_y}}\,\Phi_n(X), \tag{19}$$

where

$$X = x - \frac{\hbar k_y}{q B_z}, \tag{20}$$

and $\Phi_n(X)$ is the normalized eigenfunction of a 1D harmonic oscillator of mass m and frequency ω_c. Such a wave function is given by

$$\Phi_n(X) = N_n \exp\left(-\frac{\alpha^2 X^2}{2}\right) H_n(\alpha X), \tag{21}$$

where

$$N_n = \sqrt{\frac{\alpha}{\sqrt{\pi}\, 2^n\, n!}}, \tag{22}$$

is the normalization constant,

$$\alpha = \sqrt{\frac{m \omega_c}{\hbar}}, \tag{23}$$

is a parameter with the dimensionality of an inverse length and $H_n(x)$ are the Hermite polynomials. By using Equation (15), one sees that

$$\alpha = \sqrt{\frac{m \omega_c}{\hbar}} = \sqrt{\frac{|q|\,|B_z|}{\hbar}} = \frac{1}{l_B}, \tag{24}$$

where

$$l_B = \sqrt{\frac{\hbar}{|q||B_z|}},\qquad(25)$$

is known as the magnetic length.

By looking at the expression obtained in Equation (19), one concludes that the normalized eigenfunctions for a charged particle studied in a Landau gauge look like stripes [33]. It is straightforward to note that the probability density for such eigenfunctions, $|\Psi_{nk_y}(X,y)|^2$, depends only on the variable X (thus, x), but not y. This means that one can view such states as extended in one direction (in this case, along the y-direction), but exponentially localized around a given set of centers in the other perpendicular direction (in this case, the x-direction).

3. Results—Anisotropic Mass

Let us now consider the same quantum problem but with the assumption that the charged particle has a constant anisotropic (effective) mass of the following form:

$$m_x > 0 \; ; \; m_y > 0,\qquad(26)$$

along the respective x and y directions. This situation would apply to electrons hosted in semiconductors in which the lowest energies of the conduction band are locally approximated in parabolic form by an anisotropic dispersion relation:

$$E(\vec{k}) = E_0 + \frac{\hbar^2}{2m_x}(k_x - k_{0x})^2 + \frac{\hbar^2}{2m_y}(k_y - k_{0y})^2 + \frac{\hbar^2}{2m_z}(k_z - k_{0z})^2,\qquad(27)$$

where $E(\vec{k})$ is the energy of an electron at wavevector $\vec{k}$ in that band, E_0 is the minimum energy of that band, $m_{x,y,z}$ are the effective masses along the different axes and $\vec{k}_0 = (k_{0x}, k_{0y}, k_{0z})$ represents the wavevector of the conduction band minimum (that, in principle, may be offset from the zero value). The effective (isotropic or anisotropic) masses of conduction band electrons in common host semiconductor materials ($GaAs$, GaP, $InAs$, $AlAs$, Si, Ge, etc.) are all positive. However, if semiconductor band structures of certain exotic materials exhibit saddle points (e.g., in heterostructures, curved 2D materials, or topological insulator systems), there could be situations where the effective mass is positive in one direction ($m_x > 0$) and negative in another ($m_y < 0$). Dealing with the concept of a particle with a negative mass, either classically or quantum mechanically, is beyond the scope of this work.

The scenario envisioned by Equation (26) would lead to a starting quantum Hamiltonian

$$\hat{H} = \frac{1}{2m_x}\left[\hat{p}_x - qA_x(x,y)\right]^2 + \frac{1}{2m_y}\left[\hat{p}_y - qA_y(x,y)\right]^2.\qquad(28)$$

For the assumption of a Landau gauge, as in Equation (3), one has

$$\hat{H} = \frac{1}{2m_x}\left(-i\hbar\frac{\partial}{\partial x}\right)^2 + \frac{1}{2m_y}\left(-i\hbar\frac{\partial}{\partial y} - qB_z x\right)^2.\qquad(29)$$

The Hamiltonian in Equation (29) is the anisotropic mass counterpart to that in Equation (8), with the 2D linear momentum operators written in explicit form.

The idea behind the solution of this quantum problem is to try to identify some "new" coordinates that will allow us to see the problem of a particle with anisotropic mass in "old" coordinates as that of a "new" particle with constant "isotropic" mass in these "new" coordinates. This means that the solution of the problem will be achieved elegantly if this process comes to fruition, given that, at this juncture, one can rely on already known results.

Being inspired by this idea, the mathematical approach that we follow is centered on scaling the original coordinates, x and y. We start by scaling the variable x to γx, where γ is seen as a real positive scaling parameter whose precise value would be determined at a later stage. Let us write Equation (29) as

$$\hat{H} = \frac{1}{2\,m_x}\left[-i\hbar\gamma\frac{\partial}{\partial(\gamma x)}\right]^2 + \frac{1}{2\,m_y}\left[-i\hbar\frac{\partial}{\partial y} - \frac{q\,B_z}{\gamma}(\gamma x)\right]^2. \tag{30}$$

Since γx is going to be one of the "new" scaled coordinate variables, it immediately transpires that the scaling parameter, γ, can be factorized out of the second term in the right-hand side expression of Equation (30) if one rescales the other coordinate variable y to y/γ:

$$\hat{H} = \frac{1}{2\,m_x}\left[-i\hbar\gamma\frac{\partial}{\partial(\gamma x)}\right]^2 + \frac{1}{2\,m_y}\left[-\frac{i\hbar}{\gamma}\frac{\partial}{\partial\left(\frac{y}{\gamma}\right)} - \frac{q\,B_z}{\gamma}(\gamma x)\right]^2. \tag{31}$$

At this junction, one can check that the quantum Hamiltonian in Equation (31) can be rewritten as follows:

$$\hat{H} = \frac{\gamma^2}{2\,m_x}\left[-i\hbar\frac{\partial}{\partial(\gamma x)}\right]^2 + \frac{1}{2\,m_y\,\gamma^2}\left[-i\hbar\frac{\partial}{\partial\left(\frac{y}{\gamma}\right)} - q\,B_z(\gamma x)\right]^2. \tag{32}$$

Let us choose the value of γ, such that

$$\frac{\gamma^2}{2\,m_x} = \frac{1}{2\,m_y\,\gamma^2}. \tag{33}$$

This choice leads to

$$\gamma^2 = \sqrt{\frac{m_x}{m_y}}. \tag{34}$$

For the choice of γ^2, as in Equation (34), one has

$$\frac{\gamma^2}{m_x} = \frac{1}{m_y\,\gamma^2} = \frac{1}{\sqrt{m_x\,m_y}} \;;\; \gamma^2 = \sqrt{\frac{m_x}{m_y}}. \tag{35}$$

This means that one can use the result from Equation (35) to write the Hamiltonian in Equation (32) as

$$\hat{H} = \frac{1}{2\,m_c}\left(-i\hbar\frac{\partial}{\partial x'}\right)^2 + \frac{1}{2\,m_c}\left(-i\hbar\frac{\partial}{\partial y'} - q\,B_z\,x'\right)^2, \tag{36}$$

where the two "new" scaled coordinate variables are

$$x' = \gamma x \;;\; y' = \frac{y}{\gamma}, \tag{37}$$

and

$$m_c = \sqrt{m_x\,m_y}, \tag{38}$$

represents the effective cyclotron mass of a particle with anisotropic mass [34].

One can rewrite the quantum Hamiltonian in Equation (36) in a more compact form as follows:

$$\hat{H} = \frac{1}{2\,m_c}\left[\hat{p}_{x'}^2 + \left(\hat{p}_{y'} - q\,B_z\,x'\right)^2\right], \tag{39}$$

where $\hat{\vec{p}}' = (\hat{p}_{x'}, \hat{p}_{y'})$ is the 2D linear momentum operator with respect to the "new" primed coordinates.

A comparison of the result from Equation (39) with the original Hamiltonian for a particle with constant isotropic mass, as seen in Equation (8), indicates that the problem of a particle with anisotropic mass in variables x and y has been transformed to that of a particle with isotropic mass, $m_c = \sqrt{m_x m_y}$ in terms of "new" scaled variables x' and y'. Note that $dx\,dy = dx'\,dy'$. However, one must be careful to write

$$\int_{-\infty}^{+\infty} dx \int_0^{L_y} dy = \int_{-\infty}^{+\infty} dx' \int_0^{L_y/\gamma} dy', \qquad (40)$$

when the overall normalization condition of the wave function is applied. The solution of the quantum problem for a particle with constant isotropic mass is well known. Thus, one can immediately use such known results with the only consideration taken that all expressions must be written in terms of the "new" primed variables. The energy eigenvalues are

$$E_n = \hbar \omega_c' \left(n + \frac{1}{2} \right) \;;\; n = 0, 1, \ldots . \qquad (41)$$

where

$$\omega_c' = \frac{|q||B_z|}{m_c}. \qquad (42)$$

Note that ω_c' takes the place of the cyclotron angular frequency, ω_c, for the case of a particle with constant isotropic mass, m. Likewise, $m_c = \sqrt{m_x m_y}$ takes the place of mass, m, for the case of a particle with constant isotropic mass.

With some care, one can write the normalized eigenfunctions corresponding to the above energy levels as follows:

$$\Psi_{n k_y'}(X', y') = \frac{e^{i k_y' y'}}{\sqrt{L_y/\gamma}} \Phi_n(X'), \qquad (43)$$

where

$$X' = x' - \frac{\hbar k_y'}{q B_z}, \qquad (44)$$

and $\Phi_n(X')$ is the normalized eigenfunction of a displaced 1D harmonic oscillator of mass, m_c, and frequency, ω_c', in "new" primed coordinates. In order to have the PBC in Equation (10) still be valid, one has

$$k_y' = \gamma k_y, \qquad (45)$$

where k_y is given from Equation (17).

The displaced 1D quantum oscillator wave function in the "new" primed coordinates is written as follows:

$$\Phi_n(X') = N_n' \exp\left(-\frac{\alpha'^2 X'^2}{2} \right) H_n(\alpha' X'), \qquad (46)$$

where

$$N_n' = \sqrt{\frac{\alpha'}{\sqrt{\pi}\, 2^n\, n!}}, \qquad (47)$$

and

$$\alpha' = \sqrt{\frac{m_c \omega_c'}{\hbar}}. \qquad (48)$$

4. Discussion

Classically speaking, a charged particle experiences a magnetic force when moving through a magnetic field. The fundamental question that one must answer is what happens to the particle if this magnetic field is uniform over the motion of the charged particle. The simplest case occurs when a charged particle with fixed isotropic mass moves perpendicular to a uniform constant magnetic field. Since the magnetic force is perpendicular to the direction of travel, a charged particle follows a circular path in a magnetic field. Another way to look at this is that the magnetic force is always perpendicular to velocity, so that it does no work on the charged particle. As a result, the particle's kinetic energy and speed (magnitude of velocity) remain constant. In a nutshell, the direction of motion is affected but not the speed. The classical description above becomes more nuanced when the charged particle has a constant anisotropic mass with values $m_x \neq m_y$ along the respective x and y directions. We have investigated the classical 2D motion of a charged particle with such an anisotropic mass in the presence of a uniform constant magnetic field that is perpendicular to the plane of motion and have found that the trajectory of the particle for such a case is elliptical [34]. We also have verified that, as expected, such a trajectory becomes circular when the mass becomes isotropic ($m_x = m_y$). Overall, it was found that the resulting classical motion and trajectory of such a particle is very sensitive to the direction of the initial velocity.

The solution of the quantum counterpart to this problem is much more complicated. The main reason is that, unlike the classical scenario, the quantum Hamiltonian is given in terms of the vector potential and not the magnetic field. There are different vector potentials that can generate the same magnetic field. The freedom to choose various vector potentials, $\vec{A}(x,y)$, that lead to the same magnetic field is known as the choice of the gauge. The most common gauges used are the symmetric gauge and Landau gauge. A step-by-step solution to the quantum problem of a charged particle with constant isotropic mass in a perpendicular uniform constant magnetic field for the case of a symmetric gauge is readily available in the literature [35]. The quantum problem of a charged particle with isotropic mass is easier to solve for a Landau gauge. By providing the full details of such a solution, we reminded the reader of the peculiarities of the Landau gauge and also prepared the ground for tackling the much more difficult quantum problem for the counterpart case of a charged particle with anisotropic mass. It is shown in this work that the quantum problem of a charged particle with anisotropic mass in a perpendicular uniform constant magnetic field can be solved rather elegantly by adopting a mathematical method that rescales the original coordinates to new distorted ones. This procedure allows one to restore the mass symmetry of the stationary Schrödinger differential equation, albeit in "new" distorted coordinates.

The quantum problem of a charged particle (with or without an isotropic mass) undergoing 2D motion in a perpendicular uniform constant magnetic field leads to the physics of Landau states. The properties and the nature of Landau states is fundamental to explain a plethora of important phenomena in physics, such as the IQHE and FQHE. The unique nature of the IQHE/FQHE phases has always been fertile ground for paradigm-shifting ideas in theoretical condensed matter physics and materials science. Novel theories, phases of matter and concepts such as topological states, incompressible quantum Hall liquids or composite fermions are now well-known in the literature and all originate from studies of these two phenomena. For all these cases, the starting model assumes a standard Coulomb interaction potential between the charged particles. Obviously, a Coulomb interaction potential is isotropic, in the sense that the interaction energy of any pair of charged particles depends only on their separation distance. The same presumption is valid for many inherently anisotropic phases, such as charge density waves, liquid crystalline phases, Wigner solid phases, etc. The assumption made is that the interaction potential is isotropic (for instance, a Coulomb interaction potential) and there is no intrinsic anisotropy.

However, in a real quantum Hall sample, electrons may possess an anisotropic effective mass tensor or may interact via an effective anisotropic interaction potential (mediated from the substrate). This situation calls for a re-examination of the role played by anisotropic factors such as an anisotropic effective mass [36]. To be more specific, we consider a 2D system of charged particles with anisotropic band mass values, $m_x > 0$ and $m_y > 0$, along two respective directions labelled x and y. The charged particles interact with the usual (isotropic) Coulomb interaction potential:

$$v_C(\vec{r}_i - \vec{r}_j) = \frac{kq^2}{\sqrt{|x_i - x_j|^2 + |y_i - y_j|^2}}, \qquad (49)$$

where k is Coulomb's electric constant, q is the charge of the particles and $\vec{r}_i - \vec{r}_j = (x_i - x_j, y_i - y_j)$ is the 2D vector that separates the positions of particles i and j. The anisotropic mass of the charged particles breaks the rotational symmetry of the quantum kinetic energy operator (with or without a magnetic field). The transformation of variables in Equation (37) allows one to restore the rotational symmetry of the quantum kinetic energy in the "new" scaled coordinates. The same transformation of coordinates, when applied to the Coulomb interaction potential, would transform it into the following anisotropic Coulomb interaction potential:

$$v_\gamma(\vec{r}_i' - \vec{r}_j') = \frac{kq^2}{\sqrt{\frac{|x_i' - x_j'|^2}{\gamma^2} + \gamma^2 |y_i' - y_j'|^2}}, \qquad (50)$$

where $\gamma > 0$ is an interaction anisotropy parameter that leads to anisotropy when different from 1. This parameter (γ) is the same as the scaling parameter of the coordinates that was discussed earlier. From the perspective of Equation (50), one can view the directions of x and y (primed or unprimed) as corresponding to the two principal axes of the dielectric tensor. The potential becomes the standard isotropic Coulomb interaction potential for $\gamma = 1$. In primed variables, this anisotropic Coulomb interaction potential (for $\gamma \neq 1$) breaks the usual assumption of isotropic pair interaction potentials. It is expected that an anisotropic interaction potential of this nature can steer us towards novel conceptual frameworks [37]. The idea is to deal with the rarely tackled, but considerably more difficult, problem of understanding how anisotropic order arises in a quantum system in which the constituent particles interact with an anisotropic interaction potential. An anisotropic interaction potential, alone or in conjunction with other intrinsic degrees of anisotropy in the system, may be seen as a game changer that can lead to novel physics in the field [38,39].

The interaction/correlation effects in systems of electrons may lead to the formation of novel quantum phases of matter. Under certain conditions, one can describe various properties in terms of the underlying topology of the system. This is the case for topological insulators in general and FQHE systems in particular. In fact, FQHE liquids are the ultimate examples of a phenomenon with topological features. As already noticed, the FQHE is observed in certain 2D materials (in the presence of a large perpendicular magnetic field near absolute zero temperature). The quantum effects related to the magnetic field cause a gap to open up between energy bands in the bulk material. As a result, the electrons in the bulk become localized (they cannot move freely). This leads to bulk states of electrons that represent an insulator. However, the electrons at the edge can still move and, thus, they can conduct (this is the physics of "edge states"), while the bulk phase remains insulating. Overall, such phases are characterized by the presence of an energy band-gap within the bulk of the material, while the material's edge/boundary or surface hosts topologically protected gapless conducting modes. The non-trivial topology of the FQHE gives rise to fractionally charged elementary excitations which, in some cases, may even possess non-Abelian braiding statistics (for instance, the Pfaffian state at filling factor $\nu = 5/2$). The interaction/correlation effects between electrons are the key ingredients that lead to this sort of physics (in fact, there is no FQHE without interactions between electrons). The predominant view since the time of Laughlin's theory [13] has been that FQHE states represent isotropic quantum liquid phases that have rotational symmetry. However, recent work [40] in quantum Hall fluids has revealed the importance of a novel internal geometric degree of freedom (or metric) that has been previously overlooked. It has been pointed out that topological liquid phases arising in the FQHE regime are not required to be rotationally invariant. This means that the presence of an internal anisotropy (such as anisotropic mass and/or anisotropic interaction potential) may drastically modify our view on the remarkable topological properties of such systems. Therefore, the quantum solution of this problem for the case of a charged particle with an anisotropic mass ($m_x \neq m_y$) is very useful for analyzing situations in which the charged particles (for instance, electrons and/or holes) possess a pronounced anisotropic (effective) mass.

5. Conclusions

The quantum problem of a charged particle confined in 2D space in a uniform constant perpendicular magnetic field is the foundation of many important phenomena in physics, where the IQHE/FQHE stand out as two major discoveries from the last few decades. The basic features of all quantum Hall phenomena were initially understood by using the standard model of charged particles (electrons) with constant isotropic mass in a uniform constant perpendicular magnetic field. However, experimental sample refinements have led to an increased interest in grasping more subtle systems involving electrons that possess an effective anisotropic band mass.

Breaking any symmetry in a quantum system leads to problems that are not easy to solve in analytical form. The case of an (effective) particle with anisotropic mass is one such scenario that is important not only from a mathematical perspective, but also for real experimental applications, for instance when dealing with electrons in a semiconductor material. As a matter of fact, the concept of an effective anisotropic mass tensor is routine when it comes to studying the properties of electrons in periodic potentials, such as the ones created by the crystal structure of many semiconductor materials. The standard studies of 2D systems of electrons in a perpendicular magnetic field originally involved GaAs/AlGaAs heterostructures. In these systems, the electrons typically manifest an (effective) isotropic mass. However, the improvements in experimental samples and materials now allow one to study 2D systems of electrons for regimes that were not accessible before, including those in which the electrons manifest an (effective) anisotropic mass. Any source of internal anisotropy in a quantum system may lead to novel subtle quantum phenomena, involving scenarios that may have not been observed before. This means that the system's symmetry (or lack of it) influences the patterns of various observed quantum phenomena. In particular, this work further emphasizes the role that symmetry (or lack of it) plays in quantum problems that involve 2D systems of charged particles with anisotropic mass subject to a perpendicular magnetic field.

For a 2D system of electrons with an applied uniform constant magnetic field in the z-direction and a homogeneous electric field, E_x, in the x-direction, all states drift in the transverse y-direction (where the plane waves are). As a result, the current density in the y-direction, j_y, will be given by

$$j_y = \nu \frac{e^2}{h} E_x, \qquad (51)$$

where, for simplicity, we consider IQHE states with filling factor, $\nu = 1, 2, \ldots$. Sample details, including effects from the presence of an anisotropic (effective) mass of the electrons, are not expected to play any role in the measured Hall resistance plateaus, $R_H = |V_y/I_x| = h/(\nu e^2)$, where V_y is the Hall (transverse) voltage and I_x is the longitudinal current. However, the ranges of the magnetic field and/or the electron density where the plateaus appear will be affected by an anisotropic mass. The quantized Hall resistance shows a universal behavior, but it is known that the current distribution in real quantum Hall devices is quite complicated [41]. Therefore, an anisotropic mass of the charge carriers is expected to affect the patterns of the current distribution, too. The occurrence of such phenomena may be detected from the experimental observation of the unexpected magneto-transport anisotropy features of a 2D system of electrons in the quantum Hall regime.

Another interesting scenario where an anisotropic mass may have a profound impact arises when one discusses the effects of Earth's gravity on the quantum Hall behavior of 2D systems of electrons. To this effect, one may consider a 2D quantum Hall sample of electrons oriented in such a way that the gravitational field of the Earth acts on the 2D plane of the sample perpendicular to the magnetic field. The effect of gravity for such an orientation is to act as an effective in-plane constant homogeneous electric field due to the same linear nature of the gravitational potential. For the geometry adopted in this work, one may choose the gravitational field to act in the x-direction. The new twist that comes from the (effective) mass anisotropy of the electrons is that the resulting gravitational potential energy has the gravitational field coupled to the mass of the electrons in the x-direction, resulting in an energy term of the form, $m_x g x$, where g is the acceleration due to gravity on the Earth's surface. If the quantum Hall effect is affected by the gravitational field, the effects of the mass anisotropy should show up as a modification to the current density expression when one uses the quantum Hall effect to probe the inverse-square law of gravity, as recently suggested [42]:

$$j_y(m_x) = \left(1 + \frac{m_x g}{e E_x}\right) v \frac{e^2}{h} E_x . \tag{52}$$

This additional energy term correction due to Earth's gravity ($\propto m_x/E_x$) may lead to subtler effects than the case study of a constant isotropic mass, m, previously considered in a recent work, which takes a fresh look at the influence of gravity on the quantum Hall effect states (more precisely, on the IQHE states) of electrons for a variety of conditions [42].

Based on these considerations, one can promptly recognize the reasons why it is important to consider the 2D quantum problem of a charged particle with an anisotropic mass subject to a uniform constant perpendicular magnetic field when a Landau gauge is adopted. This problem is important to the physics of the quantum Hall effect for those situations in which the charged particles (electrons) have an effective band mass anisotropy [43–47]. It is shown that this model, despite exhibiting no axial symmetry, allows an exact analytic calculation of the energy and eigenfunctions for any value of anisotropic mass and magnetic field. The solution to the quantum problem is obtained elegantly by a scaling transformation of the original coordinates. The results of this study would be of interest to a broad audience of individuals working in quantum mechanics, as well as researchers that study the applications of quantum theory in materials science.

Funding: This research was supported, in part, by the National Science Foundation (NSF), Grant No. DMR-2001980, and the National Technology & Engineering Solutions of Sandia (NTESS) START Program.

Data Availability Statement: The data presented in this study are available upon request from the author.

Conflicts of Interest: The author declares no conflicts of interest.

References

1. Willett, R.L.; Paalanen, M.A.; Ruel, R.R.; West, K.W.; Pfeiffer, L.N.; Bishop, D.J. Anomalous sound propagation at $v = 1/2$ in a 2D electron gas: Observation of a spontaneously broken translational symmetry? *Phys. Rev. Lett.* **1990**, *65*, 112. [CrossRef] [PubMed]
2. Stanescu, T.; Martin, I.; Phillips, P. Finite-temperature density instability at high Landau level occupancy. *Phys. Rev. Lett.* **2000**, *84*, 1288. [CrossRef] [PubMed]
3. Shaji, S.; Mucha, N.R.; Majumdar, A.K.; Binek, C.; Kebede, A.; Kumar, D. Magnetic and electrical properties of Fe90Ta10 thin films. *J. Magn. Magn. Mater.* **2019**, *489*, 165446. [CrossRef]
4. Odbadrakh, K.; McNutt, N.W.; Nicholson, D.M.; Rios, O.; Keffer, D.J. Lithium diffusion at Si-C interfaces in silicon-graphene composites. *Appl. Phys. Lett.* **2014**, *105*, 053906. [CrossRef]
5. Wilson, T.E. Fabrication of robust superconducting granular aluminium/palladium bilayer microbolometers with sub-nanosecond response. *J. Low Temp. Phys.* **2008**, *151*, 201–205. [CrossRef]
6. Ciftja, O. Understanding electronic systems in semiconductor quantum dots. *Phys. Scr.* **2013**, *88*, 058302. [CrossRef]

7. Ciftja, O. Impact of an elliptical Fermi surface deformation on the energy of a spinless two-dimensional electron gas. *Phys. Scr.* **2019**, *94*, 105806. [CrossRef]
8. Gusynin, V.P.; Sharapov, S.G. Unconventional integer quantum Hall effect in graphene. *Phys. Rev. Lett.* **2005**, *95*, 146801. [CrossRef] [PubMed]
9. Herbut, I.F. Theory of integer quantum Hall effect in graphene. *Phys. Rev. B* **2007**, *75*, 165411. [CrossRef]
10. von Klitzing, K.; Dorda, G.; Pepper, M. New method for high accuracy determination of the fine structure constant based on the quantized Hall effect. *Phys. Rev. Lett.* **1980**, *45*, 494. [CrossRef]
11. Tsui, D.C.; Stormer, H.L.; Gossard, A.C. Two-dimensional magnetotransport in the extreme quantum limit. *Phys. Rev. Lett.* **1982**, *48*, 1559. [CrossRef]
12. Landau, L.D. Diamagnetismus der metalle. *Z. Phys.* **1930**, *64*, 629–637. [CrossRef]
13. Laughlin, R.B. Anomalous quantum Hall effect: An incompressible quantum fluid with fractionally charged quasiparticles. *Phys. Rev. Lett.* **1983**, *50*, 1395. [CrossRef]
14. Rezayi, E.; Read, N. Fermi-liquid-like state in a half-filled Landau level. *Phys. Rev. Lett.* **1994**, *72*, 900. [CrossRef] [PubMed]
15. Jain, J.K. Composite-fermion approach for the fractional quantum Hall effect. *Phys. Rev. Lett.* **1989**, *63*, 199. [CrossRef] [PubMed]
16. Lam, P.K.; Girvin, S.M. Liquid-solid transition and the fractional quantum-Hall effect. *Phys. Rev. B* **1984**, *30*, 473. [CrossRef]
17. Esfarjani, K.; Chui, S.T. Solidification of the two-dimensional electron gas in high magnetic fields. *Phys. Rev. B* **1990**, *42*, 10758. [CrossRef] [PubMed]
18. Zhu, X.; Louie, S.G. Wigner crystallization in the fractional quantum Hall regime: A variational quantum Monte Carlo study. *Phys. Rev. Lett.* **1993**, *70*, 335. [CrossRef] [PubMed]
19. Zhu, X.; Louie, S.G. Variational quantum Monte Carlo study of two-dimensional Wigner crystals: Exchange, correlation, and magnetic field effects. *Phys. Rev. B* **1995**, *52*, 5863. [CrossRef] [PubMed]
20. Yang, K.; Haldane, F.D.M.; Rezayi, E.H. Wigner crystals in the lowest Landau level at low-filling factors. *Phys. Rev. B* **2001**, *64*, 081301. [CrossRef]
21. de Haas, W.J.; van Alphen, P.M. The dependence of the susceptibility of diamagnetic metals upon the field. *Proc. R. Acad. Amst.* **1930**, *33*, 1106.
22. Shubnikov, L.; de Haas, W.J. Magnetic resistance increase in single crystals of bismuth at low temperatures. *Proc. R. Acad. Amst.* **1930**, *33*, 130–162.
23. Morf, R.; Halperin, B.I. Monte Carlo evaluation of trial wave functions for the fractional quantized Hall effect: Disk geometry. *Phys. Rev. B* **1986**, *33*, 2221. [CrossRef] [PubMed]
24. Fano, G.; Ortoloni, F. Interpolation formula for the energy of a two-dimensional electron gas in the lowest Landau level. *Phys. Rev. B* **1988**, *37*, 8179. [CrossRef] [PubMed]
25. Halperin, B.I.; Lee, P.A.; Read, N. Theory of the half-filled Landau level. *Phys. Rev. B* **1993**, *47*, 7312. [CrossRef] [PubMed]
26. Ciftja, O.; Ozurumba, C.; Ujeyah, F. Anisotropic quantum Hall liquids at intermediate magnetic fields. *J. Low Temp. Phys.* **2013**, *170*, 166–171. [CrossRef]
27. Ciftja, O. Anisotropic quantum Hall liquid states with no translational invariance in the lowest Landau level. *J. Low Temp. Phys.* **2016**, *183*, 85–91. [CrossRef]
28. Lilly, M.P.; Cooper, K.B.; Eisenstein, J.P.; Pfeiffer, L.N.; West, K.W. Evidence for an anisotropic state of two-dimensional electron in high Landau levels. *Phys. Rev. Lett.* **1999**, *82*, 394. [CrossRef]
29. Cooper, K.B.; Lilly, M.P.; Eisenstein, J.P.; Jungwirth, T.; Pfeiffer, L.N.; West, K.W. An investigation of symmetry-breaking mechanisms in high Landau levels. *Solid State Comm.* **2001**, *119*, 89–94. [CrossRef]
30. Jungwirth, T.; MacDonald, A.H.; Smrčka, L.; Girvin, S.M. Field-tilt anisotropy energy in quantum Hall stripe states. *Phys. Rev. B* **1999**, *60*, 15574. [CrossRef]
31. Kamilla, R.K.; Jain, J.K. Variational study of the vortex structure of composite fermions. *Phys. Rev. B* **1997**, *55*, 9824. [CrossRef]
32. Doan, Q.M.; Manousakis, E. Quantum nematic as ground state of a two-dimensional electron gas in a magnetic field. *Phys. Rev. B* **2007**, *75*, 195433. [CrossRef]
33. Ciftja, O. Interaction potential between a uniformly charged square nanoplate and coplanar nanowire. *Nanomaterials* **2023**, *13*, 2988. [CrossRef] [PubMed]
34. Ciftja, O.; Livingston, V.; Thomas, E. Cyclotron motion of a charged particle with anisotropic mass. *Am. J. Phys.* **2017**, *85*, 359–363. [CrossRef]
35. Ciftja, O. Detailed solution of the problem of Landau states in a symmetric gauge. *Eur. J. Phys.* **2020**, *41*, 035404. [CrossRef]
36. Ciftja, O. Variation of the elliptical Fermi surface for a two-dimensional electron gas with anisotropic mass. *J. Phys. Conf. Ser.* **2022**, *2164*, 012023. [CrossRef]
37. Ciftja, O. Two-dimensional finite quantum Hall clusters of electrons with anisotropic features. *Sci. Rep.* **2022**, *12*, 2383. [CrossRef] [PubMed]
38. Ciftja, O. Anisotropic magnetoresistance and piezoelectric effect in GaAs Hall samples. *Phys. Rev. B* **2017**, *95*, 075410. [CrossRef]
39. Ciftja, O. Integer quantum Hall effect with an anisotropic Coulomb interaction potential. *J. Phys. Chem. Solids* **2021**, *156*, 110131. [CrossRef]
40. Haldane, F.D.M. Geometrical description of the fractional quantum Hall effect. *Phys. Rev. Lett.* **2011**, *107*, 116801. [CrossRef] [PubMed]

41. Weis, J.; von Klitzing, K. Metrology and microscopic picture of the integer quantum Hall effect. *Philos. Trans. R. Soc. A* **2011**, *369*, 3954. [CrossRef] [PubMed]
42. Hammad, F.; Landry, A.; Mathieu, M. A fresh look at the influence of gravity on the quantum Hall effect. *Eur. Phys. J. Plus* **2020**, *135*, 449. [CrossRef]
43. Hossain, M.S.; Ma, M.K.; Chung, Y.J.; Pfeiffer, L.N.; West, K.W.; Baldwin, K.W.; Shayegan, M. Unconventional anisotropic even-denominator fractional quantum Hall state in a system with mass anisotropy. *Phys. Rev. Lett.* **2018**, *121*, 256601. [CrossRef] [PubMed]
44. Shayegan, M.; Poortere, E.P.D.; Gunawan, O.; Shkolnikov, Y.P.; Tutuc, E.; Vakili, K. Two-dimensional electrons occupying multiple valleys in AlAs. *Phys. Status Solidi B* **2006**, *243*, 3629–3642. [CrossRef]
45. Eng, K.; McFarland, R.N.; Kane, B.E. Integer quantum Hall effect on a six-valley Hydrogen-passivated Silicon (111) surface. *Phys. Rev. Lett.* **2007**, *99*, 016801. [CrossRef] [PubMed]
46. Chitta, V.A.; Desrat, W.; Maude, D.K.; Piot, B.A.; Oliveira, N.F., Jr.; Rappl, P.H.O.; Ueta, A.Y.; Abramof, E. Integer quantum Hall effect in a PbTe quantum well. *Physica E* **2006**, *34*, 124–127. [CrossRef]
47. Feldman, B.E.; Randeria, M.T.; Gyenis, A.; Wu, F.; Ji, H.; Cava, R.J.; MacDonald, A.H.; Yazdani, A. Observation of a nematic quantum Hall liquid on the surface of Bismuth. *Science* **2016**, *354*, 316–321. [CrossRef] [PubMed]

Disclaimer/Publisher's Note: The statements, opinions and data contained in all publications are solely those of the individual author(s) and contributor(s) and not of MDPI and/or the editor(s). MDPI and/or the editor(s) disclaim responsibility for any injury to people or property resulting from any ideas, methods, instructions or products referred to in the content.

Article

The GHZ Theorem Revisited within the Framework of Gauge Theory

David H. Oaknin

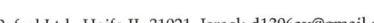

Rafael Ltd., Haifa IL-31021, Israel; d1306av@gmail.com

Abstract: The Greenberger-Horne-Zeilinger version of the Einstein-Podolsky-Rosen (EPR) paradox is widely regarded as a conclusive logical argument that rules out the possibility of reproducing the predictions of Quantum Mechanics within the framework of any physical theory sharing the notions of reality and relativistic causality that we acknowledge as a given in our classical descriptions of the macroscopic world. Thus, this renowned argument stands as a seemingly insurmountable roadblock on the path to a very desired, physically intuitive understanding of quantum phenomena and, in particular, quantum entanglement. In this paper, we notice, however, that the GHZ argument involves unaccounted spurious gauge degrees of freedom and that it can be overcome once these degrees are properly taken into account. It is then possible to explicitly build a successful statistical model for the GHZ experiment based on the usual notions of relativistic causality and physical reality. This model, thus, completes—in the EPR sense—the quantum description of the GHZ state and paves the way to a novel intuitive interpretation of the quantum formalism and a deeper understanding of the physical reality that it describes.

Keywords: quantum mechanics; EPR paradox; Bell's theorem; GHZ argument; gauge symmetries; holonomies; hidden variables; statistical physics

Citation: Oaknin, D.H. The GHZ Theorem Revisited within the Framework of Gauge Theory. *Symmetry* **2023**, *15*, 1327. https://doi.org/10.3390/sym15071327

Academic Editor: Tuong Trong Truong

Received: 10 May 2023
Revised: 14 June 2023
Accepted: 18 June 2023
Published: 29 June 2023

Copyright: © 2023 by the author. Licensee MDPI, Basel, Switzerland. This article is an open access article distributed under the terms and conditions of the Creative Commons Attribution (CC BY) license (https://creativecommons.org/licenses/by/4.0/).

1. Introduction

The inability to accommodate the seemingly trivial notions of causality and physical realism within the current interpretation of the quantum mechanical wavefunction is at the core of a long lasting debate about the foundations of quantum theory and the role played by measurements, whose origins go back to the formulation of the renowned Einstein-Podolsky-Rosen (EPR) paradox almost ninety years ago [1,2]. Solving these key issues would require developing a description of quantum phenomena in terms of a statistical model of local hidden variables. Nonetheless, according to the current wisdom, such a description is not possible in so far as we insist on keeping the notion, also seemingly trivial, that the observers' choice of their measurement settings is not constrained by the actual hidden configuration of the observed system (free-will).

Indeed, several fundamental theorems state that generic models of hidden variables that share certain intuitive features cannot fully reproduce the predictions of quantum mechanics [3–12], while carefully designed experimental tests have consistently confirmed the predictions of the quantum theory and, thus, have ruled out all these generic models of hidden variables [13–23]. The best known among these theorems is the Bell theorem [3,5–7], which proves that such generic models of hidden variables cannot reproduce the statistical correlations predicted by quantum mechanics for the outcomes of long sequences of strong polarization measurements performed along certain relative directions on pairs of entangled qubits.

The Greenberger-Horne-Zeilinger version of the Bell theorem [8] is an even more conclusive proof of the limitations of these generic models of hidden variables, since it proves that such models cannot reproduce even single outcomes of strong spin polarization measurements performed along certain relative directions on three or more entangled qubits prepared in the so-called GHZ state,

$$|\text{GHZ}\rangle \equiv \frac{|\uparrow\uparrow\cdots\uparrow\rangle + |\downarrow\downarrow\cdots\downarrow\rangle}{\sqrt{2}}. \tag{1}$$

However, in a series of recent papers [24–27], we have shown that the proof of the Bell theorem crucially relies on a subtle assumption that is not required by fundamental physical principles. Namely, we noted that the proof implicitly assumes the existence of an absolute angular frame of reference with respect to which we can define the polarization properties of the hypothetical hidden configurations of the pairs of entangled qubits as well as the orientations of the measurement devices that test them. Furthermore, we showed that such an absolute frame of reference may not exist if the hidden configurations spontaneously break the gauge rotational symmetry along an otherwise arbitrary direction.

A simple example that illustrates the absence of an absolute frame is described in Figure 3 in reference [24]. Let us first consider a Bell-type game played between three parties located at the vertices of a triangle drawn on a plane. At the start of the game, each party sets at his/her vertex a reference unit vector contained within the plane. A long sequence of unit vectors randomly oriented within the plane is then produced at the center of each of the three edges of the triangle and sent to the two parties located at their respective ends. Upon receiving a sampled random vector, each party compares its orientation to the local reference unit vector and produces a binary outcome, either $+1$ or -1, according to a deterministic response function. In this game, the affine structure of the euclidean plane allows comparing at once the relative orientations of the reference unit vectors at the three vertices, as well as the sampled random unit vectors, and, thus, it defines an "absolute frame of reference". In precise terms, the plane is equipped with an equivalence relationship that allows us to univocally define the relative orientation of vectors located at different sites. It is then straightforward to derive the Bell inequality for the pairwise correlations between the binary outcomes of the parties. However, it can be readily seen that such an "absolute frame of reference" does not exist if we consider a similar Bell-type game played between parties located on the surface of a sphere instead of a plane: a tangent vector parallel-transported over a closed-loop drawn on the sphere may acquire a non-zero geometric rotation phase due to a holonomy. Therefore, even though any two parties can calibrate and agree on a common frame of reference to describe the relative orientations of their reference unit vectors as well as the orientation of the random vectors shared between them, there does not exist a common frame of reference upon which all three parties can agree at once. In order to compare (and maybe constrain) the pairwise correlations that can be attained in the latter game, it is necessary to set the reference unit vector of one of the parties as a fixed common frame by taking advantage of the gauge degrees of freedom involved in the problem.

Gauge degrees of freedom are auxiliary degrees that may appear in the theoretical models but do not correspond to well-defined degrees of freedom in the described physical system, so that the predictions of the model cannot depend on them [28]. In fact, theoretical models that involve spurious gauge degrees of freedom may require a gauge-fixing condition in order to make physically sound predictions. In a Bell experiment, the relative orientation between the two detectors that test the pairs of entangled qubits is a well-defined physical degree of freedom that actually determines the correlation between their outcomes. On the other hand, the global orientation associated with a rigid rotation of the two detectors is a spurious gauge degree of freedom that should not play any role in the predictions of any properly defined theoretical model. Similarly, the setting of the three detectors needed to test the triplets of qubits prepared in the GHZ state is described by a single physical degree of freedom too, as we shall show later.

Following these insights, we built in [24,25,27] an explicit statistical model of local hidden variables that fully reproduces the predictions of quantum mechanics for the Bell states of two entangled qubits while complying with all the required symmetry demands and the hypothesis of 'free-will'. Thus, our model completes the description of these quantum states in the sense advocated by Einstein, Podolsky, and Rosen [1]. However, the model has been criticized because, even though it strictly complies with Einstein's causality principle, it supposedly violates Bell's definition of locality. In this respect, it is necessary to remember that Einstein's causality is a fundamental principle in modern physics that stems from the Lorentz covariance of the laws that describe the elementary building blocks of Nature and their interactions, while Bell's notion of locality arose only as a result of his intent to formulate Einstein's principle of causality in a way fit to prove his renowned theorem [3]. Therefore, wherever Einstein's causality principle and Bell's notion of locality do not agree, compliance with the former must prevail (see the discussion that precedes Equation (7) and also the discussion that follows Equation (35) for further details).

In this paper, we develop these ideas and build an explicit model of local hidden variables for the GHZ state of three entangled qubits. The paper is organized as follows. In Section 2, we review the argument put forward by Greenberger, Horne, and Zeilinger as a proof of the impossibility of reproducing the quantum mechanical predictions for the GHZ state within the framework of any model of local hidden variables. In Section 3, we introduce a simple, explicit model of hidden variables that overcomes this argument. In Section 4, we extend this model and discuss it in detail. Our conclusions are summarized in Section 5.

2. The GHZ Paradox

The Greenberger-Horne-Zeilinger spin polarization state of three entangled qubits, denoted as A, B, and C, is described by the quantum wavefunction:

$$|\Pi\rangle_\Phi = \frac{1}{\sqrt{2}} \left(|\uparrow\rangle^{(A)} |\uparrow\rangle^{(B)} |\uparrow\rangle^{(C)} + e^{i\Phi} |\downarrow\rangle^{(A)} |\downarrow\rangle^{(B)} |\downarrow\rangle^{(C)} \right),$$

where $\{|\uparrow\rangle, |\downarrow\rangle\}$ denotes a basis of single particle spin polarization eigenstates along its locally defined Z-axis. In this state, all three outcomes in every single event of a long sequence of strong spin polarization measurements performed on each one of the three qubits along their corresponding Z-axes must be consistently equal, either

$$S_Z^{(A)}(n) = S_Z^{(B)}(n) = S_Z^{(C)}(n) = +1,$$

or

$$S_Z^{(A)}(n) = S_Z^{(B)}(n) = S_Z^{(C)}(n) = -1,$$

for all $n \in \{1, \ldots, N\}$, with each one of the two possibilities happening with a probability of $1/2$. Here n labels each one of the many repetitions of the experiment, and N is the total number of repetitions.

In fact, in the GHZ state (2) the expected average values of long sequences of strong spin polarization measurements performed along any arbitrary directions $\Omega_\alpha^{(A)}$, $\Omega_\beta^{(B)}$, $\Omega_\gamma^{(C)}$ in the XY-planes orthogonal to the local Z-axes are equal to zero:

$$\langle S_{\Omega_\alpha}^{(A)}(n) \rangle_{n \in \mathbf{N}} = \langle S_{\Omega_\beta}^{(B)}(n) \rangle_{n \in \mathbf{N}} = \langle S_{\Omega_\gamma}^{(C)}(n) \rangle_{n \in \mathbf{N}} = 0, \qquad (2)$$

as well as their two-particles correlations:

$$\begin{aligned} \langle S_{\Omega_\alpha}^{(A)}(n) \cdot S_{\Omega_\beta}^{(B)}(n) \rangle_{n \in \mathbf{N}} &= \langle S_{\Omega_\beta}^{(B)}(n) \cdot S_{\Omega_\gamma}^{(C)}(n) \rangle_{n \in \mathbf{N}} \\ &= \langle S_{\Omega_\gamma}^{(C)}(n) \cdot S_{\Omega_\alpha}^{(A)}(n) \rangle_{n \in \mathbf{N}} = 0. \end{aligned} \qquad (3)$$

Notwithstanding, the three-particles correlation is non-zero, in general, and given by:

$$\langle S^{(A)}_{\Omega_\alpha}(n) \cdot S^{(B)}_{\Omega_\beta}(n) \cdot S^{(C)}_{\Omega_\gamma}(n) \rangle_{n \in \mathbf{N}}$$
$$= \cos\left(\Delta_{\Omega_\alpha^{(A)}} + \Delta_{\Omega_\beta^{(B)}} + \Delta_{\Omega_\gamma^{(C)}} + \Phi\right), \tag{4}$$

where $\Delta_{\Omega_\alpha^{(A)}}$, $\Delta_{\Omega_\beta^{(B)}}$ and $\Delta_{\Omega_\gamma^{(C)}}$ describe the relative orientations of each one of the measurement devices with respect to some implicit local reference directions labelled as X-axes, see Figure 1.

In particular, for $\Phi = 0$ the following four relationships follow:

$$\begin{array}{llllll}
S^{(A)}_X(n) & \cdot & S^{(B)}_X(n) & \cdot & S^{(C)}_X(n) & = +1, \quad n = 1, \ldots, N \\
S^{(A)}_X(m) & \cdot & S^{(B)}_Y(m) & \cdot & S^{(C)}_Y(m) & = -1, \quad m = 1, \ldots, M \\
S^{(A)}_Y(k) & \cdot & S^{(B)}_X(k) & \cdot & S^{(C)}_Y(k) & = -1, \quad k = 1, \ldots, K \\
S^{(A)}_Y(l) & \cdot & S^{(B)}_Y(l) & \cdot & S^{(C)}_X(l) & = -1, \quad l = 1, \ldots, L,
\end{array} \tag{5}$$

for any four sequences of strong measurements performed along directions (X, X, X), (X, Y, Y), (Y, X, Y) and (Y, Y, X).

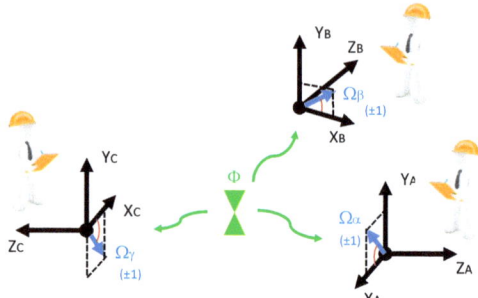

Figure 1. The GHZ argument implicitly requires the existence of an absolute frame of reference with respect to which it is possible to describe the polarization properties $(\pm 1, \pm 1, \pm 1)$ of the hypothetical hidden configurations of the triplets of qubits as well as the orientations $\Omega_\alpha^{(A)}$, $\Omega_\beta^{(B)}$, $\Omega_\gamma^{(C)}$ of the measurement devices that test them.

These four relationships (5) lie at the core of the Greenberger-Horne-Zeilinger paradox [8]. On the one hand, these relationships imply that we can gain certainty about the polarization properties of any of these three qubits, without in any sense, disturbing them. Thus, according to the notion introduced by Einstein, Podolsky, and Rosen [1], these polarization properties are *elements of reality* whose values must be set at the time when the three entangled particles are produced. On the other hand, this notion seems to be inconsistent: by multiplying the last three equations in (5) and assuming that all polarization components must take values either $+1$ or -1, we would obtain that

$$S^{(A)}_X(n) \cdot S^{(B)}_X(n) \cdot S^{(C)}_X(n) = -1, \quad n = 1, \ldots, N \tag{6}$$

which is in contradiction with the first one.

This argument is widely considered the most clear-cut evidence against the possibility of giving the wavefunction (2) a statistical interpretation within the framework of a local model of hidden configurations, in which the observers are free to choose the setting of their measurements.

3. The Paradox Revisited

The above argument crucially relies on the implicitly assumed existence of an absolute angular frame of reference, with respect to which the polarization properties of the hidden configurations of the triplets of entangled qubits, as well as the orientations of the measurement devices that test them, can be defined. In such an absolute frame of reference, all the polarization components of all possible hidden configurations must take a binary value, either $+1$ or -1, and relationships (5) immediately follow. However, as we already noticed in previous works [24–27], the existence of such an absolute angular frame of reference is not required by fundamental physical principles.

In fact, an absolute frame of reference can not be defined within the standard framework of quantum mechanics, whose predictions the models of hidden variables are aimed to reproduce. This can be readily noticed from the wavefunction that describes the GHZ state (2) in terms of the single-particle eigenstates $|\uparrow\rangle^{(A,B,C)}, |\downarrow\rangle^{(A,B,C)}$ of locally defined operators $\sigma_Z^{(A,B,C)}$. These eigenstates are defined only up to a phase (like any other normalized eigenvector of any linear operator) and, hence, the phase Φ in the wavefunction (2) is not, in principle, properly defined yet. In order to properly define this phase, it is necessary to set a reference setting (8) of the three measurement devices and experimentally obtain the threesome correlation between their outcomes. Only with respect to this reference setting of the three detectors, which we arbitrarily label as local X-axes, it is possible to properly define a subsequent rotation of any one of the devices by an angle Δ, see Figure 2. Indeed, the correlation between the binary outcomes of the three measurement devices that test the GHZ state is described by a single physical degree of freedom, the angle $\Delta + \Phi$,

$$\langle S^{(A)}_{\Omega_\alpha}(n) \cdot S^{(B)}_{\Omega_\beta}(n) \cdot S^{(C)}_{\Omega_\gamma}(n) \rangle_{n \in \mathbf{N}} = \cos(\Delta + \Phi). \tag{7}$$

while the orientations of each one of the three detectors, $\Omega_\alpha^{(A)}$, $\Omega_\beta^{(B)}$ and $\Omega_\gamma^{(C)}$, cannot be independently defined in a proper sense.

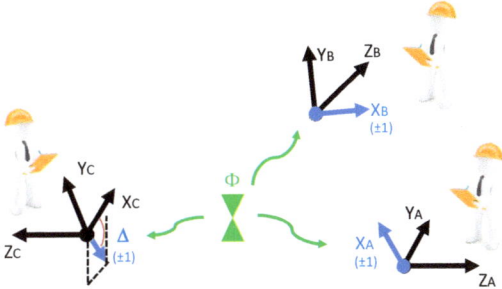

Figure 2. By symmetry considerations the orientation of one of the measurement devices, say A, can always be defined as a local X-axis for every one of the repetitions of the experiment. Moreover, the orientation of a second measurement device, say B, can also be always defined as a local X-axis since any rotation in it can be accounted for through the definition of the phase Φ that characterizes the source of the photons. In fact, as explained in Section 3, this is strictly necessary in order to properly define the quantum state (2). Thus, the experimental setting of the three measurement devices is described by a single angle Δ, while the expected correlation between their outcomes depends only on the linear combination $\Delta + \Phi$.

In the absence of an absolute frame of reference, the polarization properties of the hidden configurations may only be properly defined with respect to the reference directions set by the orientation of the corresponding measurement devices. In particular, for the reference setting of the detectors, $\Delta = 0$, whose orientations we have arbitrarily labeled as X-axes, the correlation is given by:

$$\langle S^{(A)}_X(n) \cdot S^{(B)}_X(n) \cdot S^{(C)}_X(n) \rangle_{n \in \mathbf{N}} = \cos(\Phi), \tag{8}$$

and, for $\Phi = 0$ is given by

$$\langle S_X^{(A)}(n) \cdot S_X^{(B)}(n) \cdot S_X^{(C)}(n)\rangle_{n\in \mathbf{N}} = +1. \tag{9}$$

Actually, condition (8) defines the notion of *parallel* orientations of the three measurement devices. All settings for which this condition is fulfilled are physically indistinguishable through measurements performed on triplets of entangled particles in the GHZ state, and, hence, all such sets of axes are gauge equivalent.

Moreover, the polarization properties of the hidden configurations of the triplets of entangled particles can be properly defined only with respect to the local reference directions set by the three measurement devices. That is, the actual value $s_\Omega^{(A)}(\Omega_\alpha, \omega)$ of the polarization component of, say, particle A along some direction Ω may be, in general, a function of the reference direction Ω_α set by the measurement apparatus of observer A (and, of course, also of the coordinate $\omega \in \mathcal{S}$ that labels the hidden configuration in which the system of three entangled particles occurs). This dependence does not conflict with the principle of causality, which only demands that the value of the polarization components of particle A cannot depend on the orientations of the reference directions Ω_β, Ω_γ along which observers B and C choose to test their particles. Therefore, we must not restrict our models within the constraint that all polarization components of either one of the particles must take a binary value, either $+1$ or -1: only the polarization component of each one of the particles along the reference direction set by the orientation of the corresponding measurement device must take a binary value. That is, for all possible hidden configurations of the triplet, we must have:

$$s_{\Omega_\alpha}^{(A)}(\Omega_\alpha, \omega) = \pm 1, \quad s_{\Omega_\beta}^{(B)}(\Omega_\beta, \omega) = \pm 1, \quad s_{\Omega_\gamma}^{(C)}(\Omega_\gamma, \omega) = \pm 1, \tag{10}$$

but the polarization components along any other direction must not necessarily take either one of these two values. Indeed, the only experimental access that we can have to the spin polarization components along these other directions is through weak measurements, whose outcome can have absolute values larger and smaller than one and may even be complex [29]. In fact, weak values of physical observables are complex numbers independent of the linear dimension of the Hilbert space of the described quantum system.

Therefore, it is crucial to realize that in order to obtain a meaningful description of the system, we must be careful to compare magnitudes defined with respect to the same reference directions. For example, we can state that with respect to a set of *parallel* reference directions $X^{(A)}$, $X^{(B)}$ and $X^{(C)}$ defined by condition (9), the polarization components of the particles along the orthogonal directions $Y^{(A)}$, $Y^{(B)}$, $Y^{(C)}$ take values either $+i$ or $-i$, according to the relationship:

$$\begin{aligned} s_Y^{(A)}(X,\omega) &= i\; s_X^{(A)}(X,\omega), \\ s_Y^{(B)}(X,\omega) &= i\; s_X^{(B)}(X,\omega), \\ s_Y^{(C)}(X,\omega) &= i\; s_X^{(C)}(X,\omega), \end{aligned} \tag{11}$$

with $s_X^{(A)}(X,\omega) = \pm 1, s_X^{(B)}(X,\omega) = \pm 1$ and $s_X^{(C)}(X,\omega) = \pm 1$, see Figure 3. Therefore, the four constraints (5) become trivially identical,

$$\begin{aligned} s_X^{(A)}(X,\omega) \cdot s_X^{(B)}(X,\omega) \cdot s_X^{(C)}(X,\omega) &= +1, \\ s_X^{(A)}(X,\omega) \cdot i s_X^{(B)}(X,\omega) \cdot i s_X^{(C)}(X,\omega) &= -1, \\ i s_X^{(A)}(X,\omega) \cdot s_X^{(B)}(X,\omega) \cdot i s_X^{(C)}(X,\omega) &= -1, \\ i s_X^{(A)}(X,\omega) \cdot i s_X^{(B)}(X,\omega) \cdot s_X^{(C)}(X,\omega) &= -1. \end{aligned} \tag{12}$$

In other words, the argument put forward by Greenberger, Horne, and Zeilinger as a proof of the impossibility of reproducing the predictions of quantum mechanics for the GHZ state within the framework of a model of local hidden variables can be overcome by realizing that there does not necessarily exist an absolute frame of reference with respect to which the hidden polarization properties of the entangled particles can be defined and, in consequence, allowing their actual values to depend on the reference direction with respect to which they are described.

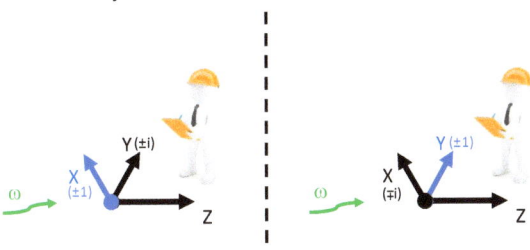

Figure 3. Two gauge-equivalent descriptions of the polarization properties of an incoming photon from a GHZ triplet, with respect to two different orientations of the measurement device that tests them.

4. A Statistical Model for the GHZ State

In this section, we build and discuss in detail an explicit statistical model of local hidden variables for the GHZ state of three entangled qubits. The model complies with the 'free-will' assumption and reproduces the quantum mechanical predictions for the average values and correlations of long sequences of strong spin polarization measurements performed on the three qubits along any three arbitrary directions.

Our statistical model consists of infinitely many possible hidden configurations continuously distributed over the unit circle $\mathcal{S}_1$, a well-defined density of probability for each one of these configurations to occur, and locally defined binary response functions that specify the outcomes that each one of these hidden configurations would produce in each one of the three measurement devices as a function of their orientations.

First, we define two sub-populations within the space of all possible hidden configurations, which we label as $\eta = \pm 1$, each one occurring with a probability of $1/2$. These two sub-populations correspond, respectively, to the two possible outcomes of the measurement performed on one of the particles, say, particle A. That is,

$$S_X^{(A)}(\eta) = s_X^{(A)}(X, \eta) = \eta. \tag{13}$$

The device that measures the polarization of particle B along an arbitrary direction orthogonal to its locally defined Z-axis fixes a reference frame of angular coordinates $\omega_B \in [-\pi, \pi)$ over the circle $\mathcal{S}_1$. We assume that the density of probability for each one of the hidden configurations to occur is given by

$$g(\omega_B) = \frac{1}{4} |\sin(\omega_B)|, \tag{14}$$

and define the outcome of the measurement on particle B as:

$$S_X^{(B)}(\omega_B, \eta) = s_X^{(B)}(X, \omega_B, \eta) = \eta \cdot S(\omega_B), \tag{15}$$

with:

$$S(x) = \text{sign}(x) = \begin{cases} +1, & \text{if } x \in (0, +\pi], \\ -1, & \text{if } x \in (-\pi, 0]. \end{cases} \tag{16}$$

Similarly, the device that measures the polarization of particle C along some other arbitrary direction Ω orthogonal to its locally defined Z-axis sets its own frame of angular coordinates $\omega_C \in [-\pi, +\pi)$ over the circle S_1. By symmetry considerations, we demand that the outcome of this measurement be described by the same response function:

$$S_\Omega^{(C)}(\omega_C, \eta) = s_\Omega^{(C)}(\Omega, \omega_C, \eta) = \eta \cdot S(\omega_C).$$

Moreover, we impose that the two sets of angular coordinates ω_B and ω_C are related by the relationship:

$$\begin{aligned} \omega_C &= \omega_B', & \text{if} \quad \eta = +1, \\ \omega_C &= \pi + \omega_B', & \text{if} \quad \eta = -1, \end{aligned} \quad (17)$$

where

$$\omega_B' = L(\omega_B; \Delta + \Phi), \quad (18)$$

and

- If $\widetilde{\Delta} \in [0, \pi)$,

$$L(\omega; \widetilde{\Delta}) = \begin{cases} q(\omega) \cdot \text{arc-cos}\left(-\cos(\widetilde{\Delta}) - \cos(\omega) - 1\right), \\ \qquad \text{if} \quad -\pi \leq \omega < \widetilde{\Delta} - \pi, \\ q(\omega) \cdot \text{arc-cos}\left(+\cos(\widetilde{\Delta}) + \cos(\omega) - 1\right), \\ \qquad \text{if} \quad \widetilde{\Delta} - \pi \leq \omega < 0, \\ q(\omega) \cdot \text{arc-cos}\left(+\cos(\widetilde{\Delta}) - \cos(\omega) + 1\right), \\ \qquad \text{if} \quad 0 \leq \omega < \widetilde{\Delta}, \\ q(\omega) \cdot \text{arc-cos}\left(-\cos(\widetilde{\Delta}) + \cos(\omega) + 1\right), \\ \qquad \text{if} \quad \widetilde{\Delta} \leq \omega < +\pi, \end{cases} \quad (19)$$

- If $\widetilde{\Delta} \in [-\pi, 0)$,

$$L(\omega; \widetilde{\Delta}) = \begin{cases} q(\omega) \cdot \text{arc-cos}\left(-\cos(\widetilde{\Delta}) + \cos(\omega) + 1\right), \\ \qquad \text{if} \quad -\pi \leq \omega < \widetilde{\Delta}, \\ q(\omega) \cdot \text{arc-cos}\left(+\cos(\widetilde{\Delta}) - \cos(\omega) + 1\right), \\ \qquad \text{if} \quad \widetilde{\Delta} \leq \omega < 0, \\ q(\omega) \cdot \text{arc-cos}\left(+\cos(\widetilde{\Delta}) + \cos(\omega) - 1\right), \\ \qquad \text{if} \quad 0 \leq \omega < \widetilde{\Delta} + \pi, \\ q(\omega) \cdot \text{arc-cos}\left(-\cos(\widetilde{\Delta}) - \cos(\omega) - 1\right), \\ \qquad \text{if} \quad \widetilde{\Delta} + \pi \leq \omega < +\pi, \end{cases} \quad (20)$$

with

$$q(\omega) = \text{sign}((\omega - \widetilde{\Delta}) \text{mod}([-\pi, \pi))),$$

and the function $y = \text{arc-cos}(x)$ is defined in its main branch, such that $y \in [0, \pi]$ while $x \in [-1, +1]$. The parameter $\widetilde{\Delta} \equiv \Delta + \Phi$ in this transformation law denotes the orientation of the measurement setting, as defined in (7) and (8).

It is straightforward to check that the density of probability (14) remains functionally invariant when described with respect to the new set of coordinates, that is,

$$g(\omega_C) = \frac{1}{4}|\sin(\omega_C)|, \quad (21)$$

since
$$|d\omega' g(\omega')| = \tfrac{1}{4}|d\omega'\,\sin(\omega')| = \tfrac{1}{4}|d(\cos(\omega'))| = \\ \tfrac{1}{4}|d(\cos(\omega))| = \tfrac{1}{4}|d\omega\,\sin(\omega)| = |d\omega\,g(\omega)|, \quad (22)$$

and
$$g(\pi+\omega) = \frac{1}{4}|\sin(\omega+\pi)| = \frac{1}{4}|\sin(\omega)| = g(\omega). \quad (23)$$

In fact, these equalities state in precise terms that the probability of each hidden configuration occuring does not depend on the orientation of the reference direction chosen by the observers to describe their particles, or, in other words, that our model complies with the requirements of 'free-will'.

We can now define a partition of the circle $\mathcal{S}_1$ into four disjoint regions,
$$\mathcal{S}_1 = \mathcal{I}_{++} \bigcup \mathcal{I}_{+-} \bigcup \mathcal{I}_{-+} \bigcup \mathcal{I}_{--}, \quad (24)$$

as follows:

$$\mathcal{I}_{++} = \{\omega_B : \omega_B \in (\widetilde{\Delta}, +\pi]\} = \\ = \begin{cases} \{\omega_C : \omega_C \in (0, -\widetilde{\Delta}+\pi]\}, & \text{if } \eta = +1 \\ \{\omega_C : \omega_C \in (-\pi, -\widetilde{\Delta}]\}, & \text{if } \eta = -1 \end{cases}$$

$$\mathcal{I}_{+-} = \{\omega_B : \omega_B \in (0, \widetilde{\Delta}]\} = \\ = \begin{cases} \{\omega_C : \omega_C \in (-\widetilde{\Delta}, 0]\}, & \text{if } \eta = +1 \\ \{\omega_C : \omega_C \in (-\widetilde{\Delta}+\pi, \pi]\}, & \text{if } \eta = -1 \end{cases}$$

$$\mathcal{I}_{--} = \{\omega_B : \omega_B \in (\widetilde{\Delta}-\pi, 0]\} = \\ = \begin{cases} \{\omega_C : \omega_C \in (-\pi, -\widetilde{\Delta}]\}, & \text{if } \eta = +1 \\ \{\omega_C : \omega_C \in (0, -\widetilde{\Delta}+\pi]\}, & \text{if } \eta = -1 \end{cases}$$

$$\mathcal{I}_{-+} = \{\omega_B : \omega_B \in (-\pi, \widetilde{\Delta}-\pi]\} = \\ \begin{cases} \{\omega_C : \omega_C \in (-\widetilde{\Delta}+\pi, \pi]\}, & \text{if } \eta = +1 \\ \{\omega_C : \omega_C \in (-\widetilde{\Delta}, 0]\}, & \text{if } \eta = -1 \end{cases}$$

where we have assumed without any loss of generality that $0 \leq \widetilde{\Delta} \leq \pi$.

In each one of these four segments, the two measurements are fully correlated or anti-correlated:

- If $\eta = +1$,
$$\begin{aligned} S_X^{(B)}(\omega_B,\eta) \cdot S_\Omega^{(C)}(\omega_C,\eta)\Big|_{\mathcal{I}_{++} \cup \mathcal{I}_{--}} &= +1, \\ S_X^{(B)}(\omega_B,\eta) \cdot S_\Omega^{(C)}(\omega_C,\eta)\Big|_{\mathcal{I}_{+-} \cup \mathcal{I}_{-+}} &= -1, \end{aligned} \quad (25)$$

- If $\eta = -1$,
$$\begin{aligned} S_X^{(B)}(\omega_B,\eta) \cdot S_\Omega^{(C)}(\omega_C,\eta)\Big|_{\mathcal{I}_{++} \cup \mathcal{I}_{--}} &= -1, \\ S_X^{(B)}(\omega_B,\eta) \cdot S_\Omega^{(C)}(\omega_C,\eta)\Big|_{\mathcal{I}_{+-} \cup \mathcal{I}_{-+}} &= +1, \end{aligned} \quad (26)$$

It is straightforward to notice that
$$\mu\!\left(\mathcal{I}_{++} \bigcup \mathcal{I}_{--}\right) - \mu\!\left(\mathcal{I}_{+-} \bigcup \mathcal{I}_{-+}\right) = \cos(\widetilde{\Delta}), \quad (27)$$

where $\mu(\cdot)$ denotes the normalized measure over the circle according to the probability density distribution (14). Hence,

- Over the sub-population of states with $\eta = +1$,

$$\langle S_X^{(B)} \cdot S_\Omega^{(C)} \rangle = \cos(\widetilde{\Delta}), \tag{28}$$

- Over the sub-population of states with $\eta = -1$,

$$\langle S_X^{(B)} \cdot S_\Omega^{(C)} \rangle = -\cos(\widetilde{\Delta}). \tag{29}$$

Therefore, over the whole population the two measurements are totally uncorrelated,

$$\langle S_X^{(B)} \cdot S_\Omega^{(C)} \rangle = 0, \tag{30}$$

since each one of the two sub-populations $\eta = +1$ and $\eta = -1$ happens with probability $1/2$. The same is true for the same reason for the correlation between the outcome of the measurement on particle A and any of the other two:

$$\langle S_X^{(A)} \cdot S_X^{(B)} \rangle = \langle S_X^{(A)} \cdot S_\Omega^{(C)} \rangle = 0. \tag{31}$$

Furthermore, the three-particles correlation is given by:

$$\langle S_X^{(A)} \cdot S_X^{(B)} \cdot S_\Omega^{(C)} \rangle = \cos(\widetilde{\Delta}), \tag{32}$$

which reproduces the quantum mechanical prediction (7) for the GHZ state.

Let us remark that in the model that we have described here, similar to quantum formalism, the orientation of two of the three measurement devices sets a reference frame with respect to which the orientation of the third device is described; see Figure 2. Therefore, it does not make sense to compare two different orientations for the reference setting since they are physically indistinguishable and, hence, the orientation of the reference setting is a spurious gauge of degree of freedom. This is the ultimate reason that allows the set of angular coordinates over the circle S_1 to acquire (due to a holonomy) a non-zero geometric phase $\alpha \neq 0, \pi$ through certain cyclic transformations (19) and (20):

$$\mathcal{L}_{-\Delta} \circ \mathcal{L}_{\Delta+\Phi} \circ \mathcal{L}_{-\Phi} = \mathcal{L}_\alpha \neq \mathbb{I}, -\mathbb{I}. \tag{33}$$

The possible appearance of a geometric phase in closed loops of gauge transformations is well-known, also in classical physics. A particularly beautiful example is the gauge theory of swimming at low Reynolds numbers described in ref. [30].

Before closing this section, let us stress that Equations (17) and (18) are coordinates transformations and do not introduce any non-local interaction between the detectors. In order to clarify this issue consider a source that produces pairs of macroscopic arrows parallel to each other and randomly oriented within a locally defined XY plane. The twin arrows are then parallel-transported in opposite directions along the Z axis towards two distant detectors, each one of them consisting of an arrow that can also be arbitrarily oriented within their local XY plane. Upon arriving at their respective detectors, the relative orientation of each one of the incoming arrows is described with respect to the orientation of the corresponding detector, and a local response is produced according to (16). Obviously, for every pair of incoming twin arrows, the following relationship must hold:

$$\omega' = \omega - \widetilde{\Delta}, \tag{34}$$

where ω and ω' are the relative angles between the orientations of the incoming arrows and their corresponding detectors, and $\widetilde{\Delta}$ is the relative angle between the two detectors. This relationship (34) does not introduce any non-local interaction between the detectors, since it is dictated by the euclidean structure of the macroscopic space, and, therefore, it is fulfilled no matter who decides how to orient the detectors or when these decisions are taken. Furthermore, the response of each one of the detectors given by (16) depends only on the orientation of the incoming arrow with respect to the local detector and does not depend either on the relative orientation between the two detectors or on the orientation of the other arrow with respect to the other detector.

Equations (17) and (18) are nothing but a non-linear generalization of the Euclidean relationship (34), and it simply means that the entangled particles might carry with them a non-euclidean metric. In this sense, it is useful to think about Equations (17) and (18) as somehow similar to the Lorentz transformation that relates, for example, the frequencies ν and ν' of a signal emitted by a source towards two detectors moving with relative velocity V,

$$\nu' = L(\nu; V). \tag{35}$$

Obviously, this non-linear relationship does not violate Einstein's principle of causality since it is dictated by the Minkowski metric of space-time, from which the very notion of causality stems.

As a last final comment, let us remind ourselves again that Bell's definition of locality arose as a result of the intent by Bell to formulate Einstein's principle of causality in a way fit to prove his renowned Bell's theorem. Therefore, wherever the notion of Bell's locality disagrees with Einstein's principle of causality, the latter must prevail.

5. Discussion

We have shown in this paper that the argument behind the renowned GHZ paradox crucially relies on an implicit assumption that is not required by fundamental physical principles and, therefore, can be overcome by giving up this unnecessary requirement. Namely, the argument put forward by Greenbereger, Horne, and Zeilinger thirty years ago [8] implicitly assumes that there exists an absolute angular frame of reference with respect to which we can define the polarization properties of the hypothetical hidden configurations of the entangled qubits, as well as the orientations of the measurement devices that test them. However, we have remarked in this and previous papers [24–27] that in order to properly define the phase Φ that characterizes the state (2) of the triplets of entangled qubits, it is necessary to fix an arbitrary reference setting of the measurement devices that test their polarizations. Only with respect to this reference setting can one properly define a subsequent relative rotation Δ of one of the devices, see Figure 2, while the orientation of the reference setting is a spurious gauge degree of freedom. In the absence of an absolute frame of reference, the polarization properties of each of the qubits can be properly defined only with respect to the orientation of the measurement device that tests them.

With these observations in mind, we have built an explicit statistical model of local hidden variables for the GHZ state of three entangled qubits that reproduces the predictions of quantum mechanics and complies with the 'free-will' assumption. The model, thus, completes—in the EPR sense—the quantum description of the GHZ state. This model closely resembles the model of hidden variables for the Bell polarization states of two entangled qubits that we recently described in [24–27].

Since these models were designed to reproduce the predictions of quantum mechanics for the Bell and GHZ experiments, they cannot be experimentally favored or disfavored against the quantum formalism through their predictions for these experiments. Further work is needed in order to develop this statistical framework and maybe find ways to test it against quantum mechanics, but this is beyond the scope of the present papers. The aim of these models at this stage is to explore the possibility that the strongly well-established quantum formalism could not be the ultimate framework for describing the fundamental building blocks of Nature and their interactions, overcoming a belief widely held by the physics community for over half a century. An underlying statistical framework would provide a physically intuitive interpretation of the quantum formalism and a better understanding of quantum phenomena.

Finally, it is worth stressing once more that the existence of an absolute frame of reference is neither demanded nor guaranteed by fundamental physical principles or any experimental evidence, and, therefore, it is at best a working assumption. However, according to the conclusions reached in this paper and [24–27], this working assumption lies at the core of the impossibility noticed by the Bell theorem, the GHZ theorem, and other renowned theorems to accommodate together within the quantum formalism some of the most fundamental physical notions. On the other hand, these difficulties can be easily overcome by lifting this working assumption. Therefore, the latter might even be considered a favored option since the apparent emergence of an absolute frame in the macroscopic world can also be easily understood [24].

Funding: This research received no external funding.

Data Availability Statement: Not applicable.

Conflicts of Interest: The author declares no conflict of interest.

References

1. Einstein, A.; Podolsky, B.; Rosen, N. Can Quantum-Mechanical Description of Physical Reality Be Considered Complete? *Phys. Rev.* **1935**, *47*, 777–780. [CrossRef]
2. Bohm, D. *Quantum Theory*; Prentice-Hall: New York, NY, USA, 1951.
3. Bell, J.S. On the Einstein Podolsky Rosen paradox. *Physics* **1964**, *1*, 195–200. [CrossRef]
4. Kochen, S; Specker, E.P. The Problem of Hidden Variables in Quantum Mechanics. *J. Math. Mech.* **1967**, *17*, 59–87. [CrossRef]
5. Clauser, J.F.; Horne, M.A.; Shimony, A.; Holt, R.A. Proposed Experiment to Test Local Hidden Variable Theories. *Phys. Rev. Lett.* **1969**, *23*, 880–884. [CrossRef]
6. Clauser, J.F.; Horne, M.A. Experimental consequences of objective local theories. *Phys. Rev. D* **1974**, *10*, 526–535. [CrossRef]
7. Fine, A. Hidden Variables, Joint Probability, and the Bell Inequalities. *Phys. Rev. Lett.* **1982**, *48*, 291–295. [CrossRef]
8. Greenberger, D.M.; Horne, M.A.; Zeilinger, A. Going beyond Bell's theorem. In *Bell's Theorem, Quantum Theory, and Conceptions of the Universe*; Kafatos, M., Ed.; Kluwer: Dordrecht, The Netherlands, 1989; pp. 69–72.
9. Klyachko, A.A.; Can, M.A.; Binicioglu, S.; Shumovsky, A.S. Simple test for hidden variables in spin-1 systems. *Phys. Rev. Lett.* **2008**, *101*, 020403. [CrossRef]
10. Leggett, A.J. Nonlocal Hidden-Variable Theories and Quantum Mechanics: An Incompatibility Theorem. *Found. Phys.* **2003**, *33*, 1469–1493. [CrossRef]
11. Cabello, A. Experimentally Testable State-Independent Quantum Contextuality. *Phys. Rev. Lett.* **2008**, *101*, 210401. [CrossRef]
12. Colbeck, R.; Renner, R. Hidden Variable Models for Quantum Theory Cannot Have Any Local Part. *Phys. Rev. Lett.* **2008**, *101*, 050403. [CrossRef]
13. Aspect, A.; Dalibard, J.; Roger, G. Experimental Test of Bell's Inequalities Using Time-Varying Analyzers. *Phys. Rev. Lett.* **1982**, *49*, 1804–1807. [CrossRef]
14. Tittel, W.; Brendel, J.; Zbinden, H.; Gisin, N. Violation of Bell Inequalities by Photons More Than 10 km Apart. *Phys. Rev. Lett.* **1998**, *81*, 3563–3566. [CrossRef]
15. Weihs, G.; Jennewein, T.; Simon, C.; Weinfurter, H.; Zeilinger, A. Violation of Bell's Inequality under Strict Einstein Locality Conditions. *Phys. Rev. Lett.* **1998**, *81*, 5039–5043. [CrossRef]
16. Matsukevich, D.N.; Maunz, P.; Moehring, D.L.; Olmschenk, S.; Monroe, C. Bell Inequality Violation with Two Remote Atomic Qubits. *Phys. Rev. Lett.* **2008**, *100*, 150404. [CrossRef] [PubMed]
17. Rowe, M.A.; Kielpinski, D.; Meyer, V.; Sackett, C.A.; Itano, W.M.; Monroe, C.; Wineland, D.J. Experimental violation of a Bell's inequality with efficient detection. *Nature* **2001**, *409*, 791–794. [CrossRef]
18. Gröblacher, S.; Paterek, T.; Kaltenbaek, R.; Brukner, C.; Żukowski, M.; Aspelmeyer, M.; Zeilinger, A. An experimental test of non-local realism. *Nature* **2007**, *446*, 871–875. [CrossRef]
19. Branciard, C.; Brunner, N.; Gisin, N.; Kurtsiefer, C.; Lamas-Linares, A.; Ling, A.; Scarani, V. Testing quantum correlations versus single-particle properties within Leggett's model and beyond. *Nat. Phys.* **2008**, *4*, 681–685. [CrossRef]
20. Amselem, E.; Rådmark, M.; Bourennane, M.; Cabello, A. State-Independent Quantum Contextuality with Single Photons. *Phys. Rev. Lett.* **2009**, *103*, 160405. [CrossRef] [PubMed]
21. Kirchmair, G.; Zähringer, F.; Gerritsma, R.; Kleinmann, M.; Gühne, O.; Cabello, A.; Blatt, R.; Roos, C.F. State-independent experimental test of quantum contextuality. *Nature* **2009**, *460*, 494–497. [CrossRef]
22. Hensen, B.; Bernien, H.; Dreau, A.E.; Reiserer, A.; Kalb, N.; Blok, M.S.; Ruitenberg, J.; Vermeulen, R.F.L.; Schouten, R.N.; Abellan, C.; et al. Loophole-free Bell inequality violation using electron spins separated by 1.3 kilometres. *Nature* **2015**, *526*, 682–686. [CrossRef]

23. Wiseman, H. Death by experiment for local realism. *Nature* **2015**, *526*, 649–650. [CrossRef] [PubMed]
24. Oaknin, D.H. The Bell theorem revisited: Geometric phases in gauge theories. *Front. Phys.* **2020**, *12*, 00142. [CrossRef]
25. Oaknin, D.H. Are models of local hidden variables for the singlet polarization state necessarily constrained by the Bell inequality? *Mod. Phys. Lett. A* **2020**, *35*, 2050229. [CrossRef]
26. Oaknin, D.H. The Franson experiment as an example of spontaneous breaking of time-translation symmetry. *Symmetry* **2022**, *14*, 380. [CrossRef]
27. Oaknin, D.H. Bypassing the Kochen-Specker theorem: An explicit non-contextual statistical model for the qutrit. *Axioms* **2023**, *12*, 90. [CrossRef]
28. Dirac, P.M. *Lectures on Quantum Mechanics*; Academic Press: New York, NY, USA, 1964.
29. Jozsa, R. Complex weak values in quantum measurement. *Phys. Rev. A* **2007**, *76*, 044103. [CrossRef]
30. Shapere, A.; Wilczek, F. Geometry of self-propulsion at low Reynolds number. *J. Fluid Mech.* **1989**, *198*, 557. [CrossRef]

Disclaimer/Publisher's Note: The statements, opinions and data contained in all publications are solely those of the individual author(s) and contributor(s) and not of MDPI and/or the editor(s). MDPI and/or the editor(s) disclaim responsibility for any injury to people or property resulting from any ideas, methods, instructions or products referred to in the content.

Article

A New Solvable Generalized Trigonometric Tangent Potential Based on SUSYQM

Lulin Xiong [1], Xin Tan [2], Shikun Zhong [1], Wei Cheng [1] and Guang Luo [1,*]

[1] College of Physics and Electronic Engineering, Chongqing Normal University, Chongqing 401331, China; lulinxiong98@outlook.com (L.X.); zhongshikun2001@outlook.com (S.Z.); c3428086519@163.com (W.C.)
[2] Chongqing Fengjie Middle School, Chongqing 404699, China; txjlja@163.com
* Correspondence: photoncn@cqnu.edu.cn

Abstract: Supersymmetric quantum mechanics has wide applications in physics. However, there are few potentials that can be solved exactly by supersymmetric quantum mechanics methods, so it is undoubtedly of great significance to find more potentials that can be solved exactly. This paper studies the supersymmetric quantum mechanics problems of the Schrödinger equation with a new kind of generalized trigonometric tangent superpotential: $A \tan npx + B \tan mpx$. We will elaborate on this new potential in the following aspects. Firstly, the shape invariant relation of partner potential is generated by the generalized trigonometric tangent superpotential. We find three shape invariance forms that satisfy the additive condition. Secondly, the eigenvalues and the eigenwave functions of the potential are studied separately in these three cases. Thirdly, the potential algebra of such a superpotential is discussed, and the discussions are explored from two aspects: one parameter's and two parameters' potential algebra. Through the potential algebra, the eigenvalue spectrums are given separately which are consistent with those mentioned earlier. Finally, we summarize the paper and give an outlook on the two-parameter shape-invariant potential.

Keywords: supersymmetric quantum mechanics; generalized trigonometric tangent superpotential; shape invariance; potential algebra

Citation: Xiong, L.; Tan, X.; Zhong, S.; Cheng, W.; Luo, G. A New Solvable Generalized Trigonometric Tangent Potential Based on SUSYQM. *Symmetry* **2022**, *14*, 1593. https://doi.org/10.3390/sym14081593

Academic Editors: Tuong Trong Truong and Eugene Oks

Received: 23 June 2022
Accepted: 27 July 2022
Published: 3 August 2022

Publisher's Note: MDPI stays neutral with regard to jurisdictional claims in published maps and institutional affiliations.

Copyright: © 2022 by the authors. Licensee MDPI, Basel, Switzerland. This article is an open access article distributed under the terms and conditions of the Creative Commons Attribution (CC BY) license (https://creativecommons.org/licenses/by/4.0/).

1. Introduction

The concept of Supersymmetry (SUSY) has permeated almost all fields of Physics: atomic and molecular physics, nuclear physics, statistical physics, and condensed matter physics [1–4]. It is even considered a necessary way to establish any unified theory [5,6]. Although SUSY has achieved great success in theoretical physics, there has been no conclusive evidence of supersymmetric partners in experiments. It was introduced by Nicolai and Witten in non-relativistic quantum mechanics [7,8]. These researchers soon found that supersymmetric quantum mechanics (SUSYQM)was of great significance and soon became a method to solve the Schrödinger equation [3,4,9,10].

The exact or quasi-exact solution of the Schrödinger equation under various potential constraints has always been a particular concern in quantum mechanics [10–14]. There are only a dozen potentials which are solvable in Schrödinger equation through SUSYQM methods. These potentials mainly include harmonic oscillator potential, Coulomb potential, Morse potential, Rosen–Morse potential, Scarf potential, Eckart potential, Pösch–Teller potential, and so on [3,14–23]. Recently, the list of these potentials has been expanded [24–26]. These precisely solvable potentials also satisfy the shape invariance condition [3,27,28], and it is found that there is a deep connection between shape invariance and SUSY. These connections need to be dealt with from the perspective of group theory. The Lie algebra is an important part of the group theory, and the potential algebra theory allows for a deep analysis of SUSYQM [29–32]. The shape invariant potentials mentioned above naturally have corresponding potential algebraic forms. Therefore, it is undoubtedly of great significance to obtain the potential algebraic form of shape invariance. The above discussion leads to the following problems: (1) How to find more solvable potentials. (2) The Riccati equation satisfied from the superpotential is only a first-order differential equation, but the solution of the equation is not easy to obtain [33]. The known solvable potential and its superpotential are consistent. Therefore, how to find more solutions to the Ricati equations is also an important problem. (3) If we can construct more solvable potentials, what exciting new results will come from these new solvable potentials?

Our group has begun tryingto promote this research from the existing superpotential. The study in [26] is our first generalization, extending the hyperbolic tangent superpotential to a linear combination of two different hyperbolic tangent, bringing positive and meaningful results. The present paper is another attempted generalization, taking the linear combination of two tangent superpotentials as our generalization potential, and the results are even more exciting.

In this paper, a superpotential with the generalized trigonometric tangent functions is proposed:

$$W(x, A, B) = A \tan npx + B \tan mpx \quad \left(-\frac{\pi}{2} < \max\{npx, mpx\} < \frac{\pi}{2}\right) \quad (1)$$

where A, B are constant coefficients, p is an arbitrary positive constant, and m and n are positive integers. The problems related to the Schrödinger equation with such superpotential are researched. Compared to the superpotential $A \tanh px + B \tanh 6px$ in [25], the superpotential in Equation (1) is undoubtedly more general. Compared with Reference [26], this article has the following differences: Firstly, the scope of the independent variable discussion is different. The potentials covered in [26] are non-periodic. The potentials studied in this paper are periodic, and we have chosen to discuss them within a period of the variable x. Secondly, the corresponding parameter binding relationship under the shape invariance constraint is completely different. Finally, the eigen-energies of these two potentials and the corresponding wave functions are not the same.

This article focuses on the following clues to illustrate our new findings. We start with a brief review of the core content of SUSYQM in the Section 2. On this basis, we proceed to study the four shape invariant algebraic relations hidden behind this new superpotential in the next section. How are the eigenvalues and potential algebras of this new potential different from other potentials? Section 4 will tell us the answer.

2. SUSYQM

For simplicity, we set $\hbar = 2m = 1$ in the steady-state Schrödinger equation $-\frac{\hbar^2}{2m}\frac{d^2\psi(x)}{dx^2} + V(x,a)\psi(x) = H\psi(x)$. The Hamiltonian of that equation is:

$$H = -\frac{d^2}{dx^2} + V(x,a) \quad (2)$$

According to the related References [10–14], the superpotential $W(x,a)$ was introduced to define the ladder operators A^+ and A^-:

$$A^{\pm}(x,a) = \mp\frac{d}{dx} + W(x,a) \quad (3)$$

The potential of the system is transformed into two partner potentials $V_{\pm}(x,a)$ to be described as:

$$V_{\pm}(x,a) = W(x,a)^2 \pm \frac{dW(x,a)}{dx} \quad (4)$$

In addition, the partner potentials $V_{\pm}(x,a)$ meet

$$V_+(x,a_0) + g(a_0) = V_-(x,a_1) + g(a_1) \quad (5)$$

where $g(a_0)$ and a_1 are functions of the additive constant a_0, and $a_1 = f(a_0)$. Equation (5) is called the shape invariance of the partner potentials. It can be rewritten as:

$$V_+(x, a_0) = V_-(x, a_1) + R(a_0) \tag{6}$$

So, it is not hard to see that

$$R(a_0) = g(a_1) - g(a_0) \tag{7}$$

The partner Hamiltonians are:

$$H_\pm = -\frac{d^2}{dx^2} + V_\pm(x, a) \tag{8}$$

That is to say:
$$H_+(x, a_0) + g(a_0) = H_-(x, a_1) + g(a_1) \tag{9}$$

The relationship between the intrinsic energies can be written as:

$$E_+(a_0) + g(a_0) = E_-(a_1) + g(a_1) \tag{10}$$

According to [3], the eigenenergy spectrum can be obtained as:

$$E_0^- = 0, E_n^+ = E_{n+1}^- \tag{11}$$

With this iterative relation, we can find all the energy levels $E_n^-(a_0)$ in turn:

$$E_n^-(a_0) = E_{n-1}^+(a_0) = g(a_n) - g(a_0)(n = 1, 2, 3 \ldots) \tag{12}$$

Not only the expression of eigenvalue $E_n^-(a_0)$, but also the expression of eigenvalue $E_n^-(a_i)(i = 0, 1, 2, \ldots)$ can be obtained:

$$E_n^-(a_i) = E_{n-1}^+(a_i) = g(a_{n+i}) - g(a_i)(n = 1, 2, 3 \ldots, i = 0, 1, 2 \ldots) \tag{13}$$

According to the superpotential and the lifting operators $A_\pm = \mp\frac{d}{dx} + W(x, a)$, we can calculate the zero-energy ground state wave function $\psi_0^-(x)$:

$$\psi_0^-(x) = N \exp\left(-\int^{(x)} W(x, a) dx\right) \tag{14}$$

where N is the normalized coefficient. According to [3], the eigenfunctions can be obtained:

$$\psi_n^+(x) = \left(E_{n+1}^-\right)^{-1/2} A^- \psi_{n+1}^-(x), \quad \psi_{n+1}^-(x) = \left(E_n^+\right)^{-1/2} A^+ \psi_n^+(x) \tag{15}$$

where $E_{n+1}^- > 0$ is required.

In SUSYQM, as long as a superpotential $W(x)$ that can be solved accurately is determined, the corresponding ascending and descending operators $A^\pm(x, a)$, partner potentials $V_\pm(x, a)$, and partner Hamiltonians $H_\pm$ can be constructed according to this superpotential $W(x)$, so as to solve the corresponding eigen energy $E_n^-(a_i)(i = 0, 1, 2, \ldots)$ and eigen wavefunction $\psi_n^-(x)$. The relationship between the superpotential and these physical quantities can be described by Figure 1.

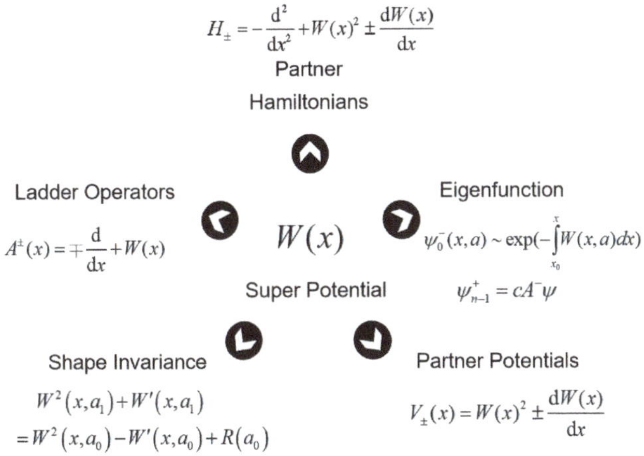

$H_{\pm} = -\frac{d^2}{dx^2} + W(x)^2 \pm \frac{dW(x)}{dx}$
Partner
Hamiltonians
Ladder Operators
$A^{\pm}(x) = \mp\frac{d}{dx} + W(x)$
$W(x)$
Super Potential
Eigenfunction
$\psi_0^-(x,a) \sim \exp(-\int_{x_0}^{x} W(x,a)dx)$
$\psi_{n-1}^+ = cA^-\psi$
Shape Invariance
$W^2(x,a_1) + W'(x,a_1)$
$= W^2(x,a_0) - W'(x,a_0) + R(a_0)$
Partner Potentials
$V_{\pm}(x) = W(x)^2 \pm \frac{dW(x)}{dx}$

Figure 1. The relationships between superpotential $W(x)$ and other physical quantities.

Figure 1 shows the importance of superpotential in SUSYQM. But the number of potentials that can be solved exactly at present is very limited. Tables A1 and A2 in Appendix A gives all the superpotentials that can be solved exactly at present and the corresponding physical quantities [3,14–26]. So, whether new superpotentials that can be solved precisely can be constructed has become the focus of research in SUSYQM. Based on this situation, this paper constructs a new superpotential, $A\tan npx + B\tan mpx$, that can be solved exactly.

3. The New Shape Invariance Derivation Idea Based on the New Solvable Potential $A\tan npx + B\tan mpx$

The generalized trigonometric tangent superpotential which we construct is given in Equation (1). The relationship between the superpotential and these parameters are shown in Figure 2.

W(x)
X
p = 0.05
p = 0.5
p = 1
A = −1.05 B = 2
A = −2.55 B = 5
A = −10.5 B = 20
A = −20.5 B = 40

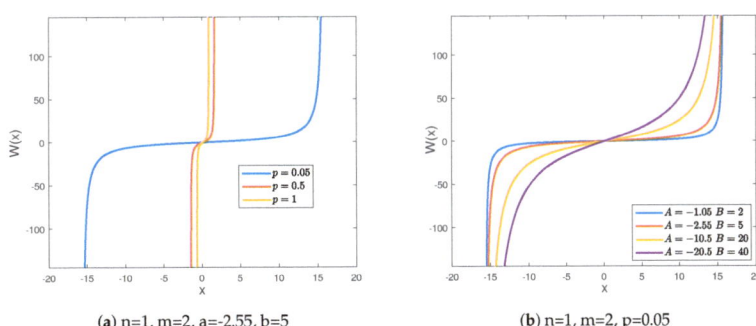

(**a**) n=1, m=2, a=-2.55, b=5

(**b**) n=1, m=2, p=0.05

Figure 2. The relationship between superpotential $W(x, A, B) = A\tan npx + B\tan mpx$ and the parameters A, B, n, p, m: (**a**) reveals the relationship between the superpotential and p; (**b**) reveals the relationship between the superpotential and A, B.

We can deduce:

$$\begin{aligned} V_+(x, A, B) &= W^2(x, A, B) + \frac{dW(x, A, B)}{dx} \\ &= A(np + A)\sec^2 npx + B(mp + B)\sec^2 mpx + 2AB\tan npx \tan mpx - A^2 - B^2 \end{aligned} \quad (16)$$

$$V_-(x,A,B) = W(x,A,B)^2 - \frac{dW(x,A,B)}{dx} \qquad (17)$$
$$= A(A-np)\sec^2 px + B(B-mp)\sec^2 mpx + 2AB\tan npx \tan mpx - A^2 - B^2$$

The figures of the partner potentials are shown in Figures 3 and 4. Figure 4 reveals the partner potentials near the origin.

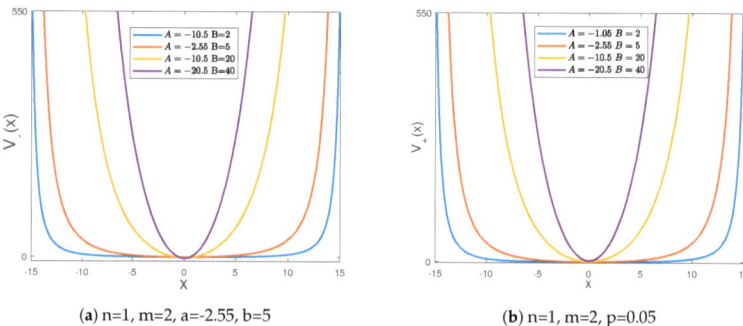

Figure 3. The Figures of the partner potentials ($n=1, m=2, p=0.05$).

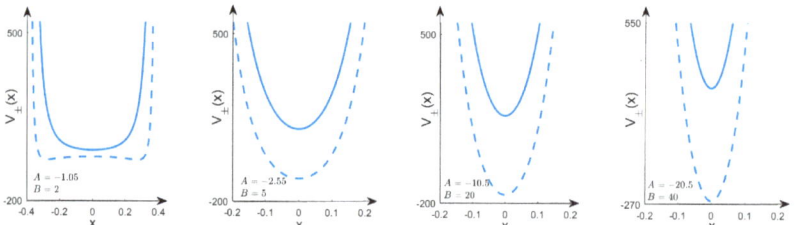

Figure 4. The partner potentials near the origin. ($n=1, m=2, p=2$). Note: the doted line is $V_-(x,A,B)$, the solid line is $V_+(x,A,B)$.

From Figure 3, it can be seen that, whatever values A and B take, the shapes of the partner potentials $V_-(x,A,B)$ and $V_+(x,A,B)$ are similar, so they conform to the shape invariance relationship described in Section 1.

Now, let us discuss the constraint relationship between A_0, A_1, B_0, and B_1. Under the condition of the shape invariance relation of $V_\pm(x,A,B)$, the independent variable x coefficient in $V_\pm(x,A,B)$ must be the same, i.e., there are:

$$A_0(np+A_0) = A_1(A_1-np) \qquad (18)$$

$$B_0(mp+B_0) = (B_1-mp)B_1 \qquad (19)$$

$$2A_0B_0 = 2A_1B_1 \qquad (20)$$

Combining Equations (18)–(20), we can obtain:

$$A_1 = A_0 + np \text{ or } A_1 = -A_0 \qquad (21)$$

$$B_1 = B_0 + mp \text{ or } B_1 = -B_0 \qquad (22)$$

It is not difficult to see that A_0, A_1, B_0 and B_1 can be combined into the following four cases which are shown in Table 1.

Table 1. The four cases of A_1, B_1.

Case 1	Case 2	Case 3	Case 4
$A_1 = A_0 + np$	$A_1 = A_0 + np$	$A_1 = -A_0$	$A_1 = -A_0$
$B_1 = B_0 + mp$	$B_1 = -B_0$	$B_1 = B_0 + mp$	$B_1 = -B_0$

As for case 4, since it does not satisfy the additivity, we do not discuss the case here. Let us analyze the wave function and energy under the other three cases in the following.

3.1. Case 1 $A_1 = A_0 + np, B_1 = B_0 + mp$

By substituting $A_1 = A_0 + np$ and $B_1 = B_0 + mp$ into Equations (16) and (17), we can obtain:

$$V_+(x, A_0, B_0) = \left(npA_0 - A_0^2\right)\sec^2 npx + \left(mpB_0 - B_0^2\right)\sec^2 mpx + 2A_0B_0 \tan npx \tan mpx - A_0^2 - B_0^2 \tag{23}$$

$$V_-(x, A_1, B_1) = \left(npA_0 - A_0^2\right)\sec^2 npx + \left(mpB_0 - B_0^2\right)\sec^2 mpx + 2(A_0 + np)(B_0 + mp) \tan npx \tan mpx - (A_0 + np)^2 - (B_0 + mp)^2 \tag{24}$$

Since the shape invariance relationship is satisfied between $V_+(x, A_0, B_0)$ and $V_-(x, A_1, B_1)$, the coefficients before independent variable x should be equal. That is to say, there is:

$$2A_0B_0 = 2(A_0 + np)(B_0 + mp) \tag{25}$$

From this formula, the binding relationship between the parameters can be further obtained as:

$$\frac{A_0}{n} = -p - \frac{B_0}{m} \tag{26}$$

Under this parameter constraint, the shape invariance relation can be written as:

$$V_+(x, A_0, B_0) = V_-(x, A_1, B_1) + (A_0 + np)^2 + (B_0 + mp)^2 - \left(A_0^2 + B_0^2\right) \tag{27}$$

It is not difficult to see the expression of $g(A_1, B_1), g(A_0, B_0)$ from the above formula that is:

$$g(A_1, B_1) = (A_0 + np)^2 + (B_0 + mp)^2 \tag{28}$$

$$g(A_0, B_0) = A_0^2 + B_0^2 \tag{29}$$

The coefficients A_k and B_k follow an additive relation and are easy to be obtained:

$$A_k = A_0 + knp \text{ and } B_k = B_0 + kmp \tag{30}$$

where $k = 0, 1, 2, \cdots$. The energy eigenvalue can be obtained as:

$$E_k^-(a_i) = g(a_{k+i}) - g(a_i)$$
$$= (A_0 + (k+i)np)^2 + (B_0 + (k+i)mp)^2 - \left((A_0 + inp)^2 + (B_0 + imp)^2\right) \tag{31}$$

note that $i = 0, 1, 2, \cdots$. When $i = 0$, there are:

$$E_k^{(-)}(a_0) = g(a_k) - g(a_0) = (A_0 + nkp)^2 + (B_0 + mkp)^2 - A_0^2 - B_0^2 \tag{32}$$

However, it is worth noting the condition that the shape invariance holds is that the ground state energy is zero, i.e., $E_0^- = 0$. According to Equation (31), there is:

$$E_0^+ = E_1 = (A_0 + np)^2 + (B_0 + mp)^2 - A_0^2 - B_0^2 \tag{33}$$

For all $k \geq 1$, we have $E_k^- \geq 0$ in Equation (31). Through $\frac{A_0}{\eta} + \frac{B_0}{m} = -p$, we can obtain:

$$k \geq -\frac{2(An + Bm)}{p(n^2 + m^2)} \qquad (34)$$

This means that the energy levels have lower limits. For example, if $A = 0.195$, $B = -0.49, n = 1, m = 2, p = 0.05$, then $k \geq 10$.

We can also find out the eigenfunctions of the Schrödinger equation:

$$\psi_k^-(x, A_0, B_0) = N_k A^+(x, A_0, B_0) A^+(x, A_1, B_1) \ldots A^+(x, A_{k-1}, B_{k-1}) e^{-\int^{(x)} W(x, A_k, B_k) dx} \qquad (35)$$

For example, the ground state wavefunction is:

$$\psi_0^{(-)}(x, A_0, B_0) = N_0 e^{-\int^{(x)} W(x, A_0, B_0) dx} = N_0 (\cos mpx)^{\frac{B_0}{np}} (\cos npx)^{\frac{A_0}{np}} \qquad (36)$$

and the first excited state wavefunction is:

$$\begin{aligned}\psi_1^-(x, A_0, B_0) &= N_1 \hat{A}^+(x, A_0, B_0) e^{-\int W(x, A_1, B_1)} \\ &= -N_1 (\cos npx)^{\frac{A_1}{np} - 1} (\cos mpx)^{\frac{B_1}{mp} - 1} (np \sin npx \cos mpx + mp \cos npx \sin mpx)\end{aligned} \qquad (37)$$

where N_k, N_0, and N_1 are the normalization coefficients. Some of the eigenfunctions and their relationships are shown in Figure 5.

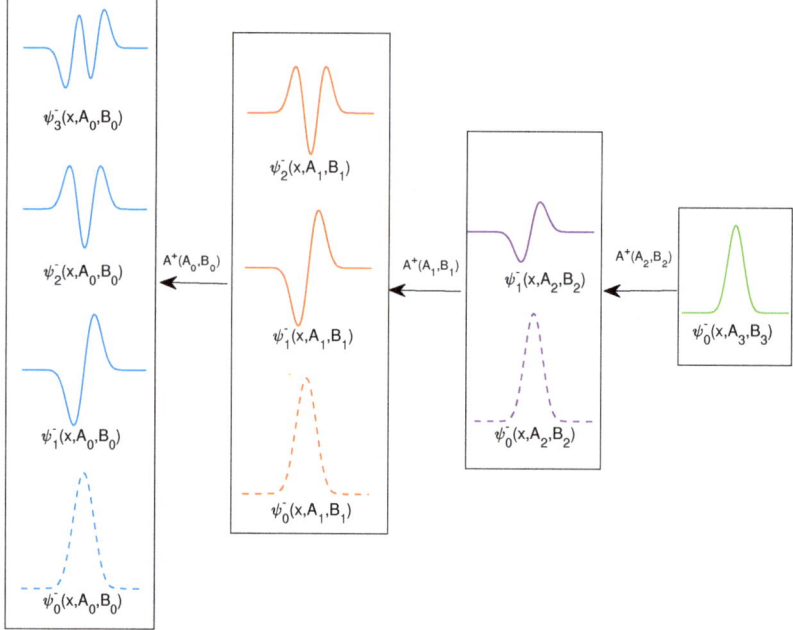

Figure 5. Some of the eigenfunctions ($n = 1, m = 2, p = 0.05$, $A_0 = -1.05$, $B_0 = 2$).

Of course, we can also obtain the eigenwave functions of the other excited states to obtain the exact solutions of the Schrödinger equation.

3.2. Case 2 $A_1 = A_0 + np$, $B_1 = -B_0$

Putting $A_1 = A_0 + np$ and $B_1 = -B_0$ into Equations (16) and (17), we have:

$$V_+(x, A_0, B_0) = A_0(np + A_0)\sec^2 npx + B_0(mp + B_0)\sec^2 mpx \\ + 2A_0 B_0 \tan npx \tan mpx - A_0^2 - B_0^2 \quad (38)$$

$$V_-(x, A_1, B_1) = (A_0 + np)A_0 \sec^2 npx + B_0(B_0 + mp)\sec^2 mpx \\ - 2(A_0 + np)B_0 \tan npx \tan mpx - (A_0 + np)^2 - B_0^2 \quad (39)$$

Analogously, the coefficients before independent variable x should be equal, that is to say:

$$2A_0 B_0 = -2(A_0 + np)B_0 \quad (40)$$

We can obtain the binding relation between the parameters corresponding to this case, which is:

$$A_0 = -\frac{np}{2} \quad (41)$$

Furthermore, the shape invariance between $V_+(x, A_0, B_0)$ and $V_-(x, A_1, B_1)$ is given by:

$$V_+(x, A_0, B_0) = V_-(x, A_1, B_1) + (A_0 + np)^2 + B_0^2 - A_0^2 - B_0^2 \quad (42)$$

In the same way, combining with Equation (5), we can obtain:

$$g(A_1, B_1) = (A_0 + np)^2 - B_0^2 \\ g(A_0, B_0) = A_0^2 - B_0^2 \quad (43)$$

Since $A_0 = -\frac{np}{2}$, substituting it into the above formula, we have:

$$E_1^-(A_0, B_0) = g(A_1, B_1) - g(A_0, B_0) = 0 \quad (44)$$

By the recurrence of energy according to the shape invariance,

$$A_k = A_0 + knp \text{ and } B_k = (-1)^k B_0 \quad (45)$$

It still needs to satisfy

$$A_k B_k = A_{k+1} B_{k+1} \Rightarrow A_k B_k = (A_k + np) B_{k+1} \quad (46)$$

Considering the Equations (41) and (45), we can obtain:

$$A_k = -np/2 = A_0 \quad (47)$$

Obviously, it can be seen that the above formula can only exist when $k = 0$; otherwise, the energy will be less than 0, which is not allowed. That is to say, only $A_0 = -np/2$ and $A_1 = np/2$ meet the requirements.

According to Equation (35), we can see that there is only a zero-energy ground state $\psi_0^-(x)$:

$$\psi_0^{(-)}(x) = N(\cos mpx)^{\frac{B_0}{mp}} (\cos npx)^{-\frac{1}{2}} \quad (48)$$

where N is the normalization constant. The figure of the ground state $\psi_0^-(x)$ is shown in Figure 6.

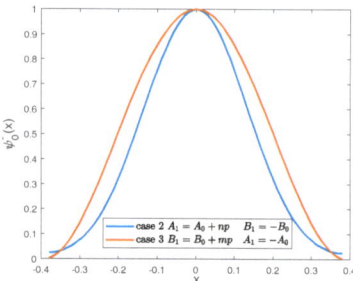

Figure 6. The figure of the ground state $\psi_0^{(-)}(x)$ in the case 2($A_0 = -2$, $B_0 = 8$) and case 3($A_0 = -2$ $B_0 = 8$) for $n = 1$, $m = 1$, $p = 2$.

3.3. Case 3 $B_1 = B_0 + mp, A_1 = -A_0$

This case is similar to the previous one. So, we have $B_0 = -mp/2$, $(A_0 \neq 0)$, and only $B_0 = -mp/2$ and $B_1 = mp/2$ meet the requirements. Since the ground state energy is zero, we can obtain:

$$E_1^- = g(A_1, B_1) - g(A_0, B_0) = 0 \qquad (49)$$

According to Equation (35), there is only a zero energy ground state $\psi_0^-(x)$:

$$\psi_0^-(x) = N'(\cos mpx)^{-\frac{1}{2}}(\cos npx)^{\frac{A_0}{np}} \qquad (50)$$

where N' is the normalization constant. The figure of the ground state $\psi_0^-(x)$ is shown in Figure 6.

From the research in the Section 3, it can be seen that the new potential $A \tan npx + B \tan mpx$ constructed in this paper can not only be precisely solved by SUSYQM but also has some special features compared with the previous potential (Appendix A); for example, it has a variety of shape invariance relationships and more rigid parameter binding relationships, which are shown in Table 2.

Table 2. The physical quantities of the new Superpotential $W(x, A, B) = A \tan npx + B \tan mpx$.

	Variation of Parms		Binding of Parms	Value of k	Eigen Energy $E_k^{(-)}$		Ground State $\psi_0^-(x)$
Case 1	$A_k = A_0 + knp$ $B_k = B_0 + kmp$	(51)	$\frac{A_0}{n} = -p - \frac{B_0}{m}$	$0, 1, 2\ldots$	$(A_0 + nkp)^2 + (B_0$ $+ mkp)^2 - A_0^2 - B_0^2$	(52)	$N_0(\cos mpx)^{\frac{B_0}{mp}}(\cos mpx)^{\frac{A_0}{np}}$
Case 2	$A_k = A_0 + knp$ $B_k = (-1)^k B_0$	(53)	$A_0 = -\frac{np}{2}$	0	0		$N(\cos mpx)^{\frac{B_0}{mp}}(\cos npx)^{-\frac{1}{2}}$
Case 3	$A_k = (-1)^n A_0$ $B_k = B_0 + kmp$	(54)	$B_0 = -\frac{mp}{2}$	0	0		$N'(\cos mpx)^{-\frac{1}{2}}(\cos npx)^{\frac{A_0}{np}}$

4. Potential Algebra of the New Superpotential $A\tan npx + B\tan mpx$

The solution and the shape invariances of Equation (5) can also be obtained by potential algebra [29–32]. Let us introduce the operators J_3, J_+ and J_- [34–37] (J_3 is a Casimir operator):

$$J_+ = e^{is\phi}\mathcal{A}^+, J_- = \mathcal{A}^- e^{-is\phi}, J_3 = k - \frac{i}{s}\partial_\phi, F(J_3) = f(\chi(sk - sJ_3)) \tag{55}$$

where s is a constant which reflects the additive step length, and k is an arbitrary constant, the function χ must satisfy the compatibility equation: $\chi(i\partial_\theta + s) = \eta(\chi(i\partial_\theta))$ in which $\eta(\chi(i\partial_\theta))$ is a function of function $\chi(i\partial_\theta)$, ϕ is an auxiliary variable, the operator $\mathcal{A}^-$ is obtained from $A^-(x, a_0)$ by introducing an auxiliary variable ϕ independent of z and replacing the parameter a_0 with an operator $\chi(i\partial_\theta)$ [34,35]:

$$x \to z, a_0 \to \chi(i\partial_\phi), a_1 \to \chi(i\partial_\phi + s), A^-(x, a_0) \to \mathcal{A}^-(z, \chi(i\partial_\phi)) \tag{56}$$

and $J_\pm$ have the characteristics of raising and lowering operators:

$$\begin{aligned}[J_+, J_-] &= J_+J_- - J_-J_+ \\ &= e^{is\phi}\mathcal{A}^+(z, \chi(i\partial_\phi))\mathcal{A}^-(z, \chi(i\partial_\phi))e^{-is\phi} - \mathcal{A}^-(z, \chi(i\partial_\phi))\mathcal{A}^+(z, \chi(i\partial_\phi)) \\ &= \mathcal{A}^+(z, \chi(i\partial_\phi + s))\mathcal{A}^-(z, \chi(i\partial_\phi + s)) - \mathcal{A}^-(z, \chi(i\partial_\phi))\mathcal{A}^+(z, \chi(i\partial_\phi))\end{aligned} \tag{57}$$

In addition, J_3 satisfies the following properties:

$$e^{\pm is\varphi}J_3 e^{\mp is\varphi} = J_3 \pm s, e^{\pm is\varphi}J_3^2 e^{\mp is\varphi} = (J_3 \pm s)^2 \tag{58}$$

For further discussion, see Reference [35]. The commutations of J_+, J_- and J_3 are satisfied with:

$$[J_3, J_\pm] = \pm J_\pm \quad [J_+, J_-] = F(J_3) \tag{59}$$

For the general algebra described in Equation (58), these operators are explicitly checked:

$$J_-J_+ + G(J_3) = J_+J_- + G(J_3 - 1), F(J_3) = G(J_3) - G(J_3 - 1) \tag{60}$$

where $G(J_3)$ is a function of J_3. Suppose $|h>$ is an arbitrary eigenstate of J_3, and $J_\pm$ plays the role of raising and lowering operators. Then, there are:

$$J_3|h> = h|h>, J_-|h> = a(h)|h-1>, J_+|h> = a(h+1)|h+1> \tag{61}$$

where $a(h)$ is a function of eigenvalue h. According to $[J_+, J_-]|h> = F(J_3)|h>$, we obtain:

$$J_+J_- - J_-J_+ = |a(h)|^2 - |a(h+1)|^2 = G(h) - G(h-1) \tag{62}$$

If $h = h_{\min}$, then $J_-|h_{\min}\rangle = 0$ and $a(h_{\min}) = 0$, we have:

$$a^2(h_{\min} + 1) = G(h_{\min} - 1) - G(h_{\min}) \tag{63}$$

By substituting Equation (63) into Equation (62), we have:

$$|a(h_{\min} + 2)|^2 = G(h_{\min} - 1) - G(h_{\min} + 1) \tag{64}$$

Repeating the above steps, we can obtain:

$$a^2(h_{\min} + k) = G(h_{\min} - 1) - G(h_{\min} + k - 1) \tag{65}$$

where k is a positive integer. If $h_{\min} + k = h$, then:

$$a^2(h) = G(h - k - 1) - G(h - 1) \tag{66}$$

From Equations (62) to (66), the expression of $G(J_3)$ is critical which can be determined by $\mathcal{H} = J_+ J_-$. If $\mathcal{H}_-$ is allowed to act on the state $\psi_n(x)$, the following relation can be obtained:

$$\mathcal{H}_- \psi_n(x) = J_+ J_- \psi_n(x) = E_n^- \psi_n(x) = (G(h-k-1) - G(h-1))\psi_n(x) \quad (67)$$

Next, we need to find the potential algebra presentation $\mathcal{H}_\pm$ and $\tilde{h}$ of H and h for this new potential $A \tan(npx) + B \tan(mpx)$. Since this new solvable potential has two parameters, it is not difficult to imagine that the potential algebra constructed should also have two parameters. According to Equation (9), we can obtain:

$$\mathcal{H}_+ \left(x, \chi_A(i\partial_{\phi A}), \chi_B(i\partial_{\phi B})\right) = \mathcal{H}_- \left(x, \chi(i\partial_{\phi A} + s_A), \chi(i\partial_{\phi B} + s_B)\right) + \tilde{h}\left(\chi(i\partial_{\phi A} + s_A), \chi(i\partial_{\phi B} + s_B)\right) - \tilde{h}\left(\chi(i\partial_{\phi A}), \chi(i\partial_{\phi B})\right) \quad (68)$$

with Equations (21) and (22), we have

$$s_A = np, s_B = mp \quad (69)$$

Let $\chi(z) = z$, then

$$i\partial_{\phi A} = np\left(k_A - J_3^A\right), i\partial_{\phi B} = mp\left(k_B - J_3^B\right) \quad (70)$$

Since parameters in need to satisfy the additivity, there are constraints similar to Equations (18)–(20), and there exist three cases:

Case (i): $\chi(i\partial_{\phi A} + s_A) = np(k_A - J_3^A + 1), B_1 = -B_0$ (the parameter A satisfies the additivity);

Case (ii): $A_1 = -A_0$; $\chi(i\partial_{\phi B} + s_B) = mp(k_B - J_3^B + 1)$ (the parameter B satisfies the additivity);

Case (iii): $\chi(i\partial_{\phi A} + s_A) = np(k_A - J_3^A + 1), \chi(i\partial_{\phi B} + s_B) = mp(k_B - J_3^B + 1)$ (both A and B satisfy the additivity).

4.1. Potential Algebra Method with One Parameter

In the above three cases, Case (i) and Case (ii) belong to the single-parameter additive shape invariance, and the discussion of Case (ii) and Case (i) is very similar. So, in this part, we only make careful calculation for Case (i) and directly give the results for Case (ii).

For Case (i), according to Equations (55), (56), and (70), we have:

$$J_3^A = k_A - \frac{i}{s_A} \partial_{\phi A}, i\partial_{\phi A} = s_A\left(k_A - J_3^A\right) = np\left(k_A - J_3^A\right), B_1 = -B_0 \quad (71)$$

and

$$J_+ J_- = e^{is_A \phi_A} \left[-\frac{d}{dx} + np\left(k_A - J_3^A\right) \tan npx - B_0 \tan mpx\right]$$
$$\left[\frac{d}{dx} + np\left(k_A - J_3^A\right) \tan npx - B_0 \tan mpx\right] e^{-is_A \phi_A}$$
$$= -\frac{d^2}{dx^2} + \left(B_0 mp + B_0^2\right) \sec^2 mpx + n^2 p^2 \left(k_A - J_3^A - np\right) \quad (72)$$
$$\left(k_A - J_3^A - np - 1\right) \sec^2 npx - 2B_0 np \left(k_A - J_3^A + np\right) \tan npx \tan mpx$$
$$- B_0^2 - n^2 p^2 \left(k_A - J_3^A - np\right)^2$$

$$\begin{aligned}
J_-J_+ &= \left[\frac{d}{dx} + np\left(k_A - J_3^A\right)\tan npx + B_0\tan mpx\right]\left[-\frac{d}{dx} + np\left(k_A - J_3^A\right)\tan npx + B_0\tan mpx\right] \\
&= -\frac{d^2}{dx^2} + \left(B_0 mp + B_0^2\right)\sec^2 mpx + n^2 p^2\left(k_A - J_3^A\right)\left(k_A - J_3^A + 1\right)\sec^2 npx + \\
&\quad | 2B_0 np\left(k_A - J_3^A\right)\tan npx \tan mpx - B_0^2 - n^2 p^2\left(k_A - J_3^A\right)^2
\end{aligned} \quad (73)$$

Furthermore, we have:

$$J_+J_- - J_-J_+ = -n^2 p^2\left[-2\left(k_A - J_3^A\right)np + n^2 p^2\right] + 2B_0 n^2 p^2 \tan npx \tan mpx \quad (74)$$

Due to the additional conditional limitations, the coefficient of the term containing the variable x can be made zero by limiting the value of k. That is, it is required that:

$$2B_0(np)^2 = 0 \quad (75)$$

$$J_+J_- - J_-J_+ = 0 = F\left(J_3^A\right) \quad (76)$$

In view of Equation (76), apparently, $G(J_3^A) = G(J_3^A - np)$ and $F(J_3) = 0$. It indicates that only a single state exists in the system, and its eigenvalue is zero. This result is the same as the shape invariance counterpart in Sections 3.2 and 3.3.

4.2. Potential Algebra Method with Two Parameters

According to Equations (55), (59), and (70), we have

$$\begin{aligned}
J_+J_- &= e^{i(s_A\phi_A + s_B\phi_B)}\mathcal{A}^+\left(z, \chi(i\partial_{\phi_A}, \partial_{\phi_B})\right)\mathcal{A}^-\left(z, \chi(i\partial_{\phi_A}, \partial_{\phi_B})\right)e^{-i(s_A\phi_A + s_B\phi_B)} \\
&= \mathcal{A}^+\left(z, \chi(i\partial_{\phi_A} + s_A, i\partial_{\phi_B} + s_B)\right)\mathcal{A}^-\left(z, \chi(i\partial_{\phi_A} + s_A, i\partial_{\phi_B} + s_B)\right) \\
&= \left[-\frac{d}{dx} + np\left(k_A - J_3^A + 1\right)\tan npx + mp\left(k_B - J_3^B + 1\right)\tan mpx\right] \\
&\quad \left[\frac{d}{dx} + np\left(k_A - J_3^A + 1\right)\tan npx + mp\left(k_B - J_3^B + 1\right)\tan mpx\right] \\
&= -\frac{d^2}{dx^2} + (np)^2\left(k_A - J_3^A + 1\right)\left(k_A - J_3^A\right)\sec^2 npx + (mp)^2\left(k_B - J_3^B + 1\right) \\
&\quad \left(k_B - J_3^B\right)\sec^2 mpx + 2mnp^2\tan npx \tan mpx - (np)^2\left(k_A - J_3^A + 1\right)^2 \\
&\quad - (mp)^2\left(k_B - J_3^B + 1\right)^2
\end{aligned} \quad (77)$$

$$\begin{aligned}
J_-J_+ &= \mathcal{A}^-\left(z, \chi(i\partial_{\phi_A}, i\partial_{\phi_B})\right)e^{-i(s_A\phi_A + s_B\phi_B)}e^{i(s_A\phi_A + s_B\phi_B)}\mathcal{A}^+\left(z, \chi(i\partial_{\phi_A}, i\partial_{\phi_B})\right) \\
&= \mathcal{A}^-\left(z, \chi(i\partial_{\phi_A}, i\partial_{\phi_B})\right)\mathcal{A}^+\left(z, \chi(i\partial_{\phi_A}, i\partial_{\phi_B})\right) \\
&= \left[\frac{d}{dx} + np\left(k_A - J_3^A\right)\tan npx + mp\left(k_B - J_3^B\right)\tan mpx\right]\left[-\frac{d}{dx} + np\left(k_A - J_3^A\right)\right. \\
&\quad \left.\tan npx + mp\left(k_B - J_3^B\right)\tan mpx\right] \\
&= -\frac{d^2}{dx^2} + m^2 p^2\left(k_B - J_3^B + 1\right)\left(k_B - J_3^B\right)\sec^2 mpx + n^2 p^2\left(k_A - J_3^A + 1\right) \\
&\quad \left(k_A - J_3^A\right)\sec^2 npx + 2mnp^2\left(k_B - J_3^B\right)\left(k_A - J_3^A\right)\tan npx \tan mpx \\
&\quad - m^2 p^2\left(k_B - J_3^B\right)^2 - n^2 p^2\left(k_A - J_3^A\right)^2
\end{aligned} \quad (78)$$

Furthermore, we have:

$$J_+J_- - J_-J_+ = (mp)^2\left(k_B - J_3^B\right)^2 + (np)^2\left(k_A - J_3^A\right)^2 - \\ \left[(mp)^2\left(k_B - \left(J_3^B - 1\right)\right)^2 + (np)^2\left(k_A - \left(J_3^A - 1\right)\right)^2\right] \quad (79)$$

Under the requirement of the shape invariance, Equation (79) must be represented only by J_3. So, we need to further rewrite the above formula as:

$$J_+J_- - J_-J_+ = (np)^2\left(2J_3^A - 2k_A - 1\right) + (mp)^2\left(2J_3^B - 2k_B - 1\right) \quad (80)$$

It is not difficult to see that if we set $k_A = -\frac{1}{2}, k_B = -\frac{1}{2}$, we obtain:

$$[J_+, J_-] = 2p^2\left(n^2 J_3^A + m^2 J_3^B\right) \quad (81)$$

Considering the function $F(J_3)$ in Equation (59)

$$J_+J_- - J_-J_+ = F(J_3) = F\left(J_3^A, J_3^B\right) = G\left(J_3^A, J_3^B\right) - G\left(J_3^A - 1, J_3^B - 1\right) \quad (82)$$

we can deduce:

$$G\left(J_3^A, J_3^B\right) = (mp)^2\left(-\frac{1}{2} - J_3^B\right)^2 + (np)^2\left(-\frac{1}{2} - J_3^A\right)^2 \quad (83)$$

and have

$$E_k^- = G(h_A - k - 1, h_B - k - 1) - G(h_A - 1, h_B - 1) \quad (84)$$

Set $-\frac{1}{2} - h_A + 1 = \frac{A}{np}, -\frac{1}{2} - h_B + 1 = \frac{B}{mp}$ and we have the energy eigenvalues

$$E_k^{(-)}(a_0) = (A_0 + nkp)^2 + (B_0 + mkp)^2 - A_0^2 - B_0^2 \quad (85)$$

This is exactly the same as Equation (32).

5. Summary and Prospect

In this paper, the Schrödinger equation with a new generalized trigonometric tangent superpotential $A \tan npx + B \tan mpx$ is solved within the framework of SUSYQM. We show that the superpotential is the new superpotential that can be solved exactly, which expands the number of exactly solvable potentials shown in Appendix A. At first, the shape invariant relation of partner potential generated by superpotential are discussed from three aspects, which are all satisfied with the additivity, and the energy spectrum and eigenfunctions are obtained. Then, we again study the three aspects with additive shape invariance from the potential algebra, and we obtain the exact same energy eigenvalues as previously. Of course, the exact solutions of the equation can be derived from the ground state wave function. Finally, the energy eigenvalues are discussed.

In conclusion, this paper studies another generalization of the existing solvable potential. Taking the linear combination of $tan mpx$ superpotential and $tan npx$ superpotential as our generalization potential, the results are still exciting. The two generalizations of our research group, including [26], actually give some important information: There are two parameters, and the relationship between the parameters is reversed by the shape invariance, with constraints between the two parameters that meet the shape invariant requirement. These are quite meaningful.

Author Contributions: L.X. (First Author): conceptualization, methodology, writing—original draft; writing—review and editing; X.T.: formal analysis, project administration; S.Z. and W.C.: validation, resources; G.L. (Corresponding Author): conceptualization, funding acquisition, resources, supervision, writing—review and editing. All authors have read and agreed to the published version of the manuscript.

Funding: This research received no external funding.

Institutional Review Board Statement: Not applicable.

Informed Consent Statement: Not applicable.

Data Availability Statement: Not applicable.

Conflicts of Interest: The authors declare no conflict of interest.

Appendix A. All Potentials That Can Be Solved Exactly

Table A1. The newly constructed potential that can be solved exactly.

Name	Superpotential	Eigenenergies	Ground State Eigenfunction
Generalized Hyperbolic Tangent 1	$A \tanh npx + B \tanh mpx$	$(A+B)^2 - (A+B-knp-kmp)^2$	$(\cosh npx)^{-\frac{A}{np}} (\cosh mpx)^{-\frac{B}{mp}}$
Generalized Hyperbolic Tangent 2	$A \tanh px + B \tanh 6px$	$(A+B)^2 - (A+B-7np)^2$	$\cosh px^{\frac{1}{2}\left(-1+\frac{B_0}{6p}\right)} \cosh 6px^{-\frac{B_0}{6p}}$
Generalized Hyperbolic Tangent 3	$\left(-\frac{b}{2}+p\right) \tanh px + b \tanh 2px$	$\left(\frac{1}{2}b+p\right)^2 - \left(\frac{1}{2}b-(n+1)p\right)^2$	$\cosh px^{-1+\frac{b_0}{2p}} \cosh 2px^{-\frac{b_0}{2p}}$
Generalized Hyperbolic Tangent 4	$\frac{1}{4}(-b+4p) \tanh px + b \tanh 4px$	$\left(\frac{3}{4}b+p\right)^2 - \left(\frac{3}{4}b-(3n+1)p\right)^2$	$\cosh px^{\frac{b}{4p}} \cosh 4px^{1-\frac{b}{4p}}$
Generalized trigonometric tangent (this paper)	$\Lambda \tan npx + B \tan mpx$	$(A_0+nkp)^2 + (B_0+mkp)^2 - A_0^2 - B_0^2$ (A1)	$N_0 (\cos mpx)^{\frac{B_0}{mp}} (\cos mpx)^{\frac{A_0}{np}}$

Table A2. Exactly solvable potentials constructed long ago.

Name	Superpotential	Eigenenergies	Ground State Eigenfunction
Harmonic oscillator	$\frac{1}{2}\omega x$	$n\omega$	$\exp\left(-\frac{1}{4}\omega x^2\right)$
3-D Oscillator	$\frac{1}{2}\omega r - \frac{\ell+1}{r}$	$2n\omega$	$r^{\ell+1}\exp\left(-\frac{\omega r^2}{4}\right)$
Coulomb	$\frac{e^2}{2(\ell+1)} - \frac{\ell+1}{r}$	$\frac{1}{4}\left[\left(\frac{e^2}{\ell+1}\right)^2 - \left(\frac{e^2}{\ell+n+1}\right)^2\right]$	$r^{\ell+1}\exp\left(-\frac{1}{2}\frac{e^2}{\ell+1}r\right)$
Morse	$A - e^{-x}$	$A^2 - (A-n)^2$	$\exp\left[-\left(Ax + \frac{B}{\alpha}e^{-\alpha x}\right)\right]$
Scarf (hyperbolic)	$A\tanh x + B\,\mathrm{sech}\,x$	$A^2 - (A-n)^2$	$(\mathrm{sech}\,\alpha x)^{A/\alpha}\exp\left[-2B\tan^{-1}(e^{\alpha x})\right]$
Scarf (trigonometric)	$A\tan x - B\sec x\,(A > B)$	$(A+n)^2 - A^2$	$\left(\cos\frac{x}{2} - \sin\frac{x}{2}\right)^{A-B}\left(\sin\frac{x}{2} + \cos\frac{x}{2}\right)^{A+B}$
Rosen–Morse (trigonometric)	$-A\cot x - \frac{B}{A}$	$(A+n)^2 - A^2 + B^2\left[\frac{1}{A^2} - \frac{1}{(A+n)^2}\right]$	$\exp\left(\frac{Bx}{A}\right)\sin^A x$
Rosen–Morse (hyperbolic)	$A\tanh\alpha x + \frac{B}{A}\,(B < A^2)$	$A^2 - (A-n)^2 - \frac{B^2}{(A-n)^2} + \frac{B^2}{A^2}$	$(\mathrm{sech}\,\alpha x)^{A/\alpha}\exp\left(-\frac{Bx}{A}\right)$
Eckart (hyperbolic)	$-A\coth r + \frac{B}{A}\,(B > A^2)$	$A^2 - (A+n)^2 + B^2\left[\frac{1}{A^2} - \frac{1}{(A+n)^2}\right]$	$(\sinh\alpha r)^{A/\alpha}\exp\left(-\frac{Br}{A}\right)$

Table A2. *Cont.*

Name	Superpotential	Eigenenergies	Ground State Eigenfunction
Eckart (trigonometric)	$-A\cot\alpha x + B\csc\alpha x (A > B)$	$(A+n\alpha)^2 - A^2$	$(\sin\alpha x)^{(A-B)/\alpha} \quad (1+\cos\alpha x)^{B/\alpha}$
Pösch–Teller (hyperbolic)	$A\coth r - B\operatorname{csch} r$ $A < B$	$A^2 - (A-n)^2$	$\dfrac{(\sinh\alpha r)^{(A/\alpha)}(B-A)}{(1+\cosh\alpha r)^{B/\alpha}}$
Pösch–Teller I (hyperbolic)	$A\tan\alpha x - B\cot\alpha x$	$(A+B+2n\alpha)^2 - (A+B)^2$	$(\sin\alpha x)^{B/\alpha}(\cos\alpha x)^{A/\alpha}$
Pösch–Teller II (hyperbolic)	$A\tanh r - B\coth r (B < A)$	$(A-B)^2 - (A-B-2n\alpha)^2$	$\dfrac{(\sinh\alpha r^\circ)^{B/\alpha}}{(\cosh\alpha r^*)^{A/\alpha}}$

89

References

1. Gendenshtin, L.; Krive, I.V. Supersymmetry in quantum mechanics. *Sov. Phys. Uspekhi* **1985**, *28*, 645. [CrossRef]
2. Junker, G. *Supersymmetric Methods in Quantum and Statistical Physics*; Springer Science & Business Media: Berlin/Heidelberg, Germany, 2012; pp. 10–19.
3. Gangopadhyaya, A.; Mallow, J.V.; Rasinariu, C. *Supersymmetric Quantum Mechanics: An Introduction*; World Scientific Publishing Company: Singapore, 2017; p. 37.
4. Cooper, F.; Khare, A.; Sukhatme, U.; Haymaker, R.W. Supersymmetry in Quantum Mechanics. *Am. J. Phys.* **2003**, *71*, 409. [CrossRef]
5. Cooper, F.; Khare, A.; Sukhatme, U. Supersymmetry and quantum mechanics. *Phys. Rep.* **1995**, *251*, 267–385. [CrossRef]
6. Beckers, J.; Debergh, N.; Nikitin, A. On supersymmetries in nonrelativistic quantum mechanics. *J. Math. Phys.* **1992**, *33*, 152–160. [CrossRef]
7. Nicolai, H. Supersymmetry and spin systems. *J. Phys. Math. Gen.* **1976**, *9*, 1497. [CrossRef]
8. Witten, E. Dynamical breaking of supersymmetry. *Nucl. Phys. B* **1981**, *188*, 513–554. [CrossRef]
9. Lahiri, A.; Roy, P.K.; Bagchi, B. Supersymmetry in quantum mechanics. *Int. J. Mod. Phys. A* **1990**, *5*, 1383–1456. [CrossRef]
10. Fernández C, D.J. SUSUSY quantum mechanics. *Int. J. Mod. Phys. A* **1997**, *12*, 171–176. [CrossRef]
11. Bagchi, B.K. *Supersymmetry in Quantum and Classical Mechanics*; CRC Press: Boca Raton, FL, USA, 2000; p. 45.
12. Ushveridze, A.G. *Quasi-Exactly Solvable Models in Quantum Mechanics*; CRC Press: Boca Raton, FL, USA, 2017; p. 82.
13. Gangopadhyaya, A.; Mallow, J.V.; Rasinariu, C.; Bougie, J. Exactness of SWKB for shape invariant potentials. *Phys. Lett. A* **2020**, *384*, 126722. [CrossRef]
14. Odake, S.; Sasaki, R. Exactly Solvable Quantum Mechanics and Infinite Families of Multi-indexed Orthogonal Polynomials. *Phys. Lett. B* **2011**, *702*, 164–170. [CrossRef]
15. Bougie, J.; Gangopadhyaya, A.; Mallow, J.V.; Rasinariu, C. Generation of a novel exactly solvable potential. *Phys. Lett. A* **2015**, *379*, 2180–2183. [CrossRef]
16. Sukumar, C. Supersymmetric quantum mechanics and its applications. In Proceedings of the AIP Conference Proceedings, Sacramento, CA, USA, 4–5 August 2004; American Institute of Physics: College Park, MD, USA, 2004; pp. 166–235.
17. Dong, S.H. *Factorization Method in Quantum Mechanics*; Springer Science & Business Media: Berlin/Heidelberg, Germany, 2007; Volume 150, p. 17.
18. Arai, A. Exactly solvable supersymmetric quantum mechanics. *J. Math. Anal. Appl.* **1991**, *158*, 63–79. [CrossRef]
19. Dutt, R.; Khare, A.; Sukhatme, U.P. Supersymmetry, shape invariance, and exactly solvable potentials. *Am. J. Phys.* **1988**, *56*, 163–168. [CrossRef]
20. Ginocchio, J.N. A class of exactly solvable potentials. I. One-dimensional Schrödinger equation. *Ann. Phys.* **1984**, *152*, 203–219. [CrossRef]
21. Khare, A.; Maharana, J. Supersymmetric quantum mechanics in one, two and three dimensions. *Nucl. Phys. B* **1984**, *244*, 409–420. [CrossRef]
22. Cooper, F.; Ginocchio, J.N.; Wipf, A. Supersymmetry, operator transformations and exactly solvable potentials. *J. Phys. A Math. Gen.* **1989**, *22*, 3707. [CrossRef]
23. Junker, G.; Roy, P. Conditionally exactly solvable potentials: A supersymmetric construction method. *Ann. Phys.* **1998**, *270*, 155–177. [CrossRef]
24. Benbourenane, J.; Eleuch, H. Exactly solvable new classes of potentials with finite discrete energies. *Results Phys.* **2020**, *17*, 103034. [CrossRef]
25. Benbourenane, J.; Benbourenane, M.; Eleuch, H. Solvable Schrodinger Equations of Shape Invariant Potentials Having Superpotential W (x, A, B) = Atanh (px) + Btanh (6px). *arXiv* **2021**, arXiv:2102.02775.
26. Zhong, S.K.; Xie, T.Y.; Dong, L.; Yang, C.X.; Xiong, L.L.; Li, M.; Luo, G. Shape invariance of solvable Schrödinger equations with a generalized hyperbolic tangent superpotential. *Results Phys.* **2022**, *35*, 105369. [CrossRef]
27. Cooper, F.; Ginocchio, J.N.; Khare, A. Relationship between supersymmetry and solvable potentials. *Phys. Rev. D* **1987**, *36*, 2458. [CrossRef]
28. Khare, A.; Sukhatme, U.P. New shape-invariant potentials in supersymmetric quantum mechanics. *J. Phys. Math. Gen.* **1993**, *26*, L901. [CrossRef]
29. Bagrov, V.G.; Samsonov, B.F. Darboux transformation, factorization, and supersymmetry in one-dimensional quantum mechanics. *Theor. Math. Phys.* **1995**, *104*, 1051–1060. [CrossRef]
30. Tian, S.F.; Zhou, S.W.; Jiang, W.Y.; Zhang, H.Q. Analytic solutions, Darboux transformation operators and supersymmetry for a generalized one-dimensional time-dependent Schrödinger equation. *Appl. Math. Comput.* **2012**, *218*, 7308–7321. [CrossRef]
31. Hall, B.C. Lie groups, Lie algebras, and representations. In *Quantum Theory for Mathematicians*; Springer: Berlin/Heidelberg, Germany, 2013; pp. 333–366.
32. Lévai, G. Solvable potentials associated with su (1, 1) algebras: A systematic study. *J. Phys. Math. Gen.* **1994**, *27*, 3809. [CrossRef]
33. Zaitsev, V.F.; Polyanin, A.D. *Handbook of Exact Solutions for Ordinary Differential Equations*; CRC Press: Boca Raton, FL, USA, 2002; p. 6.
34. Ohya, S. Algebraic Description of Shape Invariance Revisited. *Acta Polytech.* **2017**, *57*, 446–453. [CrossRef]

35. Rasinariu, C.; Mallow, J.; Gangopadhyaya, A. Exactly solvable problems of quantum mechanics and their spectrum generating algebras: A review. *Open Phys.* **2007**, *5*, 111–134. [CrossRef]
36. Su, W.C. Faddeev-Skyrme Model and Rational Maps. *Chin. J. Phys.* **2002**, *40*, 516.
37. Adams, B.; Čížek, J.; Paldus, J. Lie algebraic methods and their applications to simple quantum systems. In *Advances in Quantum Chemistry*; Elsevier: Amsterdam, The Netherlands, 1988; Volume 19, pp. 1–85.

Article

A Time-Symmetric Resolution of the Einstein's Boxes Paradox

Michael B. Heaney

Independent Researcher, 3182 Stelling Drive, Palo Alto, CA 94303, USA; mheaney@alum.mit.edu

Abstract: The Einstein's Boxes paradox was developed by Einstein, de Broglie, Heisenberg, and others to demonstrate the incompleteness of the Copenhagen Formulation of quantum mechanics. I explain the paradox using the Copenhagen Formulation. I then show how a time-symmetric formulation of quantum mechanics resolves the paradox in the way envisioned by Einstein and de Broglie. Finally, I describe an experiment that can distinguish between these two formulations.

Keywords: quantum foundations; time-symmetric; Einstein's boxes; Einstein–Podolsky–Rosen (EPR)

Citation: Heaney, M.B. A Time-Symmetric Resolution of the Einstein's Boxes Paradox. *Symmetry* **2022**, *14*, 1217. https://doi.org/10.3390/sym14061217

Academic Editor: Tuong Trong Truong

Received: 11 May 2022
Accepted: 10 June 2022
Published: 13 June 2022

Publisher's Note: MDPI stays neutral with regard to jurisdictional claims in published maps and institutional affiliations.

Copyright: © 2022 by the author. Licensee MDPI, Basel, Switzerland. This article is an open access article distributed under the terms and conditions of the Creative Commons Attribution (CC BY) license (https://creativecommons.org/licenses/by/4.0/).

1. Introduction

A grand challenge of modern physics is to resolve the conceptual paradoxes in the foundations of quantum mechanics [1]. Some of these paradoxes concern nonlocality and completeness. For example, Einstein believed the Copenhagen Formulation (CF) of quantum mechanics was incomplete. He presented a thought experiment (later known as "Einstein's Bubble") explaining his reasoning at the 1927 Solvay conference [2]. In this experiment, an incident particle's wavefunction diffracts at a pinhole in a flat screen and then spreads to all parts of a hemispherical screen capable of detecting the wavefunction. The wavefunction is then detected at one point on the hemispherical screen, implying the wavefunction everywhere else vanished instantaneously. Einstein believed that this instantaneous wavefunction collapse violated the special theory of relativity, and the wavefunction must have been localized at the point of detection immediately before the detection occurred. Since the CF does not describe the wavefunction localization before detection, it must be an incomplete theory. In an earlier paper, I analyzed a one-dimensional version of this thought experiment with a time-symmetric formulation (TSF) of quantum mechanics [3], showing that the TSF did not need wavefunction collapse to explain the experimental results.

Einstein, de Broglie, Heisenberg, and others later modified Einstein's original thought experiment to emphasize the nonlocal action-at-a-distance effects. In the modified experiment, the particle's wavefunction was localized in two boxes which were separated by a space-like interval. This modified thought experiment became known as "Einstein's Boxes." Norsen wrote an excellent analysis of the history and significance of the Einstein's Boxes thought experiment using the CF [4].

Time-symmetric explanations of quantum behavior predate the discovery of the Schrödinger equation [5] and have been developed many times over the past century [6–33]. The TSF used in this paper has been described in detail and compared to other TSFs before [3,31–33]. Note in particular that the TSF used in this paper is significantly different than the Two-State Vector Formalism (TSVF) [12,23,24]. First, the TSVF postulates that a *quantum particle* is completely described by two state vectors, written as $\langle\phi|\,|\psi\rangle$, where $|\psi\rangle$ is a retarded state vector satisfying the retarded Schrödinger equation $i\hbar\partial|\psi\rangle/\partial t = H|\psi\rangle$ and the initial boundary conditions, while $\langle\phi|$ is an advanced state vector satisfying the advanced Schrödinger equation $-i\hbar\partial\langle\phi|/\partial t = \langle\phi|H$ and the final boundary conditions. In contrast, the TSF postulates that the *transition* of a quantum particle is completely described by a complex transition amplitude density $\phi^*\psi$, defined as the algebraic product of the two wavefunctions. Second, the TSVF postulates that wavefunctions collapse upon measurement, while the TSF has no collapse postulate. The particular TSF used in this paper is a type IIB model in the classification system of Wharton and Argaman [34].

Section 2 explains the paradox associated with the CF of the Einstein's Boxes thought experiment, as described by de Broglie. Section 3 reviews a CF numerical model of the thought experiment which does not resolve the paradox. Section 4 describes a TSF numerical model of the thought experiment which resolves the paradox. Section 5 discusses the conclusions and implications.

Note that this paper only concerns a single quantum particle interfering with itself and not multiple quantum particles entangled with each other.

2. The Einstein's Boxes Paradox

The Einstein's Boxes paradox was explained by de Broglie as follows [35]:

> Suppose a particle is enclosed in a box B with impermeable walls. The associated wave ψ is confined to the box and cannot leave it. The usual interpretation asserts that the particle is "potentially" present in the whole of the box B, with a probability $|\psi|^2$ at each point. Let us suppose that by some process or other, for example, by inserting a partition into the box, the box B is divided into two separate parts B_1 and B_2 and that B_1 and B_2 are then transported to two very distant places, for example to Paris and Tokyo. The particle, which has not yet appeared, thus remains potentially present in the assembly of the two boxes and its wavefunction ψ consists of two parts, one of which, ψ_1, is located in B_1 and the other, ψ_2, in B_2. The wavefunction is thus of the form $\psi = c_1\psi_1 + c_2\psi_2$, where $|c_1|^2 + |c_2|^2 = 1$.
>
> The probability laws of [the Copenhagen Formulation] now tell us that if an experiment is carried out in box B_1 in Paris, which will enable the presence of the particle to be revealed in this box, the probability of this experiment giving a positive result is $|c_1|^2$, while the probability of it giving a negative result is $|c_2|^2$. According to the usual interpretation, this would have the following significance: because the particle is present in the assembly of the two boxes prior to the observable localization, it would be immediately localized in box B_1 in the case of a positive result in Paris. This does not seem to me to be acceptable. The only reasonable interpretation appears to me to be that prior to the observable localization in B_1, we know that the particle was in one of the two boxes B_1 and B_2, but we do not know in which one, and the probabilities considered in the usual wave mechanics are the consequence of this partial ignorance. If we show that the particle is in box B_1, it implies simply that it was already there prior to localization. Thus, we now return to the clear classical concept of probability, which springs from our partial ignorance of the true situation. But, if this point of view is accepted, the description of the particle given by ψ, though leading to a perfectly *exact* description of probabilities, does not give us a *complete* description of the physical reality, because the particle must have been localized prior to the observation which revealed it, and the wavefunction ψ gives no information about this.
>
> We might note here how the usual interpretation leads to a paradox in the case of experiments with a negative result. Suppose that the particle is charged, and that in the box B_2 in Tokyo a device has been installed which enables the whole of the charged particle located in the box to be drained off and in so doing to establish an observable localization. Now, if nothing is observed, this negative result will signify that the particle is not in box B_2 and it is thus in box B_1 in Paris. But this can reasonably signify only one thing: the particle was already in Paris in box B_1 prior to the drainage experiment made in Tokyo in box B_2. Every other interpretation is absurd. How can we imagine that the simple fact of having observed *nothing* in Tokyo has been able to promote the localization of the particle at a distance of many thousands of miles away?

3. The Conventional Formulation of Einstein's Boxes

The version of Einstein's Boxes proposed by de Broglie is experimentally impractical. We will use Heisenberg's more practical version [36], shown in Figure 1. The Conventional Formulation (CF) postulates that a single free particle wavefunction with a mass m is completely described by a retarded wavefunction $\psi(\vec{r},t)$ which satisfies the initial conditions and evolves over time according to the retarded Schrödinger equation:

$$i\hbar \frac{\partial \psi}{\partial t} = -\frac{\hbar^2}{2m}\nabla^2 \psi. \quad (1)$$

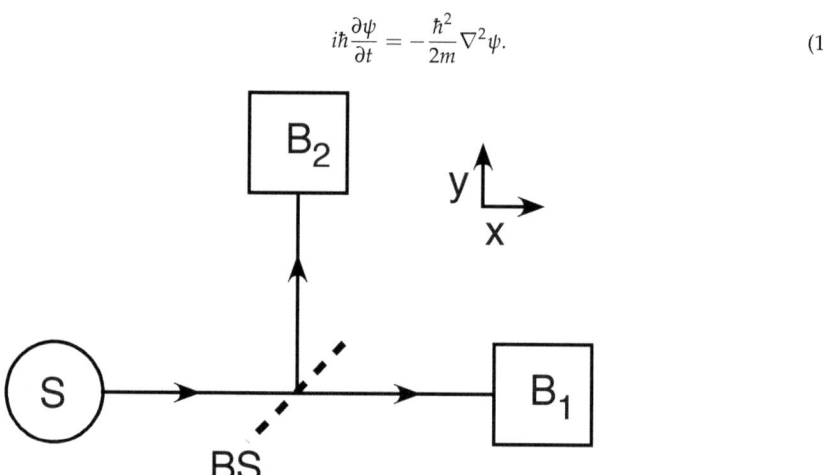

Figure 1. The modified Einstein's Boxes thought experiment. The source S can emit single-particle wavefunctions on command. Each wavefunction travels to the balanced beam splitter BS and then to box B_1 and box B_2. The two boxes are separated by a space-like interval.

The "retarded wavefunction" and "retarded Schrödinger equation" are simply the usual wavefunction and Schrödinger equation, described as retarded to distinguish them from the "advanced wavefunction" and "advanced Schrödinger equation," which will be defined below. We will use units where $\hbar = 1$ and assume the wavefunction $\psi(\vec{r},t)$ is a traveling Gaussian with an initial standard deviation $\sigma = 50$, initial momentum $k_x = 0.4$, and mass $m = 1$. We will also assume that each box contains a detector whose eigenstate is the same Gaussian as that emitted by the source. The CF assumes that a single-particle wavefunction emitted from a source S will travel to the beam splitter BS, where half of it will pass through BS and continue to box B_1 while the other half will be reflected from BS and travel to box B_2. Let us assume the two halves reach the boxes at the same time.

Figure 2 shows how the wavefunction's CF probability density $\psi^*\psi$ evolves over time, assuming the initial condition is localization in the source S. At $t = 0$, $\psi^*\psi$ is localized inside the source S. At $t = 1000$, $\psi^*\psi$ is traveling toward the beam splitter BS. At $t = 3000$, $\psi^*\psi$ has been split in half by the beam splitter, and the two halves are traveling toward box B_1 and box B_2. At $t = 4000 - \delta t$, half of $\psi^*\psi$ arrives at box B_1, while the other half arrives at box B_2. Upon a measurement at box B_2 at $t = 4000$, the CF postulates that in 50% of the runs, the half wavefunction in box B_2 collapses to zero, while simultaneously, the half wavefunction in box B_1 collapses to a full wavefunction $\phi(\vec{r},t)$, which we will assume has the same shape and size as the initial wavefunction. It was believed by de Broglie that this prediction of the CF was absurd: "How can we imagine that the simple fact of having observed *nothing* in [box B_2] has been able to promote the localization of the particle [in box B_1] at a distance of many thousands of miles away?"

The CF assumes that the probability P_c for the collapse in box B_1 is $P_c = A_c^* A_c$, where the subscript c denotes the CF and the CF transition amplitude A_c for the collapse is

$$A_c = \int_{-\infty}^{\infty} \phi^*(x,y,4000) \frac{1}{\sqrt{2}} \psi(x,y,4000) dx dy, \qquad (2)$$

where $t = 4000$ is the time of wavefunction collapse and the "quantum" factor $\frac{1}{\sqrt{2}}$ accounts for the initial wavefunction $\psi(x,y,t)$ being split in half when it reaches box B_1. Plugging in numbers gives a collapse probability $P_c = 0.43$. This probability is not $1/2$ because the evolved wavefunction at $t = 4000$ is not identical in shape to the detector eigenstate.

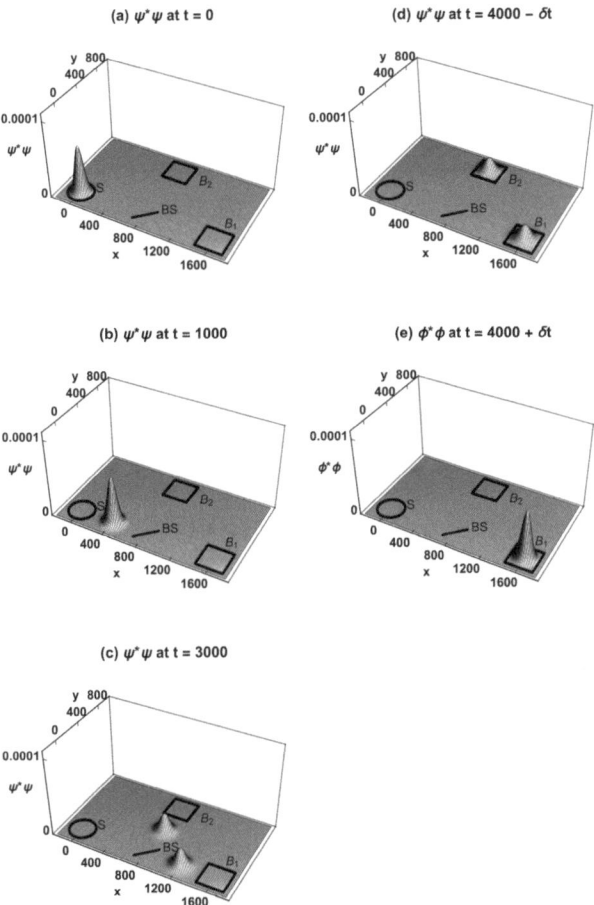

Figure 2. The Conventional Formulation (CF) explanation of the Einstein's Boxes experiment with a single-particle wavefunction emitted from source S. (**a**) The probability density $\psi^*\psi$ is localized inside S. (**b**) $\psi^*\psi$ has left S and is traveling toward the beam splitter BS. (**c**) $\psi^*\psi$ has been split in half by BS, and the two halves are traveling toward boxes B_1 and B_2. (**d**) The two halves arrive at B_1 and B_2. (**e**) A measurement at either B_1 or B_2 at $t = 4000$ causes either ψ to collapse to zero in B_2 and to a full wavefunction in B_1 (shown) or ψ to collapse to zero in B_1 and to a full wavefunction in B_2 (not shown).

4. The Time-Symmetric Formulation of Einstein's Boxes

The TSF postulates that quantum mechanics is a theory about *transitions* described by the transition amplitude density $\phi^*\psi$, where ψ is a retarded wavefunction that obeys the retarded Schrödinger equation $i\hbar \partial \psi / \partial t = H\psi$ and satisfies the initial boundary conditions, while ϕ^* is an advanced wavefunction that obeys the advanced Schrödinger equation $-i\hbar \partial \phi^* / \partial t = H\phi^*$ and satisfies the final boundary conditions. As in the TSVF, ψ can be interpreted as a retarded wavefunction from the past initial conditions, and ϕ^* can be interpreted as an advanced wavefunction from the future final conditions [3]. We will assume the same wavefunctions $\psi(\vec{r}, t)$ and $\phi(\vec{r}, t)$ as in the CF above.

An electron (e.g.) can be absorbed by a few molecules in a detector. The number of few-molecule sites in a detector is orders of magnitude larger than the number of square centimeter sites in a detector. This makes it overwhelmingly more likely that the electron will be absorbed in an area localized to a few square nanometers than much larger areas. This could explain why the transition amplitude density refocuses to a localized area at the detector. Note that there exist two unitary solutions based on the initial conditions, but time-symmetric theories also require the final conditions, which are that the particle is always found in either one or the other box. Let us then assume the final conditions are either a transition amplitude density localized in box B_1 or a transition amplitude density localized in box B_2.

Figure 3 shows the TSF explanation of the Einstein's Boxes thought experiment, assuming that the final condition is a particle transition amplitude density localized in box B_1. At $t = 0$, $|\phi^*\psi|$ is localized inside the source S. At $t = 1000$, $|\phi^*\psi|$ is traveling toward the beam splitter BS. At $t = 3000$, $|\phi^*\psi|$ has passed through the beam splitter and is traveling toward box B_1. $|\phi^*\psi|$ is zero on the path from BS to B_2 because ϕ^* is zero on this path. At $t = 4000 - \delta t$, $|\phi^*\psi|$ arrives at box B_1. Upon a measurement at box B_2 at $t = 4000$, no particle transition amplitude density is found. Upon a measurement at box B_1 at $t = 4000$, one particle's transition amplitude density is found. The one-particle transition amplitude density was localized inside box B_1 before the measurement was made.

The TSF assumes the probability P_t for the transition from localization in the source S to localization in box B_1 is $P_t = \frac{1}{2} A_t^* A_t$, where the subscript t denotes the TSF, the "classical" probability factor $\frac{1}{2}$ accounts for the fact that there are two equally likely possible final states, and the TSF amplitude A_t for the transition is

$$A_t = \int_{-\infty}^{\infty} \phi^*(x,y,t)\psi(x,y,t)dxdy, \tag{3}$$

Plugging in numbers gives a TSF transition probability $P_t = 0.43$, which is identical to the CF collapse probability result. Note that there is no transition amplitude density collapse in the TSF, so there is no need to specify the time of collapse in the integrand.

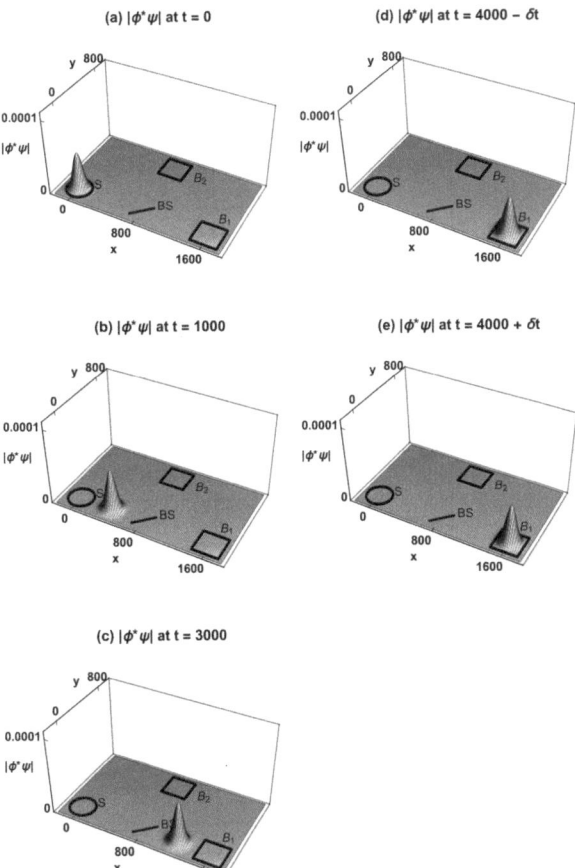

Figure 3. The time-symmetric formulation (TSF) explanation of the Einstein's Boxes experiment, with a single-particle transition amplitude density emitted from source S and detected at box B_1. (**a**) The absolute value of the transition amplitude density $|\phi^*\psi|$ is localized inside S. (**b**) $|\phi^*\psi|$ has left S and is traveling toward the beam splitter BS. (**c**) $|\phi^*\psi|$ has passed through BS and is traveling toward box B_1. $|\phi^*\psi|$ is zero on the path from BS to B_2 because ϕ^* is zero on this path. (**d**) $|\phi^*\psi|$ arrives at B_1. (**e**) Measurements at $t = 4000$ show a transition amplitude density in B_1 and no transition amplitude density in B_2. With equal probability, the final condition could have been localization in box B_2. Transition amplitude density collapse never occurs. $|\phi^*\psi|$ is normalized.

5. Discussion

To the best of my knowledge, this is the first time a TSF has been shown to resolve the Einstein's Boxes paradox. The TSF resolves the paradox in the ways that Einstein and de Broglie envisioned. The transition amplitude density $\phi^*\psi$ "localises the particle during the propagation [2]," and $\phi^*\psi$ "was already in Paris in box B_1 prior to the drainage experiment made in Tokyo in box B_2 [35]." None of the problems associated with wavefunction collapse occur. The TSF appears to give the sought-after exact description of the probabilities and a complete description of the physical reality.

One might wonder if a theory based on transition amplitude densities will be able to reproduce all of the predictions of the CF. In 1932, Dirac showed that all the experimental predictions of the CF of quantum mechanics can be formulated in terms of transition probabilities [37]. The TSF inverts this fact by postulating that quantum mechanics is a theory which experimentally predicts *only* the transition probabilities. This implies that the TSF has the same predictive power as the CF.

The TSF has the additional benefit of being consistent with the classical explanation of the Einstein's Boxes thought experiment. As the size of the "particle" becomes larger and it starts behaving more like a classical particle, it will always go to either one box or the other. There is a logical continuity between its behavior in the quantum and classical regimes, in contrast to the CF predictions.

In the TSF example above, we assumed the transition probabilities for the two boxes were the same. Now consider the case where the two transitions are not equally likely. For a very unlikely transition, the pre-experiment estimate of the TSF transition amplitude density $\phi^*\psi$ is tiny, while for a very likely transition, the pre-experiment estimate of $\phi^*\psi$ is large. However, this does not mean that $\phi^*\psi$ itself is a smaller-sized field in the event of an unlikely outcome. Before an experiment is conducted, we have classical ignorance of which transition will occur. We normalize the wavefunctions ψ and ϕ^* to unity and calculate the expected probability for each transition based on $\phi^*\psi$. After the experiment is complete, we know which of the two possible transitions actually occurred, so we renormalize the $\phi^*\psi$ of that transition to give a transition probability of one and renormalize the other $\phi^*\psi$ to zero. Note that this is an update of our classical ignorance of which transition occurred and not a physical wavefunction collapse. This may explain why the CF collapse postulate appears to work.

A central issue raised by the Einstein's Boxes paradox is the question of which elements of quantum theory should be thought of as elements of reality (ontic) and which elements are merely states of knowledge (epistemic). The TSF transition amplitude density $\phi^*\psi$ and the wavefunctions ψ and ϕ^* should be thought of as elements of reality, with the understanding that $\phi^*\psi$ is the TSF equivalent of a real particle wavefunction while ψ and ϕ^* are the TSF equivalents of virtual particle wavefunctions. For multiple particles, $\phi^*\psi$ lives in a higher dimensional configuration spacetime, which should be thought of as the stage for reality [32]. The CF concept of a superposition of paths after the beam splitter then becomes just a state of knowledge in the TSF. In reality, only one path is taken; we just do not know in advance which one. Since the TSF assumes that the sources and sinks are randomly emitting ψ and ϕ^* wavefunctions, it is a probabilistic theory. In analogy with the classical theory of special relativity, the TSF transition amplitude density can be thought of as a quantum worldtube. The higher dimensional configuration spacetime is then the quantum equivalent of Minkowski spacetime.

Finally, the CF predicts a rapid oscillating motion of a free particle's wavefunction in empty space. Schrödinger discovered the possibility of this rapid oscillating motion in 1930, naming it zitterbewegung [38]. This prediction of the CF is inconsistent with Newton's first law, since it implies a free particle's wavefunction does not move with a constant velocity. The TSF predicts zitterbewegung will never occur [3]. Direct measurements of zitterbewegung are beyond the capability of current technology, but future technological developments should allow measurements to confirm or deny its existence. Given the technology, one possible way to test for zitterbewegung would be to hold an electron in the ground state in a parabolic potential and then turn off the potential while looking for radiation at the zitterbewegung frequency of 10^{21} s^{-1}. This could distinguish between the CF and the TSF.

Funding: This research received no external funding.

Institutional Review Board Statement: Not applicable.

Informed Consent Statement: Not applicable.

Data Availability Statement: Not applicable.

Acknowledgments: I thank Travis Norsen and David A. Fotland for helpful conversations.

Conflicts of Interest: The author declares no conflict of interest.

References

1. Smolin, L. *The Trouble with Physics: The Rise of String Theory, the Fall of a Science, and What Comes Next*; Houghton Mifflin Company: New York, NY, USA, 2006; pp. 3–17.
2. Bacciagaluppi, G.; Valentini, A. *Quantum Theory at the Crossroads: Reconsidering the 1927 Solvay Conference*; Cambridge University Press: Cambridge, UK, 2009.
3. Heaney, M.B. A symmetrical interpretation of the Klein-Gordon equation. *Found. Phys.* **2013**, *43*, 733–746. [CrossRef]
4. Norsen, T. Einstein's Boxes. *Am. J. Phys.* **2005**, *73*, 164. [CrossRef]
5. Tetrode, H.M. Über den Wirkungszusammenhang der Welt. Eine Erweiterung der klassischen Dynamik. *Z. Phys.* **1922**, *10*, 317–328. Available online: http://www.mpseevinck.ruhosting.nl/seevinck/Translation_Tetrode.pdf (accessed on 20 January 2021). [CrossRef]
6. Lewis, G.N. The nature of light. *Proc. Natl. Acad. Sci. USA* **1926**, *12*, 22. [CrossRef]
7. Eddington, A.S. *The Nature of the Physical World: Gifford Lectures Delivered at the University of Edinburgh, January to March 1927*; The Macmillan Co.: New York, NY, USA, 1928; pp. 216–217.
8. Costa de Beauregard, O. Mécanique Quantique. *Compt. Rend.* **1953**, *236*, 1632–1634.
9. Watanabe, S. Symmetry of physical laws. Part III. prediction and retrodiction. *Rev. Mod. Phys.* **1955**, *27*, 179–186. [CrossRef]
10. Watanabe, S. Symmetry in time and Tanikawa's method of superquantization in regard to negative energy fields. *Prog. Theor. Phys.* **1956**, *15*, 523–535. [CrossRef]
11. Sciama, D. Determinism and the Cosmos. In *Determinism and Freedom in the Age of Modern Science*; Hook, S., Ed.; Collier Books: New York, NY, USA, 1961; pp. 90–91.
12. Aharonov, Y.; Bergmann, P.G.; Lebowitz, J.L. Time symmetry in the quantum process of measurement. *Phys. Rev.* **1964**, *134*, B1410–B1416. [CrossRef]
13. Davidon, W.C. Quantum physics of single systems. *Il Nuovo C. B* **1976**, *36*, 34–40. [CrossRef]
14. Roberts, K.V. An objective interpretation of Lagrangian quantum mechanics. *Proc. R. Soc. Lond. A* **1978**, *360*, 135–160.
15. Rietdijk, C.W. Proof of a Retroactive Influence. *Found. Phys.* **1978**, *8*, 615–628. [CrossRef]
16. Cramer, J.G. The transactional interpretation of quantum mechanics. *Rev. Mod. Phys.* **1986**, *58*, 647–687. [CrossRef]
17. Hokkyo, N. Variational formulation of transactional and related interpretations of quantum mechanics. *Found. Phys. Lett.* **1988**, *1*, 293–299. [CrossRef]
18. Sutherland, R.I. Density formalism for quantum theory. *Found. Phys.* **1998**, *28*, 1157–1190. [CrossRef]
19. Pegg, D.T.; Barnett, S.M. Retrodiction in quantum optics. *J. Opt. B Quantum Semiclass. Opt.* **1999**, *1*, 442–445. [CrossRef]
20. Wharton, K.B. Time-symmetric quantum mechanics. *Found. Phys.* **2007**, *37*, 159–168. [CrossRef]
21. Hokkyo, N. Retrocausation acting in the single-electron double-slit interference experiment. *Stud. Hist. Philos. Sci. B Stud. Hist. Philos. Mod. Phys.* **2008**, *39*, 762–766. [CrossRef]
22. Miller, D.J. Quantum mechanics as a consistency condition on initial and final boundary conditions. *Stud. Hist. Philos. Sci. B Stud. Hist. Philos. Mod. Phys.* **2008**, *39*, 767–781. [CrossRef]
23. Aharonov, Y.; Vaidman, L. The Two-State Vector Formalism: An Updated Review. In *Time in Quantum Mechanics*; Muga, G., Sala Mayato, R., Egusquiza, I., Eds.; Springer: Berlin/Heidelberg, Germany, 2008; pp. 399–447.
24. Aharonov, Y.; Popescu, S.; Tollaksen, J. A time-symmetric formulation of quantum mechanics. *Phys. Today* **2010**, *63*, 27–32. [CrossRef]
25. Wharton, K.B. A novel interpretation of the Klein-Gordon equation. *Found. Phys.* **2010**, *40*, 313–332. [CrossRef]
26. Gammelmark, S.; Julsgaard, B.; Mølmer, K. Past quantum states of a monitored system. *Phys. Rev. Lett.* **2013**, *111*, 160401. [CrossRef]
27. Price, H. *Time's Arrow and Archimedes' Point: New Directions for the Physics of Time*; Oxford University Press: New York, NY, USA, 1997.
28. Corry, R. Retrocausal models for EPR. *Stud. Hist. Philos. Mod. Phys.* **2015**, *49*, 1–9. [CrossRef]
29. Schulman, L.S. *Time's Arrows and Quantum Measurement*; Cambridge University Press: New York, NY, USA, 1997.
30. Drummond, P.D.; Reid, M.D. Retrocausal model of reality for quantum fields. *Phys. Rev. Res.* **2020**, *2*, 033266-1–033266-15. [CrossRef]
31. Heaney, M.B. A symmetrical theory of nonrelativistic quantum mechanics. *arXiv* **2013**, arXiv:1310.5348.
32. Heaney, M.B. A Time-Symmetric Formulation of Quantum Entanglement. *Entropy* **2021**, *23*, 179. [CrossRef] [PubMed]
33. Heaney, M.B. Causal Intuition and Delayed-Choice Experiments. *Entropy* **2021**, *23*, 23. [CrossRef]
34. Wharton, K.B.; Argaman, N. Colloquium: Bell's theorem and locally mediated reformulations of quantum mechanics. *Rev. Mod. Phys.* **2020**, *92*, 021002. [CrossRef]
35. de Broglie, L. *The Current Interpretation of Wave Mechanics: A Critical Study*; Elsevier: Amsterdam, The Netherlands, 1964; pp. 28–29.
36. Heisenberg, W. *The Physical Principals of the Quantum Theory*; Dover Publications: Mineola, NY, USA, 1949; p. 39.
37. Dirac, P.A.M. Relativistic quantum mechanics. *Proc. R. Soc. Lond. Ser. A* **1932**, *136*, 453–464.
38. Schrödinger, E. Über die kräftefreie Bewegung in der relativistischen Quantenmechanik. *Sitz. Preuss. Akad. Wiss. Phys. Math. Kl.* **1930**, *24*, 418–428.

Article

Global Quantum Information-Theoretic Measures in the Presence of Magnetic and Aharanov-Bohm (AB) Fields

Collins Okon Edet [1,2,3,*], Emmanuel Benjamin Ettah [1], Syed Alwee Aljunid [3], Rosdisham Endut [3], Norshamsuri Ali [3,*], Akpan Ndem Ikot [2] and Muhammad Asjad [4]

[1] Department of Physics, Cross River University of Technology, Calabar PMB 1123, Nigeria; emmanuelettah@crutech.edu.ng
[2] Theoretical Physics Group, Department of Physics, University of Port Harcourt, East/West Road, Choba PMB 5323, Nigeria; akpan.ikot@uniport.edu.ng
[3] Faculty of Electronic Engineering Technology, Universiti Malaysia Perlis, Kangar 01000, Perlis, Malaysia; syedalwee@unimap.edu.my (S.A.A.); rosdisham@unimap.edu.my (R.E.)
[4] Department of Applied Mathematics and Sciences, Khalifa University, Abu Dhabi P.O. Box 127788, United Arab Emirates; muhammad.asjad@ku.ac.ae
* Correspondence: collinsokonedet@gmail.com or collinsokonedet@crutech.edu.ng (C.O.E.); norshamsuri@unimap.edu.my (N.A.)

Citation: Edet, C.O.; Ettah, E.B.; Aljunid, S.A.; Endut, R.; Ali, N.; Ikot, A.N.; Asjad, M. Global Quantum Information-Theoretic Measures in the Presence of Magnetic and Aharanov-Bohm (AB) Fields. *Symmetry* 2022, 14, 976. https://doi.org/10.3390/sym14050976

Academic Editor: Tuong Trong Truong

Received: 18 March 2022
Accepted: 27 April 2022
Published: 10 May 2022

Publisher's Note: MDPI stays neutral with regard to jurisdictional claims in published maps and institutional affiliations.

Copyright: © 2022 by the authors. Licensee MDPI, Basel, Switzerland. This article is an open access article distributed under the terms and conditions of the Creative Commons Attribution (CC BY) license (https://creativecommons.org/licenses/by/4.0/).

Abstract: The global quantum information-theoretical analysis of the class of Yukawa potential (CYP) in the presence of magnetic and Aharonov–Bohm (AB) fields has been examined both analytically and numerically in this research piece. The energy equation and wave function for the CYP are obtained by solving the Schrodinger equation in the presence of external magnetic and AB fields using the functional analysis technique. The probability density is used to calculate the Tsallis, Rényi, and Onicescu information energy entropies numerically. The influence of the screening parameter (β), magnetic ($\vec{B}$), and AB (ξ) fields on the global information-theoretical measurements for the CYP is explored. Atomic and molecular physics, quantum chemistry, and physics are specific areas where these research findings will find application.

Keywords: magnetic and AB fields; Onicescu information energy; Rényi entropy; Shannon entropy; class of Yukawa potential; Tsallis entropy

1. Introduction

Rényi, Tsallis, and Shannon information entropies, and Onicescu information energy, are all global quantum information-theoretic measures (GQITM). These measures are focused on quantifying the spread of the probability distribution that characterizes the permitted quantum mechanical states of a system [1–5]. The importance of these global measures is to study the uncertainty of the probability distribution [6–13].

These theoretical techniques have been widely used in atomic and molecular systems, and they provide excellent insight into density functionals and electron correlation, which assists in the study of atomic structure and dynamics [14–18]. Quantum information theory (QIT) has acquired a lot of traction recently and has piqued the interest of many scholars. It has also proven to be incredibly useful in a variety of domains ranging from physics, chemistry, biology, medicine, computer science, neural networks, image recognition, linguistics, and other social sciences [19–22].

This is because QIT has a connection to current quantum communications, computing, and density functional techniques, which are the underlying theories and building blocks for a number of technological advances [18–20]. The quantification of information is a sub-discipline of applied mathematics, physics, and engineering. Nonetheless, these metrics, as well as the uncertainty relations that go with them, are essential factors in identifying a variety of atomic and molecular processes [21,22]. In quantum physics, they are commonly utilized to study quantum entanglement [22,23], quantum revivals [23,24], and atomic ionization characteristics [25]. This study has been done by several scholars for various quantum mechanical systems [25–38].

Olendski [39] studied the Shannon quantum information entropies, Fisher informations, and Onicescu energies and complexities both in the position and momentum spaces for the azimuthally symmetric two-dimensional nano-ring that is placed in uniform magnetic and Aharonov–Bohm fields. Olendski [40] calculated the one-parameter functionals of the Rényi and Tsallis entropies both in the position and momentum spaces for the azimuthally symmetric 2D nano-ring that is placed into the combination of the transverse uniform magnetic field and the Aharonov–Bohm (AB) flux and whose potential profile is modeled by the superposition of the quadratic and inverse quadratic dependencies on the radius r.

We are interested in investigating information-theoretical measures for the CYP in the presence of magnetic and Aharonov–Bohm fields in the current work. Onate and Ojonubah were the first to propose this potential [41]. Since it is a generalization of the Yukawa, Hellmann, Coulomb, and inverse quadratic Yukawa potentials, this atomic model is important [41,42]. CYP has a wide range of applications in physics, including high-energy physics, atomic and solid-state physics, and many more [43,44]. The CYP is expressed as [41]:

$$V(r) = -\frac{\tilde{A}e^{-2\beta r}}{r^2} - \frac{\tilde{B}}{r} + \frac{\tilde{C}e^{-\beta r}}{r} \qquad (1)$$

where r is the interparticle distance, $\tilde{A}$, $\tilde{B}$, and $\tilde{C}$ are the potential parameters, and β is the screening parameter which characterizes the range of the interaction [41].

In this study, we are looking for answers to the following questions: what happens to information-entropies when magnetic and Aharonov–Bohm fields have an all-encompassing effect? What happens when a lone effect occurs? As a result, we are interested in using information-theoretical measurements to investigate this spreading in both position and momentum spaces.

GITM are measures of uncertainty and information of a probability distribution and are useful in identifying strong variations on the distribution over a small region in a system; thus, they identify the local changes in the probability density, giving a good description of the quantum system [9,45].

The Shannon entropy is extended by the Rényi entropy. It is a single-parameter entropy measure family that has some important link with Shannon entropy. In the position space, Rényi entropy is defined as [2,27,28,46]:

$$R_p[\Xi_n] = \frac{1}{1-p} \ln\left[\int (\Xi(r))^p dr\right] = \frac{1}{1-p} \ln \Upsilon_p[\Xi_n], p > 0, p \neq 1. \qquad (2)$$

For the momentum space coordinate, the associated Rényi entropy is given as:

$$R_q[X_n] = \frac{1}{1-p} \ln\left[\int (X(\rho))^p d\rho\right] = \frac{1}{1-p} \ln \Upsilon_p[X_n], p > 0, p \neq 1. \qquad (3)$$

where $X_n = X(\rho) = |\Psi(\rho)|^2$. The parameter's permissible range of values is governed by the integral's convergence condition in the definition, with the crucial condition $p > 0$. In the limit $p \to 0$, the Rényi entropy changes to the Shannon entropy [34].

As p approaches zero, the Rényi entropy increasingly weighs all events with nonzero probability more equally, regardless of their probabilities. In the limit for $p \to 0$, the Rényi entropy is just the logarithm of the size of the support of Ξ_n. The limit for $p \to 1$ is the Shannon entropy. As p approaches infinity, the Rényi entropy is increasingly determined by the events of highest probability [34].

Onicescu proposed a better measure of dispersion distribution in an attempt to establish a generalization to the Shannon entropy [5]. Onicescu information energy is described as [5]:

$$E[\Xi_n] = \int \Xi^2(r)dr = \int (\Xi(r))^p dr = \Upsilon_p[\Xi_n], p = 2 \tag{4}$$

For the momentum space coordinate, the equivalent Onicescu energy is given as:

$$E[X_n] = \int X^2(\rho)d\rho = \int (X(\rho))^p d\rho = \Upsilon_q[X_n], p = 2 \tag{5}$$

The probability distribution is more concentrated and the information content is smaller as the Onicescu information energy increases. The energy product of Onicescu can thus be calculated as $E_{\rho\gamma} = E_\rho E_\gamma$.

In the position and momentum space coordinates, the Tsallis entropy is defined as [4]:

$$T_p[\Xi_n] = \frac{1}{p-1}\left(1 - \left[\int (\Xi(r))^p dr\right]\right) = \frac{1}{p-1}(1 - \Upsilon_p[\Xi_n]), p > 0, p \neq 1 \tag{6}$$

and:

$$T_p[X_n] = \frac{1}{p-1}\left(1 - \left[\int (X(\rho))^p d\rho\right]\right) = \frac{1}{p-1}(1 - \Upsilon_p[\gamma_n]), p > 0, p \neq 1 \tag{7}$$

where $X_p[\Xi_n]$ is the entropic moments. In the limit $p \to 1$, the Tsallis entropy also changes to the Shannon entropy. In Equations (2)–(7): p is a non-negative dimensionless coefficient, which can be construed as a factor describing the reaction of the system to its deviation from the equilibrium; $\Xi(r)$ is the position space probability density; and $X(\rho)$ is the momentum space probability density.

The following is how this article is structured: the normalized wave function and probability density for the CYP in the presence of magnetic and Aharonov–Bohm fields are presented in the next section. The numerical findings and explanations of the Rényi entropy, Tsallis entropy, and Onicescu information energy, as well as their respective uncertainty relations, are presented in Section 3. A final remark is made in Section 4.

2. The Model Formulation

In cylindrical coordinates, the Hamiltonian operator of a charged particle moving in the class of Yukawa potential (CYP) under the combined influence of AB and external magnetic fields may be expressed [47–49] as:

$$\left[\frac{1}{2\mu}\left(i\hbar\vec{\nabla} - \frac{e}{c}\vec{A}\right)^2 - \frac{\tilde{A}e^{-2\beta r}}{r^2} - \frac{\tilde{B}}{r} + \frac{\tilde{C}e^{-\beta r}}{r}\right]\psi(r,\varphi) = E_{nm}\psi(r,\varphi), \tag{8}$$

where E_{nm} denotes the energy level, μ is the effective mass of the system, and the vector potential which is denoted by "$\vec{A}$" is given as: $\vec{A} = \left(0, \frac{Be^{-\beta r}}{(1-e^{-\beta r})} + \frac{\phi_{AB}}{2\pi r}, 0\right)$ [47,48].

Equation (8) cannot be solved analytically, so Greene and Aldrich approximation scheme have to be employed in order to obtain the eigen solutions [46]. The energy is obtained as follows using the functional analysis approach (FAA):

$$E_{nm} = \frac{\hbar^2\beta^2\eta_m}{2\mu} - \tilde{B}\beta - \frac{\hbar^2\beta^2}{8\mu}\left[\frac{\frac{2\mu\tilde{B}}{\hbar^2\beta} - \frac{2\mu\tilde{C}}{\hbar^2\beta} - \frac{2\mu\tilde{A}}{\hbar^2} + \left(\frac{\mu\omega_c}{\hbar\beta}\right)^2 - \eta_m - (n+\sigma_m)^2}{(n+\sigma_m)}\right]^2 \tag{9}$$

where m is the magnetic quantum number:

$$\sigma_m = \frac{1}{2} + \sqrt{(m+\xi)^2 - \frac{2\mu\tilde{A}}{\hbar^2} + \left(\frac{\mu\omega_c}{\hbar\beta}\right)^2 + \frac{2\mu\omega_c}{\hbar\beta}(m+\xi)} \tag{9a}$$

and:

$$\eta_m = (m+\xi)^2 - \frac{1}{4} \tag{9b}$$

The normalized wave function $\chi_{nm}(s)$ that corresponds to the two lowest lying states $n = 0, 1$ are presented as follows:

$$\chi_0(s) = \sqrt{\frac{\beta\Gamma(1+2\lambda+2\sigma_m)}{\Gamma(2\lambda)\Gamma(1+2\sigma_m)}} \left(e^{-\beta r}\right)^{\lambda} \left(1 - e^{-\beta r}\right)^{\sigma_m} \tag{10}$$

and:

$$\chi_1(s) = \frac{1}{1+2\lambda}\sqrt{\frac{\beta\lambda(1+2\lambda)\Gamma(3+2\lambda+2\sigma_m)}{(1+\sigma_m)(1+2\lambda+2\sigma_m)\Gamma(1+2\lambda)\Gamma(1+2\sigma_m)}} \left(e^{-\beta r}\right)^{1+\lambda} \times \\ \left(1 - e^{-\beta r}\right)^{\sigma_m}\left(e^{-\beta r}(1+2\lambda) - 2\sigma_m - (1+2\lambda)\right) \tag{11}$$

where $\lambda = \sqrt{-\frac{2\mu E_{nm}}{\hbar^2\beta^2} - \frac{2\mu \tilde{C}}{\hbar^2\beta} + \eta_m}$.

The normalized momentum-space $\chi_{nm}(\rho)$ wave function for the two lowest lying states $n = 0, 1$, are obtained as [39,40]:

$$\chi_0(\rho) = \frac{1}{\sqrt{2\pi}}\int_0^{\infty} \chi_0(r)e^{-i\rho r}dr \tag{12}$$

$$\chi_0(\rho) = \sqrt{\frac{\beta\Gamma(1+2\lambda+2\sigma_m)}{\Gamma(2\lambda)\Gamma(1+2\sigma_m)}} \frac{\Gamma\left(\frac{i\rho}{\beta}+\lambda\right)\Gamma(1+\sigma_m)}{\sqrt{2\pi}\beta\Gamma\left(1+\frac{i\rho}{\beta}+\lambda+\sigma_m\right)} \tag{13}$$

$$\chi_1(\rho) = \frac{1}{\sqrt{2\pi}}\int_0^{\infty} \chi_1(r)e^{-i\rho r}dr \tag{14}$$

$$\chi_1(\rho) = \sqrt{\frac{\beta(1+2\lambda)\Gamma(3+2\lambda+2\sigma_m)}{2(1+\sigma_m)(1+2\lambda+2\sigma_m)\Gamma(2\lambda)\Gamma(1+2\sigma_m)}} \left(\frac{(-2i\rho\sigma_m + \alpha(1+2\lambda+\sigma_m))\Gamma(1+\sigma_m)\Gamma\left(\frac{i\rho}{\beta}+\lambda\right)}{\sqrt{2\pi}\beta^2(1+2\lambda)\Gamma\left(2+\frac{i\rho}{\beta}+\lambda+\sigma_m\right)}\right) \tag{15}$$

Full details of the solutions can be found in ref. [50]. We point out here that Edet and Ikot [50] have recently treated one of these global information entropies known as the Shannon entropy. In a bid to broaden the scope of our application, we will in the next section consider other global entropies.

In the absence of magnetic and AB fields, if we set $m = \ell + \frac{1}{2}$, $-\tilde{A} = \tilde{A}$, $\tilde{C} = -\tilde{C}$, $\tilde{B} = 0$, and $\beta = 0$, we recover the Kratzer–Feus potential:

$$V(r) = \frac{\tilde{A}}{r^2} - \frac{\tilde{C}}{r} \tag{16}$$

with energy:

$$E_{n\ell} = -\frac{\mu}{2\hbar^2}\frac{\tilde{C}^2}{\left(n+\frac{1}{2}+\sqrt{\ell+\frac{1}{4}+\frac{2\mu\tilde{A}}{\hbar^2}}\right)^2} \tag{17}$$

The above expressions (16) and (17) are in agreement with Ref. [51].

3. Global Information-Theoretic Measures for the CYP

In general, the derivation of these information entropies is difficult and time-consuming, particularly the analytical formulation for the Tsallis and Rényi entropies and Onicescu information energy in momentum space. This is due to the Fourier transform's intricate computation; as a result, we find the numerical result.

Figure 1a–d displays the plot of Tsallis entropies in position and momentum space, which reveals that the CYP's position Tsallis entropies diminish as the potential parameter increases, whereas the momentum space expands when the potential parameter β is amplified. In the position space with rising magnetic and AB fields, Tsallis entropy is likewise shown to decrease. In the momentum space, the opposite is the case.

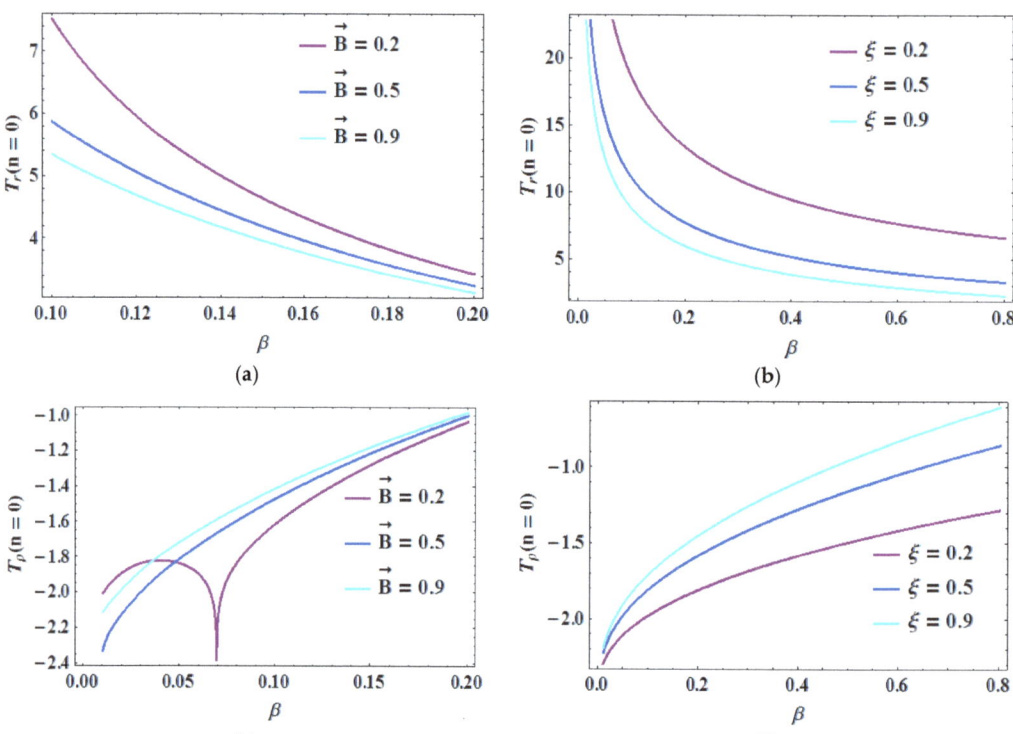

Figure 1. Position space Tsallis entropies $T_r(r)$ versus: (**a**) screening parameter (β) with varying magnetic field; (**b**) screening parameter (β) with varying AB field. Momentum space Tsallis entropies $T_\rho(\rho)$ versus (**c**) screening parameter (β) with varying magnetic field; (**d**) screening parameter (β) with varying AB field.

The Rényi entropies (RE) in position and momentum space are shown in Figure 2a–d. RE increases with rising potential parameter β and decreases with the increasing magnetic and AB fields in position space. RE gets larger with the screening parameter β and is inversely proportional to magnetic and AB fields in momentum space. This behaves similarly to the Shannon entropies in position space seen in Figure 1a–d in Ref. [51].

The Onicescu information energy (OIE) in position and momentum space is shown in Figure 3a–d. The OIE in position space increases as the screening parameter β upsurges and declines as the magnetic and AB fields rise. The OIE reduces as the screening parameter rises and upsurges as the magnetic and AB fields grow in momentum space. This highlights the fact that the greater the system's OIE, the more concentrated the probability distribution is and the smaller the information content. According to the definition of the Shannon entropy, more localized distributions and position space probability density correspond to the smaller value of the RE, which means that the delocalization of the probability density increases with increasing quantum number.

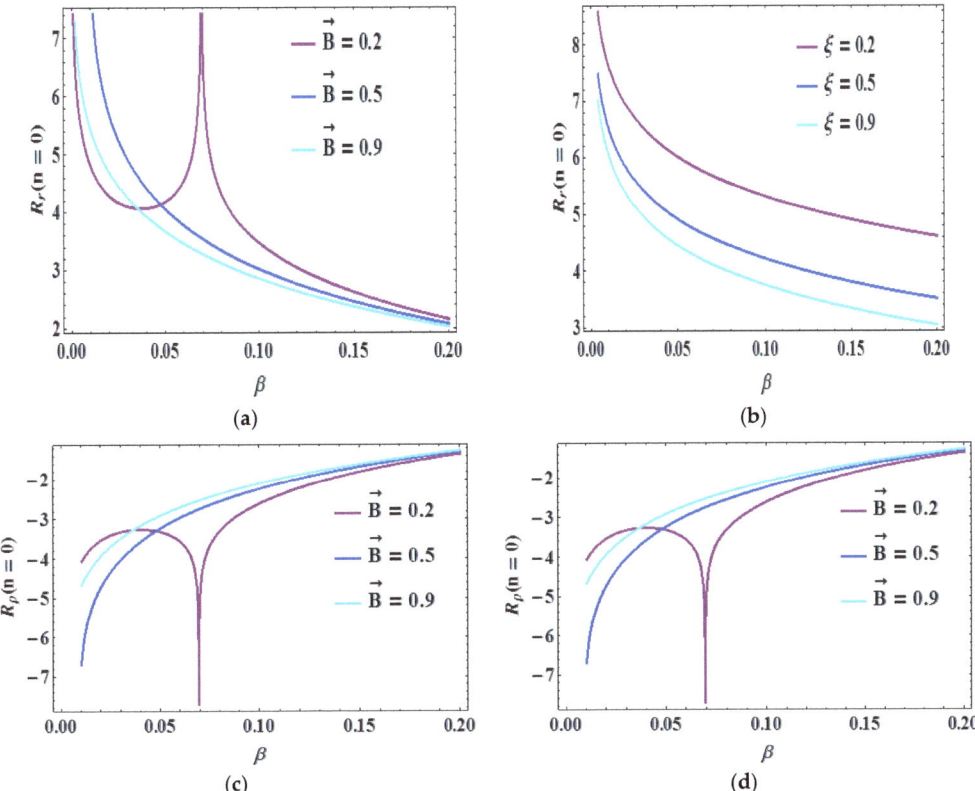

Figure 2. Position space Rényi entropies $R_r(r)$ versus: (**a**) screening parameter (β) with varying magnetic field; (**b**) screening parameter (β) with varying AB field. Momentum space Rényi entropies $R_\rho(\rho)$ versus (**c**) screening parameter (β) with varying magnetic field; (**d**) screening parameter (β) with varying AB field.

The numerical findings in Tables 1 and 2 demonstrate that the position-space Tsallis entropy reduces as the potential parameter, magnetic, and AB fields rise, whereas the momentum-space Tsallis information entropy grows as the potential parameter β, magnetic, and AB fields increase. This is consistent with what we observed in Figure 1. The single influence of these fields is examined in Table 3. The Tsallis entropy in the position space grows as the potential parameter β increases when just the magnetic field is present, and a similar condition is observed in the momentum space. This contradicts our findings for the all-inclusive impact in momentum space. This finding is also confirmed when only the AB field is functioning.

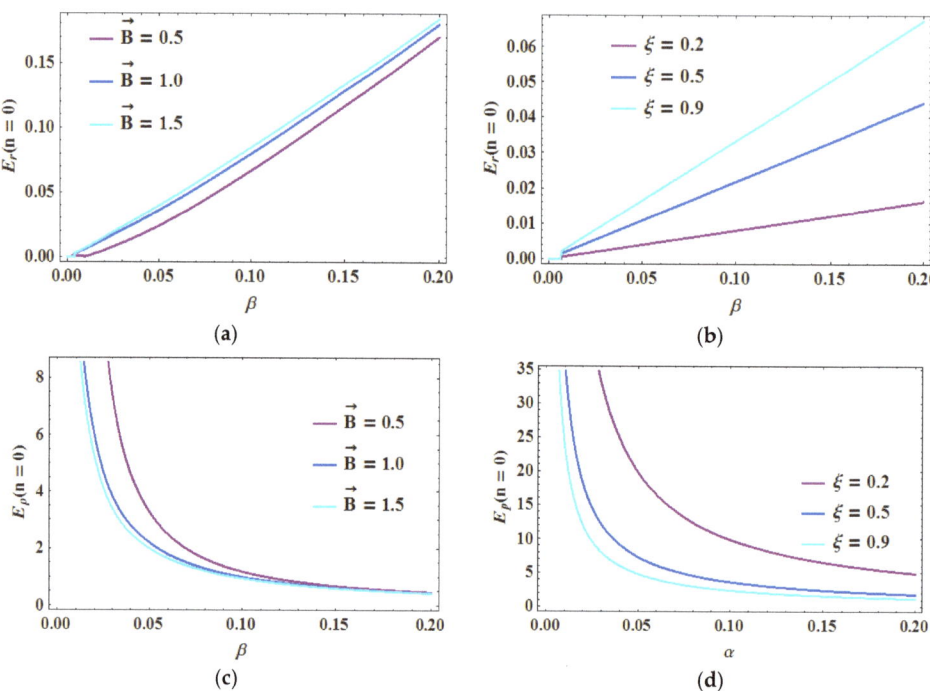

Figure 3. Position space Onicescu information energy $E_r(r)$ versus: (**a**) screening parameter (α) with varying magnetic field; (**b**) screening parameter (β) with varying AB field. Momentum space Onicescu information energy $E_\rho(\rho)$ versus (**c**) screening parameter (β) with varying magnetic field; (**d**) screening parameter (β) with varying AB field.

Table 1. Numerical results of the Tsallis entropy for CYP of $\widetilde{A}=1$, $\widetilde{B}=2$ and $\widetilde{C}=-1$ for $m=0$ in the presence of AB and magnetic fields with varying β and $\vec{B}$.

$n=0$	β	T_r	T_ρ	$T_r T_\rho$	$\vec{B}$	T_r	T_ρ	$T_r T_\rho$
	0.1	4.85597	−1.36078	−6.60789	0.1	10.3199	−1.76531	−18.2178
	0.2	3.00089	−0.97275	−2.9191	0.2	7.51722	−1.6137	−12.1305
	0.3	2.12788	−0.68065	−1.44834	0.3	6.62342	−1.54307	−10.2204
	0.4	1.58914	−0.437	−0.69446	0.4	6.15692	−1.5002	−9.23659
	0.5	1.21291	−0.22413	−0.27185	0.5	5.86766	−1.47134	−8.63333
	0.6	0.930544	−0.03313	−0.03083	0.6	5.67091	−1.45071	−8.22686
	0.7	0.708313	0.141231	0.100036	0.7	5.52895	−1.43534	−7.93591
	0.8	0.527399	0.30238	0.159475	0.8	5.42214	−1.42353	−7.71857
	0.9	0.376339	0.452694	0.170367	0.9	5.33924	−1.41423	−7.55093
$n=1$	0.1	0.331132	−1.13946	−0.37731	0.1	4.20379	−1.38461	−5.8206
	0.2	−0.22603	−0.66985	0.151406	0.2	3.13651	−1.302	−4.08373
	0.3	−0.4842	−0.31568	0.152849	0.3	2.56567	−1.25414	−3.21771
	0.4	−0.64587	−0.0206	0.013304	0.4	2.19632	−1.22327	−2.68669
	0.5	−0.76184	0.236612	−0.18026	0.5	1.93233	−1.2021	−2.32285
	0.6	−0.85165	0.466783	−0.39753	0.6	1.73169	−1.18699	−2.05549
	0.7	−0.92462	0.676366	−0.62538	0.7	1.57271	−1.17587	−1.84931
	0.8	−0.98589	0.869596	−0.85733	0.8	1.44287	−1.16752	−1.68457
	0.9	−1.03857	1.04943	−1.08991	0.9	1.33436	−1.16112	−1.54936

Table 2. Numerical results of the Tsallis entropy for CYP of $\widetilde{A} = 1$, $\widetilde{B} = 2$ and $\widetilde{C} = -1$ for $m = 0$ in the presence of AB and magnetic fields with varying ξ.

	ξ	T_r	T_ρ	$T_r T_\rho$
$n = 0$	0.1	75.1126	−2.36279	−177.476
	0.2	18.4411	−1.97689	−36.4561
	0.3	14.1096	−1.8945	−26.7308
	0.4	12.1546	−1.84339	−22.4055
	0.5	10.9773	−1.80543	−19.8188
	0.6	10.1662	−1.77482	−18.0432
	0.7	9.56125	−1.74892	−16.7219
	0.8	9.08577	−1.72634	−15.6851
	0.9	8.69795	−1.70622	−14.8406
$n = 1$	0.1	4.34999	−1.49708	−6.51228
	0.2	4.03499	−1.48134	−5.97721
	0.3	3.75866	−1.46679	−5.51319
	0.4	3.51378	−1.45323	−5.10635
	0.5	3.29489	−1.44051	−4.74632
	0.6	3.09776	−1.42851	−4.42517
	0.7	2.91908	−1.41713	−4.13671
	0.8	2.75619	−1.4063	−3.87602
	0.9	2.60694	−1.39595	−3.63916

Table 3. Numerical results of the Tsallis entropy for CYP of $\widetilde{A} = 1$, $\widetilde{B} = 2$ and $\widetilde{C} = -1$ for $m = 0$ in the absence of AB and magnetic fields.

	β	T_r	T_ρ	$T_r T_\rho$	$n = 0, \xi = 0$	β	T_r	T_ρ	$T_r T_\rho$
$\vec{B} = 0, n = 0$	0.1	10.0506	−0.77122	−7.75125		0.1	18.5869	−0.34567	−6.425
	0.2	3.68752	−1.26202	−4.65372		0.2	13.49	−0.62472	−8.42742
	0.3	2.28039	−1.41008	−3.21553		0.3	11.0811	−0.77215	−8.55619
	0.4	1.59787	−1.49039	−2.38145		0.4	9.57162	−0.87116	−8.33842
	0.5	1.1687	−1.5445	−1.80506		0.5	8.49297	−0.9456	−8.03094
	0.6	0.864823	−1.5848	−1.37057		0.6	7.65982	−1.00545	−7.70155
	0.7	0.634176	−1.61665	−1.02524		0.7	6.9822	−1.05578	−7.37164
	0.8	0.450886	−1.64281	−0.74072		0.8	6.41051	−1.09948	−7.04822
	0.9	0.300389	−1.66492	−0.50012		0.9	5.91497	−1.13834	−6.73327
$\vec{B} = 0, n = 1$	0.1	7.63677	−0.94899	−7.24723	$n = 0, \xi = 0$	0.1	4.71345	−1.03892	−4.89688
	0.2	1.34222	−1.40999	−1.89252		0.2	3.07693	−1.2298	−3.78402
	0.3	0.489068	−1.52626	−0.74644		0.3	2.32993	−1.32995	−3.09868
	0.4	0.067563	−1.59329	−0.10765		0.4	1.87899	−1.39616	−2.62338
	0.5	−0.19899	−1.63956	0.326249		0.5	1.56944	−1.44481	−2.26753
	0.6	−0.38826	−1.67449	0.650143		0.6	1.34038	−1.48278	−1.9875
	0.7	−0.53219	−1.70232	0.905953		0.7	1.16234	−1.51362	−1.75935
	0.8	−0.6467	−1.72533	1.11577		0.8	1.01907	−1.53937	−1.56873
	0.9	−0.74081	−1.74485	1.2926		0.9	0.90080	−1.5613	−1.40643

The numerical results in Tables 4 and 5 demonstrate that the position-space Rényi entropy decreases as the potential parameter, magnetic, and AB fields increase, but the momentum-space Rényi information entropy increases as the potential parameter β, magnetic, and AB fields increase. This is consistent with what we saw in Figure 2.

When we looked at the lone influence of these fields on the Rényi entropy in Table 6, we saw something intriguing. When just the magnetic field is active, we find that the Rényi entropy in the position space grows as the potential parameter rises, but the opposite is true in the momentum space.

Table 4. Numerical results of the Rényi entropy for CYP of $\widetilde{A} = 1$, $\widetilde{B} = 2$ and $\widetilde{C} = -1$ for $m = 0$ in the presence of AB and magnetic fields with varying β and $\vec{B}$.

$n = 0$	β	R_r	R_ρ	$R_r + R_\rho$	$\vec{B}$	R_r	R_ρ	$R_r + R_\rho$
	0.01	5.03916	−4.31242	0.726733	0.1	4.08677	−3.06149	1.02528
	0.02	4.34122	−3.61373	0.727489	0.2	3.47004	−2.59247	0.877566
	0.03	3.9311	−3.20287	0.728235	0.3	3.23639	−2.40078	0.835604
	0.04	3.63891	−2.90994	0.728971	0.4	3.10517	−2.29122	0.813953
	0.05	3.41138	−2.68168	0.729699	0.5	3.02021	−2.22009	0.800121
	0.06	3.2248	−2.49438	0.730416	0.6	2.96073	−2.17044	0.790281
	0.07	3.0665	−2.33538	0.731125	0.7	2.91691	−2.13409	0.78282
	0.08	2.92894	−2.19712	0.731823	0.8	2.88343	−2.1065	0.776922
	0.09	2.80723	−2.07472	0.732513	0.9	2.85713	−2.08501	0.772117
$n = 1$	0.01	2.86638	−3.88188	−0.91592	0.1	2.50324	−2.01771	1.03444
	0.02	2.18509	−3.18133	−0.89902	0.2	2.0905	−1.83909	0.507536
	0.03	1.79113	−2.76872	−0.88264	0.3	1.84131	−1.74117	0.27248
	0.04	1.51464	−2.47415	−0.86674	0.4	1.66793	−1.67997	0.123167
	0.05	1.30236	−2.24434	−0.85132	0.5	1.53782	−1.63886	0.014201
	0.06	1.1306	−2.05558	−0.83635	0.6	1.43544	−1.60992	−0.07122
	0.07	0.986711	−1.89521	−0.82182	0.7	1.35222	−1.58885	−0.1411
	0.08	0.863154	−1.75565	−0.80777	0.8	1.28293	−1.57312	−0.1999
	0.09	0.755069	−1.63204	−0.79445	0.9	1.22415	−1.56115	−0.25036

Table 5. Numerical results of the Rényi entropy for CYP of $\widetilde{A} = 1$, $\widetilde{B} = 2$ and $\widetilde{C} = -1$ for $m = 0$ in the presence of AB and magnetic fields with varying ζ.

$n = 0$	ζ	R_r	R_ρ	$R_r + R_\rho$
	0.1	8.46512	−7.25642	1.2087
	0.2	5.31356	−3.91064	1.40292
	0.3	4.73423	−3.545	1.18924
	0.4	4.42115	−3.34237	1.07878
	0.5	4.2118	−3.20189	1.00991
	0.6	4.05662	−3.09406	0.962565
	0.7	3.93427	−3.00635	0.927922
	0.8	3.83372	−2.93228	0.901435
	0.9	3.7486	−2.86809	0.880507
$n = 1$	0.1	2.57713	−2.28344	0.293695
	0.2	2.46699	−2.24452	0.222474
	0.3	2.36656	−2.20906	0.157495
	0.4	2.27435	−2.17646	0.097891
	0.5	2.18922	−2.14626	0.042958
	0.6	2.11021	−2.11809	−0.00788
	0.7	2.03658	−2.09168	−0.0551
	0.8	1.96769	−2.0668	−0.09911
	0.9	1.90302	−2.04327	−0.14025

This is in contrast to what we saw in the overall impact. The Rényi entropy in the position space reduces as the potential parameter β grows when just the AB field is active, but the opposite is true in the momentum space. This supports our observation of the all-encompassing influence. However, we may deduce that the magnetic is necessary to produce a rising Rényi entropy with regard to the potential parameter β. This finding is comparable to what the Shannon entropy shows [50]. It is important to realize that the conjugates of position and momentum space information entropies have an inverse relationship with each other. A strongly localized distribution in the position space corresponds to widely delocalized distribution in the momentum space.

Table 6. Numerical results of the Rényi entropy for CYP of $\widetilde{A} = 1$, $\widetilde{B} = 2$ and $\widetilde{C} = -1$ for $m = 0$ in the absence of AB and magnetic fields.

$\vec{B} = 0, n = 0$	β	R_r	R_ρ	$R_r + R_\rho$	$\xi = 0, n = 0$	β	R_r	R_ρ	$R_r + R_\rho$
	0.01	2.68063	−1.87887	0.801764		0.01	7.62858	−6.22541	1.40317
	0.02	2.73321	−1.9305	0.802702		0.02	6.93615	−5.53222	1.40392
	0.03	2.79274	−1.98819	0.804547		0.03	6.53136	−5.1267	1.40466
	0.04	2.86121	−2.05353	0.807671		0.04	6.2443	−4.83893	1.40538
	0.05	2.94157	−2.12888	0.812688		0.05	6.02175	−4.61567	1.40608
	0.06	3.03855	−2.21788	0.820669		0.06	5.83997	−4.43321	1.40676
	0.07	3.16036	−2.32666	0.833696		0.07	5.68632	−4.2789	1.40743
	0.08	3.32321	−2.46682	0.856396		0.08	5.55325	−4.14517	1.40808
	0.09	3.56648	−2.66486	0.901613		0.09	5.43589	−4.02717	1.40871
$\vec{B} = 0, n = 1$	0.01	1.7365	−1.70708	0.029415	$\xi = 0, n = 1$	0.01	4.95852	−4.64268	0.315837
	0.02	1.80368	−1.76118	0.042503		0.02	4.2703	−3.94798	0.322324
	0.03	1.89214	−1.82383	0.068313		0.03	3.86973	−3.54097	0.328754
	0.04	2.01026	−1.89798	0.112277		0.04	3.5869	−3.25177	0.335129
	0.05	2.17217	−1.98861	0.183554		0.05	3.36856	−3.02711	0.341448
	0.06	2.40431	−2.10526	0.299049		0.06	3.19101	−2.8433	0.347714
	0.07	2.76485	−2.27025	0.494601		0.07	3.04159	−2.68767	0.353926
	0.08	3.42965	−2.56125	0.868406		0.08	2.91276	−2.55267	0.360085
	0.09	6.79656	−4.58251	2.21405		0.09	2.79963	−2.43344	0.366193

The numerical results in Tables 7 and 8 demonstrate that the position-space Onicescu information energy surges as the potential parameter β, magnetic, and AB fields rise, whereas the momentum space Onicescu information energy information entropy reduces as the potential parameter β, magnetic, and AB fields rise. When we looked at the single influence of these fields on the Onicescu information energy in Table 9, we discovered something interesting. When just the magnetic field remains operational, the Onicescu information energy in the position space drops as the potential parameter β rises, although in the momentum space the opposite is the case.

Table 7. Numerical results of the Onicescu information energy for CYP of $\widetilde{A} = 1$, $\widetilde{B} = 2$ and $\widetilde{C} = -1$ for $m = 0$ in the presence of AB and magnetic fields with varying β and $\vec{B}$.

$n = 0$	β	E_r	E_ρ	$E_r E_\rho$	$\vec{B}$	E_r	E_ρ	$E_r E_\rho$
	0.01	0.008762	9.07971	0.079559	0.1	0.024594	3.19723	0.078632
	0.02	0.017609	4.51712	0.079541	0.2	0.043439	1.80812	0.078544
	0.03	0.026536	2.99677	0.079523	0.3	0.054213	1.45001	0.07861
	0.04	0.035542	2.23696	0.079505	0.4	0.061476	1.27996	0.078687
	0.05	0.044623	1.78135	0.079488	0.5	0.066716	1.18048	0.078757
	0.06	0.053776	1.47782	0.079472	0.6	0.070659	1.11549	0.078819
	0.07	0.063	1.2612	0.079455	0.7	0.073716	1.06996	0.078873
	0.08	0.072291	1.09888	0.079439	0.8	0.076143	1.03648	0.078921
	0.09	0.081648	0.972761	0.079424	0.9	0.078105	1.01097	0.078962
$n = 1$	0.01	0.007422	5.59089	0.041495	0.1	0.040827	0.912806	0.037267
	0.02	0.014928	2.77409	0.041412	0.2	0.050152	0.737849	0.037004
	0.03	0.022514	1.83575	0.041331	0.3	0.056263	0.660556	0.037165
	0.04	0.030177	1.36701	0.041252	0.4	0.060563	0.617627	0.037405
	0.05	0.037912	1.08607	0.041175	0.5	0.063724	0.590905	0.037655
	0.06	0.045716	0.899033	0.041101	0.6	0.06612	0.573083	0.037893
	0.07	0.053587	0.765634	0.041028	0.7	0.067982	0.560632	0.038113
	0.08	0.061521	0.665751	0.040958	0.8	0.069456	0.551639	0.038314
	0.09	0.069516	0.588203	0.04089	0.9	0.070642	0.544983	0.038499

Table 8. Numerical results of the Onicescu information energy for CYP of $\widetilde{A} = 1$, $\widetilde{B} = 2$ and $\widetilde{C} = -1$ for $m = 0$ in the presence of AB and magnetic fields with varying ζ.

	ζ	E_r	E_ρ	$E_r E_\rho$
$n=0$	0.1	0.0000933	806.906	0.075303
	0.2	0.008064	9.86189	0.079521
	0.3	0.013689	5.807	0.079491
	0.4	0.018143	4.38037	0.079472
	0.5	0.021898	3.62857	0.079458
	0.6	0.025189	3.15415	0.079449
	0.7	0.028147	2.82241	0.079441
	0.8	0.030854	2.57461	0.079436
	0.9	0.033363	2.38079	0.079431
$n=1$	0.1	0.031595	1.14182	0.036076
	0.2	0.033139	1.09123	0.036162
	0.3	0.034608	1.04845	0.036285
	0.4	0.036014	1.01157	0.036431
	0.5	0.037364	0.979267	0.03659
	0.6	0.038666	0.950587	0.036756
	0.7	0.039925	0.924839	0.036924
	0.8	0.041145	0.901502	0.037092
	0.9	0.04233	0.88018	0.037258

This contrasts our findings in the case of the comprehensive impact. When just the AB field is present, we notice that the Onicescu information energy in the position space grows as the potential parameter β increases, but the opposite is true in the momentum space. This is consistent with our findings in the case of the all-inclusive effect. However, we could deduce that the AB field is necessary to acquire a rising Onicescu information energy in position space with regard to the potential parameter.

Table 9. Numerical results of the Onicescu information energy for CYP of $\widetilde{A} = 1$, $\widetilde{B} = 2$ and $\widetilde{C} = -1$ for $m = 0$ in the absence of AB and magnetic fields.

	β	E_r	E_ρ	$E_r E_\rho$		β	E_r	E_ρ	$E_r E_\rho$
$\vec{B}=0, n=0$	0.01	0.092604	0.841859	0.07796	$\zeta=0, n=0$	0.01	0.000797	99.8309	0.079572
	0.02	0.087888	0.887062	0.077962		0.02	0.001593	49.9446	0.079566
	0.03	0.082858	0.940966	0.077966		0.03	0.002388	33.3145	0.079561
	0.04	0.077453	1.00672	0.077974		0.04	0.003182	24.9986	0.079555
	0.05	0.071589	1.08934	0.077985		0.05	0.003976	20.0082	0.079549
	0.06	0.065144	1.19742	0.078004		0.06	0.004769	16.6806	0.079544
	0.07	0.057925	1.34718	0.078035		0.07	0.005561	14.3032	0.079538
	0.08	0.049601	1.57429	0.078087		0.08	0.006353	12.5197	0.079533
	0.09	0.039507	1.97911	0.078189		0.09	0.007144	11.1321	0.079527
$\vec{B}=0, n=1$	0.01	0.058928	0.642116	0.037839	$\zeta=0, n=1$	0.01	0.002969	12.0965	0.035914
	0.02	0.055738	0.678174	0.0378		0.02	0.005944	6.04393	0.035927
	0.03	0.052193	0.722853	0.037728		0.03	0.008926	4.02654	0.035941
	0.04	0.04821	0.780259	0.037616		0.04	0.011914	3.01793	0.035954
	0.05	0.043661	0.858218	0.037471		0.05	0.014907	2.41282	0.035969
	0.06	0.038324	0.974424	0.037344		0.06	0.017907	2.00947	0.035984
	0.07	0.031739	1.18204	0.037517		0.07	0.020913	1.72141	0.035999
	0.08	0.022607	1.76184	0.03983		0.08	0.023924	1.50541	0.036015
	0.09	0.00187	39.5402	0.073944		0.09	0.026941	1.33744	0.036031

4. Conclusions

The GQITM was investigated in both the position and momentum spaces for the CYP in the ground and first excited states in this research. The wave function and energy equations are obtained by solving the Schrodinger equation with the CYP in the presence of magnetic and AB fields using the functional analytical method [50]. The probability density is evaluated by squaring the CYP's wave function given in terms of hypergeometric functions. Numerical results at $p = 2$ for the Rényi entropy, Tsallis entropy, and Onicescu information energy have also been produced. The effects of magnetic and AB fields on these entropies have been well investigated. Our results show that these fields and potential parameters are relevant for the manipulation of the behavior of the quantum system. The findings obtained in this study will find possible applications in quantum information processing, quantum chemistry, etc. The present study can be extended to the investigation of the information entropies of heavy mesons such as charmonium and bottomonium in the presence of magnetic and AB fields [52–55].

Author Contributions: C.O.E. and S.A.A.: Conceived and designed the study, acquired, analyzed and interpreted the data, and handled the review. R.E. and N.A. handled the computational analysis. A.N.I., M.A. and E.B.E.: revised the manuscript for important intellectual content, and contributed in general recommendations and data interpretation. All authors have read and agreed to the published version of the manuscript.

Funding: The research was carried out under Long Term Research Grant Scheme project LRGS/1/2020/UM/01/5/2 provided by Ministry of Higher Education of Malaysia (MOHE).

Institutional Review Board Statement: Not applicable.

Informed Consent Statement: Not applicable.

Data Availability Statement: The datasets used and/or analyzed during the current study are available from the corresponding authors on reasonable request.

Acknowledgments: C.O.E. acknowledges eJDS (ICTP).

Conflicts of Interest: The authors declare no conflict of interest.

References

1. Shannon, C.E. A mathematical theory of communication. *Bell Syst. Tech. J.* **1948**, *27*, 379. [CrossRef]
2. Renyi, A. Measures of Information and Entropy. In Proceedings of the 4th Symptom on Mathematics, Statistics and Probability, Berkeley University Press, Berkeley, CA, USA, 30 July 1960.
3. Kullberg, S.; Leibler, R.A. On information and sufficiency. *Ann. Math Stat.* **1951**, *22*, 79.
4. Tsallis, C.J. Possible generalization of Boltzmann-Gibbs statistics. *Stat. Phys.* **1988**, *54*, 479. [CrossRef]
5. Onicescu, C.R.O. Theorie de l'information energie informationelle. *Acad. Sci. Paris. A* **1966**, *263*, 25.
6. Sun, G.H.; Aoki, M.A.; Dong, S.H. Quantum information entropies of the eigenstates for the Pöschl—Teller-like potential. *Chin. Phys. B* **2013**, *22*, 050302. [CrossRef]
7. Sun, G.H.; Dong, S.H.; Saad, N. Quantum information entropies for an asymmetric trigonometric Rosen–Morse potential. *Ann. Phys.* **2013**, *525*, 934. [CrossRef]
8. Dong, S.; Sun, G.H.; Dong, S.H.; Draayer, J.P. Quantum information entropies for a squared tangent potential well. *Phys. Lett. A* **2014**, *378*, 124. [CrossRef]
9. Navarro, G.Y.; Sun, G.H.; Dytrych, T.; Launey, K.D.; Dong, S.H.; Draayer, J.P. Quantum information entropies for position-dependent mass Schrödinger problem. *Ann. Phys.* **2014**, *348*, 153. [CrossRef]
10. Torres, R.V.; Sun, G.H.; Dong, S.H. Quantum information entropy for a hyperbolical potential function. *Phys. Scr.* **2015**, *90*, 035205. [CrossRef]
11. Song, X.D.; Sun, G.H.; Dong, S.H. Shannon information entropy for an infinite circular well. *Phys. Lett. A* **2015**, *379*, 1402. [CrossRef]
12. Sun, G.H.; Dong, S.H.; Launey, K.D.; Dytrych, T.; Draayer, J.P. Shannon information entropy for a hyperbolic double-well potential. *Int. J. Quantum Chem.* **2015**, *115*, 891. [CrossRef]
13. Sun, G.H.; Dusan, P.; Oscar, C.N.; Dong, S.H. Shannon information entropies for position-dependent mass Schrödinger problem with a hyperbolic well. *Chin. Phys. B* **2015**, *24*, 100303.
14. Omugbe, E.; Osafile, O.E.; Okon, I.B.; Eyube, E.S.; Inyang, E.P.; Okorie, U.S.; Jahanshir, A.; Onate, C.A. Non-relativistic bound state solutions with α-deformed Kratzer-type potential using the super-symmetric WKB method: Application to theoretic-information measures. *Eur. Phys. J. D* **2022**, *76*, 72. [CrossRef]
15. Song, X.D.; Dong, S.H.; Zhang, Y. Quantum information entropy for one-dimensional system undergoing quantum phase transition. *Chin. Phys. B* **2016**, *25*, 050302. [CrossRef]
16. Shi, Y.J.; Sun, G.H.; Jing, J.; Dong, S.H. Shannon and Fisher entropy measures for a parity-restricted harmonic oscillator. *Laser Phys.* **2017**, *27*, 125201. [CrossRef]

17. Solaimani, M.; Sun, G.H.; Dong, S.H. Shannon information entropies for rectangular multiple quantum well systems with constant total lengths. *Chin. Phys. B* **2018**, *27*, 040301. [CrossRef]
18. Najafizade, S.A.; Hassanabadi, H.; Zarrinkamar, S. Nonrelativistic Shannon information entropy for Killingbeck potential. *Can. J. Phys.* **2016**, *94*, 1085. [CrossRef]
19. Najafizade, S.A.; Hassanabadi, H.; Zarrinkamar, S. Information Theoretic Global Measures of Dirac Equation With Morse and Trigonometric Rosen–Morse Potentials. *Few-Body Syst.* **2017**, *58*, 149. [CrossRef]
20. Panahi, H.; Najafizade, A.; Hassanabadi, H. Study of the Shannon Entropy in the Quantum Model Obtained from SO (2, 2). *J. Korean Phys. Soc.* **2019**, *75*, 87. [CrossRef]
21. Najafizade, A.; Panahi, H.; Hassanabadi, H. Study of information entropy for involved quantum models in complex Cayley-Klein space. *Phys. Scr.* **2020**, *95*, 085207. [CrossRef]
22. Zare, S.; Hassanabadi, H. Properties of Quasi-Oscillator in Position-Dependent Mass Formalism. *Adv. High Energy Phys.* **2016**, *2016*, 4717012. [CrossRef]
23. Romera, E.; de los Santos, F. Fractional revivals through Rényi uncertainty relations. *Phys. Rev. A* **2008**, *78*, 013837. [CrossRef]
24. Romera, E.; Nagy, A. Rényi information of atoms. *Phys. Lett. A* **2008**, *372*, 4918. [CrossRef]
25. Najafizade, S.A.; Hassanabadi, H.; Zarrinkamar, S. Nonrelativistic Shannon information entropy for Kratzer potential. *Chin. Phys. B* **2016**, *25*, 040301. [CrossRef]
26. Ghafourian, M.; Hassanabadi, H. Shannon information entropies for the three-dimensional Klein-Gordon problem with the Poschl-Teller potential. *J. Korean Phys. Soc.* **2016**, *68*, 1267. [CrossRef]
27. Amadi, P.O.; Ikot, A.N.; Ngiangia, A.T.; Okorie, U.S.; Rampho, G.J.; Abdullah, H.Y. Shannon entropy and Fisher information for screened Kratzer potential. *Int. J. Quantum Chem.* **2020**, *120*, e26246. [CrossRef]
28. Ikot, A.N.; Rampho, G.J.; Amadi, P.O.; Sithole, M.J.; Okorie, U.S.; Lekala, M.I. Shannon entropy and Fisher information-theoretic measures for Mobius square potential. *Eur. Phys. J. Plus* **2020**, *135*, 6. [CrossRef]
29. Yahya, W.A.; Oyewumi, K.J.; Sen, K.D. Information and complexity measures for the ring-shaped modified Kratzer potential. *Indian J. Chem.* **2014**, *53A*, 1307.
30. Yahya, W.A.; Oyewumi, K.J.; Sen, K.D. Position and momentum information-theoretic measures of the pseudoharmonic potential. *Int. J. Quantum Chem.* **2014**, *115*, 1543. [CrossRef]
31. Hassanabadi, H.; Zare, S.; Alimohammadi, M. Investigation of the information entropy for the X (3) model. *Eur. Phys. J. Plus* **2017**, *132*, 498. [CrossRef]
32. Sun, G.H.; Dong, S.H. Quantum information entropies of the eigenstates for a symmetrically trigonometric Rosen–Morse potential. *Phys. Scr.* **2013**, *87*, 045003. [CrossRef]
33. Isonguyo, C.N.; Oyewumi, K.J.; Oyun, O.S. Quantum information-theoretic measures for the static screened Coulomb potential. *Int. J. Quantum Chem.* **2018**, *118*, e25620. [CrossRef]
34. Patil, S.H.; Sen, K.D. Net information measures for modified Yukawa and Hulthén potentials. *Int. J. Quantum Chem.* **2007**, *107*, 1864. [CrossRef]
35. Falaye, B.J.; Serrano, F.A.; Dong, S.H. Fisher information for the position-dependent mass Schrödinger system. *Phys. Lett. A* **2016**, *380*, 267. [CrossRef]
36. Serrano, F.A.; Falaye, B.J.; Dong, S.H. Information-theoretic measures for a solitonic profile mass Schrödinger equation with a squared hyperbolic cosecant potential. *Phys. A* **2016**, *446*, 152. [CrossRef]
37. Jiao, L.G.; Zan, L.R.; Zhang, Y.Z.; Ho, Y.K. Benchmark values of S hannon entropy for spherically confined hydrogen atom. *Int. J. Quantum Chem.* **2017**, *117*, e25375. [CrossRef]
38. Pooja; Sharma, A.; Gupta, R.; Kumar, A. Quantum information entropy of modified Hylleraas plus exponential Rosen Morse potential and squeezed states. *Int. J. Quantum Chem.* **2017**, *117*, 25368. [CrossRef]
39. Olendski, O. Quantum information measures of the Aharonov–Bohm ring in uniform magnetic fields. *Phys. Lett. A* **2019**, *383*, 1110. [CrossRef]
40. Olendski, O. Rényi and tsallis entropies of the aharonov–bohm ring in uniform magnetic fields. *Entropy* **2019**, *21*, 1060. [CrossRef]
41. Hamzavi, M.; Thylwe, K.E.; Rajabi, A.A. Approximate bound states solution of the Hellmann potential. *Commun. Theor. Phys.* **2013**, *60*, 1. [CrossRef]
42. Ahmadov, A.I.; Demirci, M.; Aslanova, S.M.; Mustamin, M.F. Arbitrary ℓ-state solutions of the Klein-Gordon equation with the Manning-Rosen plus a Class of Yukawa potentials. *Phys. Lett. A* **2020**, *384*, 126372. [CrossRef]
43. Purohit, K.R.; Rai, A.K.; Parmar, R.H. Rotational vibrational partition function using attractive radial potential plus class of Yukawa potential. *AIP Conf. Proc.* **2020**, *2220*, 120004.
44. Bialynicki-Birula, I.; Mycielski, J. Uncertainty relations for information entropy in wave mechanics. *Commun. Math. Phys.* **1975**, *44*, 129. [CrossRef]
45. Guerrero, A.; Sanchez-Moreno, J.; Dehesa, J.S. Information-theoretic lengths of Jacobi polynomials. *J. Phys. A Math. Theor.* **2010**, *43*, 305203. [CrossRef]
46. Greene, R.L.; Aldrich, C. Variational wave functions for a screened Coulomb potential. *Phys. Rev. A* **1976**, *14*, 2363. [CrossRef]
47. Edet, C.O.; Amadi, P.O.; Ettah, E.B.; Ali, N.; Asjad, M.; Ikot, A.N. The magnetocaloric effect, thermo-magnetic and transport properties of LiH diatomic molecule. *Mol. Phys.* **2022**, e2059025. [CrossRef]

48. Edet, C.O.; Ikot, A.N. Analysis of the impact of external fields on the energy spectra and thermo-magnetic properties of N_2, I_2, CO, NO and HCl diatomic molecules. *Mol. Phys.* **2021**, *119*, 23. [CrossRef]
49. Edet, C.O.; Ikot, A.N. Effects of Topological Defect on the Energy Spectra and Thermo-magnetic Properties of $$ CO $$ CO Diatomic Molecule. *J. Low Temp. Phys.* **2021**, *203*, 84. [CrossRef]
50. Edet, C.O.; Ikot, A.N. Shannon information entropy in the presence of magnetic and Aharanov–Bohm (AB) fields. *Eur. Phys. J. Plus* **2021**, *136*, 432. [CrossRef]
51. Bayrak, O.; Boztosun, I.; Ciftci, H. Exact analytical solutions to the Kratzer potential by the asymptotic iteration method. *Int. J. Quantum Chem.* **2007**, *107*, 540. [CrossRef]
52. Inyang, E.P.; Inyang, E.P.; Ntibi, J.E.; Ibekwe, E.E.; William, E.S. Approximate solutions of D-dimensional Klein–Gordon equation with Yukawa potential via Nikiforov–Uvarov method. *Indian J. Phys.* **2021**, *95*, 2733–2739. [CrossRef]
53. Akpan, I.O.; Inyang, E.P.; William, E.S. Approximate solutions of the Schrödinger equation with Hulthén-Hellmann Potentials for a Quarkonium system. *Rev. Mex. Fís.* **2021**, *67*, 482–490.
54. Ibekwe, E.E.; Okorie, U.S.; Emah, J.B.; Inyang, E.P.; Ekong, S.A. Mass spectrum of heavy quarkonium for screened Kratzer potential (SKP) using series expansion method. *Eur. Phys. J. Plus* **2021**, *136*, 1–11. [CrossRef]
55. Abu-Shady, M.; Edet, C.O.; Ikot, A.N. Non-relativistic quark model under external magnetic and Aharanov–Bohm (AB) fields in the presence of temperature-dependent confined Cornell potential. *Can. J. Phys.* **2021**, *99*, 1024–1031. [CrossRef]

Article

Dynamical Symmetries of the 2D Newtonian Free Fall Problem Revisited

Tuong Trong Truong

Laboratoire de Physique Théorique et Modélisation, CNRS UMR 8089, CY Cergy Paris Université, 95302 Cergy-Pontoise, France; truong@cyu.fr

Abstract: Among the few exactly solvable problems in theoretical physics, the 2D (two-dimensional) Newtonian free fall problem in Euclidean space is perhaps the least known as compared to the harmonic oscillator or the Kepler–Coulomb problems. The aim of this article is to revisit this problem at the classical level as well as the quantum level, with a focus on its dynamical symmetries. We show how these dynamical symmetries arise as a special limit of the dynamical symmetries of the Kepler–Coulomb problem, and how a connection to the quartic anharmonic oscillator problem, a long-standing unsolved problem in quantum mechanics, can be established. To this end, we construct the Hilbert space of states with free boundary conditions as a space of square integrable functions that have a special functional integral representation. In this functional space, the free fall dynamical symmetry algebra is shown to be isomorphic to the so-called Klink's algebra of the quantum quartic anharmonic oscillator problem. Furthermore, this connection entails a remarkable integral identity for the quantum quartic anharmonic oscillator eigenfunctions, which implies that these eigenfunctions are in fact *zonal* functions of an underlying symmetry group representation. Thus, an appropriate representation theory for the 2D Newtonian free fall quantum symmetry group may potentially open the way to exactly solving the difficult quantization problem of the quartic anharmonic oscillator. Finally, the initial value problem of the acoustic Klein–Gordon equation for wave propagation in a sound duct with a varying circular section is solved as an illustration of the techniques developed here.

Keywords: super-integrability; dynamical symmetry algebras; integral transforms

1. Introduction

In an elementary freshman physics course, the problem of particle motion under constant gravitational acceleration on the surface of the earth in Euclidean space, usually considered as one dimensional, is frequently dubbed as a free fall problem [1,2]. This is not the more general relativistic problem in which an inertial particle subject to no force moves along a space-time geodesic [3].

To the best knowledge of the author, up to now, no comprehensive report on the dynamical symmetries of this 2D Newtonian free fall problem has been found in the literature besides the seminal work of T. Iwai and S. G. Rew [4]. As the one-dimensional free fall problem, which has been thoroughly treated classically and quantum mechanically, displays limited interesting features, we introduce an extra space dimension to allow for new specific features to emerge. This is how a body of dynamical symmetries comes about, and how a connection to the quartic anharmonic oscillator problem is generated, which is one of the most challenging theoretical problems as it is generally thought of as a non-trivial quantum field theory in zero space dimension. These issues were not considered in [4].

This paper is divided into two main sections and a short section on an application.

Section 2 is devoted to the classical physics of the 2D Newtonian free fall problem. For the convenience of the reader, some basic concepts on the symmetries of a system are recalled in the framework of Hamiltonian mechanics before a derivation of the dynamical symmetry algebra is undertaken along the lines of [4]. Then, we prove a theorem which states that the dynamical symmetry algebra of the free fall problem may be obtained as a special limit of the the dynamical symmetry algebra of the Kepler–Coulomb problem, which concerns the dynamics of a particle moving in the inverse separation distance potential field. Both problems are known to be super-integrable, and now they are shown to be connected at the classical level. Hence, one may access the free fall dynamical symmetry algebra from the well-known Kepler–Coulomb dynamical symmetry algebra. In addition, this connection goes even further when a higher order integral of motion in the free fall problem is shown to be an outgrowth of a Kepler–Coulomb integral of motion at zero coupling constant. Furthermore, a passage to parabolic coordinates reveals a connection between the free fall problem and the quartic anharmonic oscillator.

In Section 3, we apply the Schrödinger quantization to the system, which is in fact not the quantization scheme adopted by T. Iwai and S. G. Rew in [4]. For this, we introduced a Hilbert space of square integrable wave functions that verify free boundary conditions in $\mathbb{R}^2$ and are suitable for obtaining an image representation of the free fall dynamical symmetry algebra as Klink's so-called one-variable algebra of the quartic anharmonic oscillator. Thus, the Schrödinger quantization does naturally lead to the quartic anharmonic oscillator besides the passage to parabolic coordinates. This is due to a particular integral representation of the free fall wave functions, which takes the form of an Airy-Fourier transform. Finally, by performing a passage to parabolic coordinates, one obtains an integral identity for the eigenfunctions of the quartic anharmonic oscillator problem, which turns out to be the characteristic integral equation for *zonal* functions in the representation of some underlying group. As the only natural group arising from the 2D free fall problem is its own dynamical symmetry group, we suspect that a comprehensive development of its representation theory would reveal that the eigenfunctions of the quartic anharmonic oscillator problem are just the corresponding zonal functions of this group. We defer this tantalizing investigation to a future work. In short, the present results on quantizing the 2D Newtonian free fall dynamical symmetry algebra appear as a necessary intermediate step in the search for an exact solution to the quantum quartic anharmonic oscillator problem.

The last section, Section 4, uses the functional techniques of Section 3 to solve the initial problem of sound wave propagation in a cylindrical duct with a particular varying circular section.

The paper ends with a short conclusion and perspectives on possible further research.

2. The Classical Two-Dimensional Free Fall Problem

2.1. Generalities on Classical Symmetries, Canonical Structure, and Integrals of Motion

As generally admitted, the classical state of a non-moving physical system is described by a set of functions in a coordinate system. Under a coordinate transformation, these functions may take different forms. But if they remain *invariant*, the system is said to admit a *symmetry*.

Now, for a physical system in motion, its states are specified by functions of time t and dynamical coordinates in *phase space*. If under a transformation of dynamical coordinates its state functions remain invariant, we say that the system admits a *dynamical symmetry*. The set of symmetry transformations may have a group structure, which is called *dynamical symmetry group*. Such a dynamical symmetry group (or equivalently, the algebra of its generators) is of highest interest in the search for solutions of its equation of motion (The term *dynamical symmetry* was coined by A. O. Barut [5]. Earlier, such symmetry was known as *hidden* or *accidental* [6]).

In this paper, we are concerned with the dynamical symmetries of a system in the framework of Hamilton's *canonical* formalism of mechanics, with the time-independent Hamiltonian function H. We now recall some useful main points of this framework.

- A system with n degrees of freedom is described by $2n$ canonical variables (q_i, p_i) with $i = 1, \ldots, n$ in phase space, verifying the following fundamental Poisson bracket commutation relations:

$$\{q_i, q_j\} = \{p_i, p_j\} = 0, \quad \{q_i, p_j\} = \delta_{ij}, \tag{1}$$

where the Poisson bracket between two functions in phase space ϕ and ψ is defined by the following:

$$\{\phi, \psi\} = \sum_{i=1}^{n} \left(\frac{\partial \phi}{\partial q_i} \frac{\partial \psi}{\partial p_i} - \frac{\partial \phi}{\partial p_i} \frac{\partial \psi}{\partial q_i} \right),$$

and the canonical equations of motion:

$$\dot{q}_i = \frac{\partial H}{\partial p_i}, \quad \dot{p}_i = -\frac{\partial H}{\partial q_i}, \tag{2}$$

the solutions of which depend on the $2n$ initial conditions $(q_i(t), p_i(t))|_{t=0} = (q_i^0, p_i^0)$.

Consequently a dynamical symmetry exists if it originates from a *canonical* coordinate transformation in phase space, leaving *both* the Hamiltonian H and the fundamental Poisson bracket relations invariant. In differential geometry, this is called a *symplectic* structure. Non-trivial dynamical symmetry exists only for $n > 1$.

- Canonical transformations (see Chapters 10–12 of [7]) are parts of a wider class of coordinate transforms in phase space called *contact transformations*. A differentiable mapping $(q_1, \ldots, q_n, p_1, \ldots, p_n) \to (Q_1, \ldots, Q_n, P_1, \ldots, P_n)$ is called a contact transformation if the differential form $\sum_{i=1}^{n} (P_i dQ_i - p_i dq_i)$ is the exact differential dW of a function $W(q_1, \ldots, q_n, p_1, \ldots, p_n)$.

This can be equivalently expressed by requiring the fundamental Poisson brackets in the variables $(Q_1, \ldots, Q_n, P_1, \ldots, P_n)$ to be valid:

$$\begin{aligned}
\{Q_i, Q_j\} = \{P_i, P_j\} &= 0, \quad (i, j = 1, \ldots, n), \\
\{Q_i, P_j\} &= 0, \quad (i, j = 1, \ldots, n; i < j, i > j), \\
\{Q_i, P_i\} &= 1, \quad (i = 1, \ldots, n).
\end{aligned} \tag{3}$$

The Poisson bracket of two functions (ϕ, ψ) is *invariant* under the contact transformations $\{\phi, \psi\}_{(Q,P)} = \{\phi, \psi\}_{(q,p)}$. The lower index (q, p) (resp. (Q, P)) refers to the variables in the partial derivatives of the respective Poisson brackets. A contact transformation $(q_1, \ldots, q_n, p_1, \ldots, p_n) \to (Q_1, \ldots, Q_n, P_1, \ldots, P_n)$, which *preserves* the equations of motion of the system is a *canonical* transformation if $\int \sum_{i=1}^{n} P_i dQ_i$ is an *integral invariant* of the system (*Jacobi's theorem*). Consequently, a canonical transformation implements a dynamical symmetry.

An *infinitesimal* canonical transformation of parameter τ has the following form:

$$Q_i = q_i + \frac{\partial F_\tau}{\partial p_i} \Delta \tau, \quad P_i = p_i - \frac{\partial F_\tau}{\partial q_i} \Delta \tau, \tag{4}$$

where F_τ is an arbitrary function of $(q_1, \ldots, q_n, p_1, \ldots, p_n)$ and $\Delta \tau$ is a small increment of the parameter τ. F_τ is called the *the generating function* of the contact transformation. The variation Δf of an arbitrary function f in phase space under an infinitesimal contact transformation of generator F_τ is given by $\Delta f = \{f, F_\tau\} \Delta \tau$. F_τ can be an *integral of the motion*, a function which takes a constant value and does not explicitly depend on time t, see [7]. A system of integrals of motion F_i is said to be in involution when $\{F_i, F_j\} = 0$ for all $(i, j = 1, \ldots, m)$. If F and F' are two integrals of the motion, their Poisson bracket is also an integral of the motion (*Poisson's theorem*). The set of all F_j forms a *Lie* algebra with respect to the Poisson bracket.

The determination of all canonical transformations for a dynamical system is at the core of finding the dynamical symmetries of this system.

- For a system with an n degree of freedom, with a time-independent Hamiltonian H, total energy is *conserved* and $H = E$. If there exists n functionally *independent* integrals of motion F_i, where $i = 1, \ldots, n$ in involution (or $\{F_i, F_j\} = 0$ for all (i,j)), the system is said to be Liouville *integrable* and its solution can be given up to *quadratures*. If there are further m functionally independent integrals of motion F_m, where $m = 1, \ldots, (n-1)$, the system is called *super-integrable* [8]. For $m = 1$ (resp. $m = (n-1)$) it is called *minimal* super-integrable (resp. *maximal* super-integrable). These m extra integrals of motion usually build a Lie algebra. Maximal super-integrable systems are also known as *exactly solvable* systems, and their properties can be derived algebraically. Each integral of motion F_i may originate from Noether's conservation law or from a coordinate variable separation. The Kepler–Coulomb (or inverse distance potential) and the isotropic harmonic oscillator problems are known to be super-integrable systems in two dimensions [9].

2.2. Statement of the Classical Two-Dimensional Free Fall Problem

As it is widely known, the motion of a particle under constant force in two-dimensional space occurs along parabolas with a symmetry axis parallel to the direction of the constant force. To simplify the writing, the particle is assumed to have a unit mass and moves in a two-dimensional configuration space $(v, u) \in \mathbb{R}^2$, with Ov being the horizontal axis and the gravitational constant set is also equal to one. Let $(\dot{v}, \dot{u})$ be the time derivatives of the Cartesian coordinates (v, u). From the Lagrangian $L = \left(\frac{1}{2}(\dot{u}^2 + \dot{v}^2) - u\right)$, one gets the conjugate momenta and the Hamiltonian:

$$p_u = \frac{\partial L}{\partial \dot{u}} = \dot{u}, \quad p_v = \frac{\partial L}{\partial \dot{v}} = \dot{v}, \quad \text{and} \quad H = \frac{1}{2}p_v^2 + \left(\frac{1}{2}p_u^2 + u\right). \tag{5}$$

Note that H is the sum of the v-free motion part and of the u-free fall part.

2.3. Dynamical Symmetries

As in two dimensions, the maximal number of possible symmetries is three. Iwai and Rew [4] were the first to obtain dynamical symmetries by considering linear inhomogeneous transformations $S(\rho, \sigma, \tau)$ in the phase space dependent on three parameters (ρ, σ, τ):

$$\begin{aligned} V &= v + \tau p_v + \sigma p_u + \rho, & P_v &= p_v - \sigma, \\ U &= u + \sigma p_v - \tau p_u - \frac{1}{2}(\sigma^2 + \tau^2), & P_u &= p_u + \tau. \end{aligned} \tag{6}$$

Proposition 1. $S(\rho, \sigma, \tau)$ *is a canonical transformation.*

Proof. By the simple substitution of the (V, U, P_v, P_u) in terms of the (v, u, p_v, p_u), as given by Equation (6), it appears that the Hamiltonian function form is invariant:

$$H = \frac{1}{2}(P_u^2 + P_v^2) + U = \frac{1}{2}(p_u^2 + p_v^2) + u. \tag{7}$$

Moreover, using the expression of the Poisson bracket and Equation (6), one can also check the following:

$$\begin{aligned} \{V, U\} &= \{P_v, P_u\} &= 0, \\ \{V, P_u\} &= \{U, P_v\} &= 0, \\ \{V, P_v\} &= \{U, p_u\} &= 1, \end{aligned} \tag{8}$$

Hence, $S(\rho, \sigma, \tau)$ is a bona fide canonical transformation. □

Proposition 2. *For all triplet $(\rho, \sigma, \tau) \in \mathbb{R}^3$, the canonical transforms form a group—the dynamical symmetry group of the free fall problem in two dimensions.*

Proof. *Composition law.* Let another canonical transform $S(\rho', \sigma', \tau')$ with parameters (ρ', σ', τ') act on the previous one $S(\rho, \sigma, \tau)$ and compute the resulting product $S(\rho', \sigma', \tau') S(\rho, \sigma, \tau)$, which is expressed by the set of new parameters $(\overline{V}, \overline{U}, \overline{P}_v, \overline{P}_v)$:

$$\begin{aligned}
\overline{V} &= V + \tau' P_v + \sigma' P_u + \rho', \\
\overline{U} &= U + \sigma' P_v - \tau' P_u - \frac{1}{2}(\sigma'^2 + \tau'^2), \\
\overline{P}_v &= P_v - \sigma', \\
\overline{P}_u &= P_u + \tau'.
\end{aligned} \qquad (9)$$

Then, the substitution of Equation (6) into Equation (9) yields the following equations:

$$\begin{aligned}
\overline{V} &= (v + \tau p_v + \sigma p_u + \rho) + \tau'(p_v - \sigma) + \sigma'(p_u + \tau) + \rho' \\
&= v + (\tau + \tau')p_v + (\sigma + \sigma')p_u + (\rho + \rho') - (\sigma\tau' - \sigma'\tau), \\
\overline{U} &= (u + \sigma p_v - \tau p_u - \frac{1}{2}(\sigma^2 + \tau^2)) + \sigma'(p_v - \sigma) - \tau'(p_u + \tau) - \frac{1}{2}(\sigma'^2 + \tau'^2) \\
&= u + (\sigma + \sigma')p_v - (\tau + \tau')p_u - \frac{1}{2}(\sigma + \sigma')^2 - \frac{1}{2}(\tau + \tau')^2, \\
\overline{P}_v &= (p_{v'} - \sigma) - \sigma' = p_v - (\sigma + \sigma'), \\
\overline{P}_u &= (p_{u'} + \tau') + \tau' = p_u + (\tau + \tau').
\end{aligned} \qquad (10)$$

Hence, we conclude that the composition of two operations is as follows:

$$S(\rho', \sigma', \tau') \cdot S(\rho, \sigma, \tau) = S((\rho + \rho') - (\sigma\tau' - \sigma'\tau), \sigma + \sigma', \tau + \tau'). \qquad (11)$$

Because of the extra term $(\sigma\tau' - \sigma'\tau)$, the previous formula does not correspond to the additive structure of a group operation with respect to the parameters (ρ, σ, τ). But since the v variable is a *cyclic* variable, this term is physically *irrelevant* '; in this way, the group structure is *physically* restored.

Iwai and Rew proposed to represent each element of the group with fifth-order, upper-triangular matrices, as can be checked by the matrix multiplication rule:

$$\begin{pmatrix}
1 & 0 & \tau & \sigma & \rho \\
0 & 1 & \sigma & -\tau & -\frac{1}{2}(\sigma^2 + \tau^2) \\
0 & 0 & 1 & 0 & -\sigma \\
0 & 0 & 0 & 1 & \tau \\
0 & 0 & 0 & 0 & 1
\end{pmatrix}$$

- The *inverse transform* $S^{-1}(\rho, \sigma, \tau)$ is clearly given by $S(-\rho, -\sigma, -\tau)$ since the extra term $(\sigma\tau' - \sigma'\tau)$ vanishes at $(\sigma + \sigma') = 0$ and $(\tau + \tau') = 0$.

-The *identity* matrix is obviously the neutral group element. □

2.4. Infinitesimal Iwai–Rew Canonical Linear Transforms and Integrals of Motion

The infinitesimal form of the Iwai–Rew canonical linear transformations is written in the form of Equation (8), with infinitesimal $(\Delta\rho, \Delta\sigma, \Delta\tau)$ as:

$$\begin{aligned}
V &= v + p_v \Delta\tau + p_u \Delta\sigma + \Delta\rho + \ldots, \\
U &= u + p_v \Delta\sigma - p_u \Delta\tau + \ldots, \\
P_v &= p_v - \Delta\sigma + \ldots, \\
P_u &= p_u + \Delta\tau + \ldots.
\end{aligned} \qquad (12)$$

Proposition 3. *The infinitesimal transformations in phase space defined by Equation (12) are canonical transformations: H and the Poisson brackets* $\{V, P_v\}, \{V, P_u\}, \{V, U\}, \{U, P_u\}, \{U, P_v\}, \{P_u, P_v\}$ *are invariant.*

Proof. By working out the Poisson brackets using Equation (12), they will appear to remain in the canonical form of Equation (8), while the form of the Hamiltonian in the variables (V, U, P_v, P_u) is the same as the form in variables (v, u, p_v, p_u). □

We now determine the three generating functions $(F_\rho, F_\sigma, F_\tau)$ from the three subgroups of canonical transformations, respectively defined by the following equations:

$$
\begin{aligned}
F_\rho &: \quad V = v + \Delta\rho, U = u, P_v = p_v, P_u = p_u, \\
F_\sigma &: \quad V = v + p_u\Delta\sigma, U = u + p_v\Delta\sigma, P_v = p_v - \Delta\sigma, P_u = p_u, \\
F_\tau &: \quad V = v + p_v\Delta\tau, U = u - p_u\Delta\tau, P_v = p_v, P_u = p_u + \Delta\tau.
\end{aligned}
\quad (13)
$$

Then, partial derivatives of $(F_\rho, F_\sigma, F_\tau)$ with respect to (v, u, p_v, p_u) can be deduced as follows:

$$
\begin{aligned}
F_\rho &: \quad \frac{\partial F_\rho}{\partial v} = 0, \frac{\partial F_\rho}{\partial u} = 0, \frac{\partial F_\rho}{\partial p_v} = 1, \frac{\partial F_\rho}{\partial p_u} = 0, \\
F_\sigma &: \quad \frac{\partial F_\sigma}{\partial v} = 1, \frac{\partial F_\sigma}{\partial u} = 0, \frac{\partial F_\sigma}{\partial p_v} = p_u, \frac{\partial F_\sigma}{\partial p_u} = p_v \\
F_\tau &: \quad \frac{\partial F_\tau}{\partial v} = 1, \frac{\partial F_\tau}{\partial u} = -1, \frac{\partial F_\sigma}{\partial p_v} = p_v, \frac{\partial F_\sigma}{\partial p_u} = -p_u.
\end{aligned}
\quad (14)
$$

Therefore, we obtain the exact differentials $(dF_\rho, dF_\sigma, dF_\tau)$ in terms of (dv, du, dp_v, dp_u) (Schwarz's theorem is trivially verified for all pairs of variables). Their integration yields the sought integrals of motion:

$$
F_\rho = p_v, \quad F_\sigma = (p_v p_u + v), \quad F_\tau = \frac{1}{2}p_v^2 - \left(\frac{1}{2}p_u^2 + u\right). \quad (15)
$$

Proposition 4. *The free fall system in two dimensions is* super-integrable *since its has three integrals of the motion:*

$$
\{F_\rho, H\} = \{F_\sigma, H\} = \{F_\tau, H\} = 0. \quad (16)
$$

Proof. Compute the three Poisson brackets and observe that they are zero. □

Proposition 5. *The three generating functions $(F_\rho, F_\sigma, F_\tau)$ build a Lie algebra structure, the dynamical symmetry algebra $\mathfrak{A}_{FF}$ of the two-dimensional free fall with respect to the Poisson bracket:*

$$
\{F_\rho, F_\sigma\} = -I, \quad \{F_\sigma, F_\tau\} = 2F_\rho, \quad \{F_\tau, F_\rho\} = 0. \quad (17)
$$

Proof. Work out the Poisson brackets and use the expressions of the $(F_\rho, F_\sigma, F_\tau)$ from Equation (15). They are identical to those of Iwai–Rew [4]. If we make the substitutions $F_\rho \to F_3, \quad F_\sigma \to F_2, \quad F_\tau \to F_1$ in our Equation (15), we recover the Iwai–Rew commutation relations given by their Equation (2.21). □

Remark 1. *Observe that the classical trajectory data (two initial position coordinates, two initial momentum coordinates) can be used to compute the values of $\hat{H}, \hat{F}_\rho, \hat{F}_\sigma, \hat{F}_\tau$ and vice versa.*

Proposition 6. *Origins of the integrals of motion*
(a) F_ρ *is due to translational invariance of H in the Ov direction;*
(b) F_σ *is due the separability of H in a $\pi/4$-rotated coordinate system;*
(c) F_τ *is due to the manifest separability of H in v and u variables.*

Proof. (a) For this case, the proof is trivial because v is a *cyclic* variable in H.

(b) Consider the change of variables $\xi = (u+v)$, $\zeta = (u-v)$, from which one deduces $u = \frac{1}{2}(\xi + \zeta), v = \frac{1}{2}(\xi - \zeta)$. Then, since $p_v = \dot{v}$, $p_u = \dot{u}$, we have $p_v = \frac{1}{2}(p_\xi + p_\zeta)$, $p_u = \frac{1}{2}(p_\xi - p_\zeta)$. Substitution into the expressions of H and F_σ yields the following equation:

$$H = \frac{1}{2}\left\{\left(\frac{1}{2}p_\xi^2 + \xi\right) + \left(\frac{1}{2}p_\zeta^2 + \zeta\right)\right\}, \text{ and } F_\sigma = \frac{1}{2}\left\{\left(\frac{1}{2}p_\zeta^2 + \zeta\right) - \left(\frac{1}{2}p_\xi^2 + \xi\right)\right\}. \quad (18)$$

where H and F_σ are both separable in the new coordinate system obtained by a $\pi/4$ rotation of the (v, u) coordinate system around the origin. In fact, they are the sum and difference of the one-dimensional free fall Hamiltonian in the ξ and ζ directions. Hence, $\{F_\sigma, H\} = 0$. In reference [10], it was claimed that at the quantum level, F_σ is due to separability in *translated parabolic* coordinates. But so far, no proof has appeared in print.

(c) As H is the sum of a free-motion Hamiltonian in the v-direction, and the Hamiltonian is that of a one-dimensional free fall in the u-direction, it is then clear that F_τ, which is the difference of these two Hamiltonians, should verify $\{F_\tau, H\} = 0$. Since the total energy is conserved as $H = E$, it may be simpler to write $F_\tau = (p_v^2 - E)$. □

Remark 2. *We notice that v-parity is also a dynamical symmetry and a* discrete canonical transformation *since* $v \to -v$ *(which also implies* $p_v \to -p_v$*) leaves H as well as* $\{v, p_v\} = 1$ *and all other Poisson brackets invariant.*

2.5. The Free Fall Problem as a Special Limiting Case of the Kepler–Coulomb Problem

In this section, we show that the free fall problem can emerge from a special limit of the Kepler–Coulomb (KC) problem or the problem of the inverse distance potential, occurring either in gravitational or electrical interaction. This problem is known to be super-integrable and in two dimensions, and the set of its three integrals of motion makes up the components of the so-called Runge–Lenz vector [11]. Here, we adopt the notations of [12], with the particle mass set equal to one.

The idea is simple. The inverse distance potential κ/r arising from a source of strength κ is rotation-symmetric, where r is the distance from the source to the observation site. If the source recedes to infinity in the u-direction, the potential at the observation site tends toward zero. But one may compensate this potential decrease by taking a source strength $\kappa(r)$, which increases with the distance. Thus, we may choose an increasing functional dependence so that in the limit of infinite source-observation separation, a linear potential appears.

In our Cartesian coordinate system, if the source of strength κ is placed at the coordinate origin, the Hamiltonian of the inverse distance problem is the following:

$$H_{KC} = \frac{1}{2}(p_v^2 + p_u^2) - \frac{\kappa}{\sqrt{u^2 + v^2}}, \quad (19)$$

and the three integrals of motion are given by [12]:

$$\begin{aligned}
L_w &= (vp_u - up_v), \\
K_v &= (vp_u - up_v)p_u - v\frac{\kappa}{\sqrt{u^2 + v^2}}, \\
K_u &= -(vp_u - up_v)p_v - u\frac{\kappa}{\sqrt{u^2 + v^2}},
\end{aligned} \quad (20)$$

where L_w is the angular momentum around Ow, which is orthogonal to the plane Ovu. They verify the Poisson bracket relations of the dynamical symmetry algebra $\mathfrak{A}_{KC}$

$$\{L_w, K_v\} = K_u, \quad \{K_u, L_w\} = K_v, \quad \{K_v, K_u\} = -2H_{KC}L_w. \quad (21)$$

Now, if the source is no longer at the origin O but situated on the Ou axis at a distance l from the origin, the Hamiltonian and the components of the Runge–Lenz vector have new expressions H'_{KC} and (L'_w, K'_v, K'_u), which are deduced from the previous expressions in which u is replaced by $(u+l)$ and κ by $\kappa(l)$:

$$H'_{KC} = \frac{1}{2}(p_v^2 + p_u^2) - \frac{\kappa(l)}{\sqrt{(u+l)^2 + v^2}}, \qquad (22)$$

and

$$\begin{aligned} L'_w &= (vp_u - up_v) - lp_v, \\ K'_v &= (vp_u - up_v - lp_v)p_u - v\frac{\kappa(l)}{\sqrt{(u+l)^2 + v^2}}, \\ K'_u &= -(vp_u - up_v - lp_v)p_v - (u+l)\frac{\kappa(l)}{\sqrt{(u+l)^2 + v^2}}. \end{aligned} \qquad (23)$$

Now, as $l \to \infty$, the asymptotic behaviour of the inverse distance potential is as follows:

$$\frac{\kappa(l)}{\sqrt{(u+l)^2 + v^2}} \sim \frac{\kappa(l)}{l}\left(1 - \frac{u}{l} + \mathcal{O}(\frac{1}{l^2})\right). \qquad (24)$$

Hence, if the source strength increases as $\kappa(l) = l^2$, then the inverse distance potential reaches the limiting form of a linear potential in u:

$$-\frac{\kappa(l)}{\sqrt{(u+l)^2 + v^2}} \to -l + u + \mathcal{O}(\frac{1}{l}). \qquad (25)$$

In this limit, the inverse distance potential problem tends toward the free fall problem up to a negative infinite constant:

$$H'_{KC} \sim H - l + \mathcal{O}\left(\frac{1}{l}\right), \qquad (26)$$

and the components of the Runge–Lenz vector take the following asymptotic forms:

$$\begin{aligned} L'_w &\sim -lp_v + (vp_u - up_v), \\ K'_v &\sim -l(p_u p_v + v) + (vp_u - up_v)p_u + uv - v\mathcal{O}(\frac{1}{l}), \\ K'_u &\sim -l^2 + lp_v^2 - \left[(vp_u - up_v)p_v - u^2\right] + \mathcal{O}(\frac{1}{l}), \end{aligned} \qquad (27)$$

Theorem 1. *The dynamical symmetry algebra of the free fall problem $\mathfrak{A}_{FF}$ is a contraction of the dynamical symmetry algebra of the Kepler–Coulomb problem $\mathfrak{A}_{KC}$.*

Proof. We now rewrite the Poisson brackets of the Kepler–Coulomb problem when the potential source is at a large distance l from the origin, and then replace the generators (L'_w, K'_v, K'_u) by their asymptotic expansions for $l \to \infty$. Therefore:

(a) $\{L'_w, K'_v\} = K'_u$ becomes

$$\{(vp_u - up_v) - l F_\rho, -l F_\sigma + (vp_u - up_v)p_u + uv - v\mathcal{O}(\frac{1}{l})\} =$$
$$-l^2 + l(F_\tau + H) - \left[(vp_u - up_v)p_v - u^2\right] + \mathcal{O}(\frac{1}{l}). \qquad (28)$$

Extracting the leading order in l^2 on the left-hand side and on the right-hand side, we obtain $\{F_\rho, F_\sigma\} = -I$, as expected; see Equation (17).

(b) $\{K'_u, L'_w\} = K'_v$ becomes

$$\{-l^2 + l(F_\tau + H) - \left[(vp_u - up_v)p_v - u^2\right] + \mathcal{O}(\tfrac{1}{l}), (vp_u - up_v) - l F_\rho\} =$$

$$-l F_\sigma + (vp_u - up_v)p_u + uv - v\mathcal{O}(\tfrac{1}{l}). \tag{29}$$

Collecting terms of leading order in l^2 on both sides of this equation, we get $\{F_\tau, F_\rho\} = 0$ since there is no term in l^2 on the right-hand side.

(c) $\{K'_v, K'_u\} = -2H'_{KC} L'_w$ becomes

$$\{-l F_\sigma + (vp_u - up_v)p_u + uv - v\mathcal{O}(\tfrac{1}{l}), -l^2 + l(F_\tau - H) - \left[(vp_u - up_v)p_v - u^2\right] + \mathcal{O}(\tfrac{1}{l})\} =$$

$$-2\left(H - l + \mathcal{O}\left(\tfrac{1}{l}\right)\right)\left((vp_u - up_v) - l F_\rho\right). \tag{30}$$

Equating terms of order d^2 on both sides of this equation yields precisely $\{F_\sigma, F_\tau\} = 2F_\rho$. Hence, we reproduce all the Poisson brackets of $\mathfrak{A}_{FF}$. □

2.6. *A "Higher" Order Integral of the Motion*

We now raise the question whether there exists a "higher" order integral of the motion as a construct of the dynamical symmetry algebra generators. What comes to mind is a weighted sum of squares of the $(F_\rho, F_\sigma, F_\tau)$, an object similar to the square of the angular momentum in a rotation algebra. Instead of a tedious systematic search, a more astute way of finding such an integral of the motion would start by observing that the symmetry algebra of the Kepler–Coulomb problem does have a non-trivial limit when the source strength is turned off, i.e., $\kappa = 0$. Then, the generators take the following form:

$$L_w = (vp_u - up_v), \quad K_v(0) = (vp_u - up_v)p_u, \quad K_u(0) = -(vp_u - up_v)p_v. \tag{31}$$

They fulfil the same Poisson bracket relations as those with $\kappa \neq 0$:

$$\{L_w, K_v(0)\} = K_u(0), \quad \{K_u(0), L_w\} = K_v(0), \quad \{K_v(0), K_u(0)\} = -2H_0 L_w, \tag{32}$$

where H_0 is the Hamiltonian for free particle motion in two dimensions.

Now, if we allow a linear potential to "grow" in the u-direction, rotational symmetry disappears. Hence, L_w must be discarded as a possible integral of motion in the presence of a linear potential u. Thus, from the two remaining $(K_v(0), K_u(0))$, only one can survive under a modified form as a "higher" order integral of the motion, because of Poisson's theorem in Section 2.1.

Let F^2 be this hypothetical "higher" integral of the motion. As the introduced linear potential is in the u direction, we may assume F^2 to be of the simple form:

$$F^2 = K_u(0) + h(v, u, p_v, p_u), \tag{33}$$

where $h(v, u, p_v, p_u)$ is an unknown function in phase space.

Proposition 7. *There exists a second order integral of the motion F^2 given by the following equation:*

$$F^2 = K_u(0) - \frac{1}{2}v^2 = -(vp_u - up_v)p_v - \frac{1}{2}v^2. \tag{34}$$

Proof. The function $h(v, u, p_v, p_u)$ is determined by the condition $\{F^2, H\} = 0$. Since from explicit computation one gets the following:

$$\{F^2, H\} = \frac{\partial h}{\partial v}p_v + \frac{\partial h}{\partial u}p_u + \frac{\partial h}{\partial p_u} + vp_v = 0, \tag{35}$$

it is obvious that one should require that $h(v, u, p_v, p_u) = -\frac{1}{2}v^2$. □

Remark 3. *The same search procedure for another quadratic integral of the motion does not work with $K_v(0)$ because the commutativity of the Poisson bracket with H leads to impossible conditions having to be satisfied.*

Proposition 8. *The expression of F^2 in terms of the generators (F_r, F_s, F_t) is as follows:*

$$F^2 = HF_\rho^2 - \frac{1}{2}F_\sigma^2 - \frac{1}{2}(F_\tau + H)^2, \tag{36}$$

Proof. Substitute the expressions $(H, F_\rho, F_\sigma, F_\tau)$ into the expression of F^2. □

Corollary 1. *The Poisson brackets of F^2 with the generators $(F_\rho, F_\sigma, F_\tau)$ are easily obtained:*

$$\{F^2, F_\rho\} = -F_\sigma, \quad \{F^2, F_\sigma\} = 2F_\rho F_\tau, \quad \{F^2, F_\tau\} = -2F_\rho F_\sigma. \tag{37}$$

Proposition 9. *The infinitesimal transformation generated by F^2 with parameter z is canonical.*

Proof. The infinitesimal canonical transform generated by F^2 with the parameter z is given by the following equations:

$$\begin{aligned} V &= v + (vp_u - 2up_v)\Delta z + \ldots \\ U &= u + vp_v \Delta z + \ldots \\ P_v &= p_v - (p_v p_u + v)\Delta z + \ldots \\ P_u &= p_u + p_v^2 \Delta z \ldots \end{aligned} \tag{38}$$

We can check that the Hamiltonian remains invariant if the terms of order $(\Delta z)^2$ are ignored:

$$H(v', u', p_{v'}, p_{u'}) = H(v, u, p_v, p_u) + \left((p_v p_u + v)^2 + p_v^4\right)(\Delta z)^2 + \ldots \tag{39}$$

Next, we can verify that the six canonical Poisson brackets are preserved, but the details are not presented here. Moreover, it does possess the additive abelian group property with respect to z, i.e., $S(z) \cdot S(z') = S(z + z')$, as can be checked explicitly. □

2.7. Passage to Parabolic Coordinates and the Physical Meaning of F^2

The issue here is to understand how F^2 arises as a dynamical symmetry. The comprehensive work of Miller et al. [13] on quantum separability has revealed that parabolic coordinates do play a central role. Following this indication, we make a passage to parabolic coordinates (x, y) from our Cartesian coordinates (v, u), as defined by $u = \frac{1}{2}(x^2 - y^2)$, $v = xy$. The Lagrangian L then changes to a new expression:

$$L = \frac{1}{2}(\dot{u}^2 + \dot{v}^2) - u = \frac{1}{2}(x^2 + y^2)(\dot{x}^2 + \dot{y}^2) - \frac{1}{2}(x^2 - y^2). \tag{40}$$

From the following definitions of conjugate momenta:

$$p_u = \frac{\partial L}{\partial \dot{u}} = \dot{u}, \qquad p_v = \frac{\partial L}{\partial \dot{v}} = \dot{v},$$
$$p_x = \frac{\partial L}{\partial \dot{x}} = (x^2 + y^2)\dot{x}, \qquad p_y = \frac{\partial L}{\partial \dot{y}} = (x^2 + y^2)\dot{y}, \qquad (41)$$

we deduce a relation between the (v, u) and the (x, y) momenta:

$$p_u = \frac{xp_x - yp_y}{(x^2 + y^2)}, \qquad p_v = \frac{yp_x + xp_y}{(x^2 + y^2)}. \qquad (42)$$

This allows for the acquisition of new expressions of H and F^2 in parabolic coordinates:

$$H = \frac{1}{2}\frac{p_x^2 + p_y^2}{(x^2 + y^2)} + \frac{1}{2}(x^2 - y^2), \qquad (43)$$

$$F^2 = \frac{x^2 y^2}{2(x^2 + y^2)}\left\{\left(\frac{p_x^2}{x^2} + x^2\right) - \left(\frac{p_y^2}{y^2} - y^2\right)\right\}. \qquad (44)$$

As such, these expressions do not show any obvious x and y variable separation. The three integrals of motion $(F_\rho, F_\sigma, F_\tau)$ also do not display any obvious separation into an x part and a y part when re-expressed in the parabolic coordinates. Next, we recall that the initial conditions (v_0, u_0, p_v^0, p_v^0) fully determine the four integrals of motion $(H, F_\rho, F_\sigma, F_\tau)$. Hence, F^2 has a fixed value because it is a construct of these four integrals of motion.

Proposition 10. *For $H = E$, F^2 takes a constant value in the range of energy values $E_+(E)$ of a confining quartic oscillator with an angular frequency square $(-E)$.*

Proof. As total energy is conserved, $H = E$ implies that the following relation must be verified for all $(x, y) \neq (0, 0)$:

$$\frac{1}{(x^2 + y^2)}\left\{\left(\frac{1}{2}p_x^2 - Ex^2 + \frac{1}{2}x^4\right) + \left(\frac{1}{2}p_y^2 - Ey^2 - \frac{1}{2}y^4\right)\right\} = 0. \qquad (45)$$

This means that for a given E, the sum of the Hamiltonians of a confining quartic x-oscillator and a non-confining quartic y-oscillator must be equal to zero for all $(x, y) \neq (0, 0)$. Since these one dimensional quartic oscillators are time-independent, their respective Hamiltonians have fixed values, i.e.:

$$\left(\frac{1}{2}p_x^2 - Ex^2 + \frac{1}{2}x^4\right) = E_+(E), \qquad \left(\frac{1}{2}p_y^2 - Ey^2 - \frac{1}{2}y^4\right) = E_-(E), \qquad (46)$$

however, subjected to the condition $E_+(E) + E_-(E) = 0$. Note that for a given E, $E_+(E)$ takes all real values above the minimum value of the quartic x-oscillator Hamiltonian polynomial in phase space.

On the other hand, F^2 may be rewritten as follows:

$$F^2 = \frac{x^2 y^2}{(x^2 + y^2)}\left\{\frac{1}{x^2}\left(\frac{1}{2}p_x^2 - Ex^2 + \frac{1}{2}x^4\right) + \frac{1}{y^2}\left(\frac{1}{2}p_y^2 - Ey^2 - \frac{1}{2}y^4\right)\right\}$$

$$= \frac{x^2 y^2}{(x^2 + y^2)}\left(\frac{E_+(E)}{x^2} - \frac{E_-(E)}{y^2}\right). \qquad (47)$$

Since $E_+(E) + E_-(E) = 0$, one gets $F^2 = E_+(E)$. This is due to this *new* aspect of separation of variables called the Stäckel separation of variables. □

2.8. Third Order Integrals of the Motion

In [13], it is shown that third order super-integrable systems separable in parabolic coordinates admit second order integrals, while third order integrals are reducible, i.e., they are Poisson brackets of second order integrals of motion. Here, we have:

$$F^{3\sigma} = \{F_\sigma, F^2\} = p_v^3 - 2up_v - p_v p_u^2, \quad F^{3\tau} = \{F_\tau, F^2\} = -(p_u p_v^2 + 2vp_v). \tag{48}$$

However, these third order reducible integrals of motion do not generate one parameter infinitesimal canonical transformations because they do not leave H functionally invariant and are consequently uninteresting.

3. Schrödinger Quantization of the Two-Dimensional Free Fall Problem

3.1. Quantization

The Hamiltonian form of the free fall problem lends itself nicely to its quantization. The classical canonical variables are put into one-to-one correspondence with their quantum counterparts, as self-adjoint operators in a Hilbert space of states $\mathfrak{H}$:

$$(v, p_v, u, p_u) \leftrightarrow (\widehat{Q}_v, \widehat{P}_v, \widehat{Q}_u, \widehat{P}_u). \tag{49}$$

They build a direct product of v and u Heisenberg algebras (here $\hbar = 1$ for ease of writing):

$$[\widehat{Q}_v, \widehat{P}_v] = i, \quad [\widehat{Q}_u, \widehat{P}_u] = i, \quad \text{and} \quad [\widehat{O}_v, \widehat{O}_u] = 0, \tag{50}$$

where $\widehat{O}_v$ (respectively $\widehat{O}_u$) means $(\widehat{Q}_v, \widehat{P}_v)$ (respectively $\widehat{O}_u = (\widehat{Q}_u, \widehat{P}_u)$).

The quantum Hamiltonian, the dynamical symmetry algebra generators, and the quadratic integral of motion are then given by the following equation:

$$\widehat{H} = \frac{1}{2}\left(\widehat{P}_v^2 + \widehat{P}_u^2\right) + \widehat{Q}_u, \quad \widehat{F}_\rho = \widehat{P}_v, \quad \widehat{F}_\sigma = \widehat{P}_v \widehat{P}_u + \widehat{Q}_v, \quad \widehat{F}_\rho = \widehat{P}_v^2. \tag{51}$$

They fulfil the quantum commutation relations deduced from their Poisson brackets classical counterparts:

$$[\widehat{F}_\rho, \widehat{F}_\sigma] = -iI, \quad [\widehat{F}_\sigma, \widehat{F}_\tau] = 2i\widehat{F}_\rho, \quad [\widehat{F}_\tau, \widehat{F}_\rho] = 0, \tag{52}$$

where I is the identity operator. The quadratic integral of motion $\widehat{F}^2$ is as follows:

$$\widehat{F}^2 = \widehat{Q}_u \widehat{P}_v^2 - \frac{1}{2}(\widehat{P}_v \widehat{Q}_v + \widehat{Q}_v \widehat{P}_v)\widehat{P}_u - \frac{1}{2}\widehat{Q}_v^2. \tag{53}$$

3.2. Schrödinger Representation

It is convenient to work with the Schrödinger coordinate representation. The Hilbert space $\mathfrak{H}$ of states $|\psi\rangle \in \mathfrak{H}$ is represented by the square integrable functions $\psi(v, u) = \langle v, u | \psi \rangle$, where $|v, u\rangle$ for $(v, u) \in \mathbb{R}^2$ is a continuous set of complete and total set in $\mathfrak{H}$. The canonical dynamical operators by differential operators in (v, u) are as follows:

$$(\widehat{Q}_v, \widehat{P}_v, \widehat{Q}_u, \widehat{P}_u) = \left(v, -i\frac{\partial}{\partial v}, u, -i\frac{\partial}{\partial u}\right). \tag{54}$$

Consequently, the differential operator representation of the Hamiltonian, the dynamical symmetry generators, and the quadratic integral of motion are as follows:

$$\begin{aligned}
\hat{H} &= -\frac{1}{2}\left(\frac{\partial^2}{\partial v^2} + \frac{\partial^2}{\partial u^2}\right) + u, \\
\hat{F}_\rho &= -i\frac{\partial}{\partial v}, \hat{F}_\sigma = \left(-\frac{\partial^2}{\partial v \partial u} + v\right), \hat{F}_\tau = -\left(\frac{\partial^2}{\partial v^2} + \hat{H}\right), \\
\hat{F}^2 &= \left(v\frac{\partial^2}{\partial v \partial u} - u\frac{\partial^2}{\partial v^2} + \frac{1}{2}\frac{\partial}{\partial u} - \frac{1}{2}v^2\right).
\end{aligned} \quad (55)$$

All Poisson brackets from the classical $(H, F_\rho, F_\sigma, F_\tau, F^2)$ now become quantum commutators between $(\hat{H}, \hat{F}_\rho, \hat{F}_\sigma, \hat{F}_\tau, \hat{F}^2)$, as can be easily checked. Let $\hat{\mathfrak{A}}_{FF}$ be the algebra generated by $(\hat{H}, \hat{F}_\rho, \hat{F}_\sigma, \hat{F}_\tau)$.

Corollary 2. $\hat{F}^2$ *can be expressed in terms of* $(\hat{H}, \hat{F}_\rho, \hat{F}_\sigma, \hat{F}_\tau)$ *as*

$$\hat{F}^2 = \hat{H}\hat{F}_\rho^2 - \frac{1}{2}\hat{F}_\sigma^2 - \frac{1}{2}\left(\hat{F}_\tau + \hat{H}\right)^2. \quad (56)$$

Proof. Use the expressions in Equation (55) and substitute in (56). □

As usual, the space of relevant wave functions $\mathcal{L}^2(v, u)$ are generated by the eigenfunctions of the stationary Hamiltonian operator $\hat{H}$. At this step, to determine this functional space, boundary conditions should be specified. As pointed out in the previous section on the classical mechanics of free fall motion, we are concerned with global motion along parabolic trajectories in $\mathbb{R}^2$, with concavity turned downward and not with the bouncing motion on a horizontal line $v =$ constant or the billiard motion inside a two-dimensional box. Both have overly restrictive boundary conditions, rending the wave functions uninteresting. Therefore, only stationary wave functions with free boundary conditions on the v-axis and on the negative u-axis solutions of $\hat{H}\psi_E(v, u) = E\psi_E(v, u)$ are considered here.

Since $\hat{H}$ is separable in the v and u parts, elementary solutions are of the product form:

$$e^{ikv} \operatorname{Ai}\left(\gamma u + \frac{k^2}{\gamma^2} - \gamma E\right), \quad (57)$$

where $\gamma^3 = 2$ and $\operatorname{Ai}(x)$ is the first Airy function which decreases at $x \to \infty$; see [14].

Hence, the spectrum of $\hat{H}$ is real and continuous. An arbitrary stationary eigenfunction $\psi_E(u, v)$ with an eigenvalue E is given by an integral on k:

$$\psi_E(u, v) = \int_\mathbb{R} dk\, e^{ikv}\, \widetilde{\psi}_E(k) \operatorname{Ai}\left(\gamma u + \frac{k^2}{\gamma^2} - \gamma E\right), \quad (58)$$

where $\widetilde{\psi}_E(k)$, yet to be determined, appears as a density amplitude representing the relative distribution of the v-motion with respect to the u-motion parts for a fixed E in $\psi_E(u, v)$. Both $\psi_E(u, v)$ and $\widetilde{\psi}_E(k)$ describe two aspects of the same quantum state $|\psi_E\rangle$.

Proposition 11. *As* $\psi_E(u, v)$ *represents a probability amplitude in* $\mathcal{L}^2(v, u)$, $\widetilde{\psi}_E(k)$ *is a square integrable function on* $\mathbb{R}$, *i.e.*, $\widetilde{\psi}_E(k) \in \mathcal{L}^2(k)$.

Proof. Let us compute the overlap integral $\langle \psi_E | \psi_{E'} \rangle$ between two eigenstates of energies E and E':

$$\langle \psi_E | \psi_{E'} \rangle = \int_{\mathbb{R}^2} \frac{du\, dv}{2\pi}\, \psi_E^*(v, u)\, \psi_{E'}(v, u) = \int_{\mathbb{R}^2} \frac{du\, dv}{2\pi}$$

$$\int_\mathbb{R} dk\, e^{-ikv}\, \widetilde{\psi}_E^*(k) \operatorname{Ai}\left(\gamma u + \frac{k^2}{\gamma^2} - \gamma E\right) \int_\mathbb{R} dk'\, e^{ik'v}\, \widetilde{\psi}_{E'}(k') \operatorname{Ai}\left(\gamma u + \frac{k'^2}{\gamma^2} - \gamma E'\right). \quad (59)$$

The 2π factor is just for convenience. Integration over dv yields $2\pi\,\delta(k-k')$. Then, after k'-integration, one may consider the remaining u-integration.

This u-integral is readily given in [14] by equation 3.108 on page 57 as follows:

$$\frac{1}{|\alpha\beta|}\int_{-\infty}^{\infty} du\, \text{Ai}\left(\frac{u+a}{\alpha}\right) \text{Ai}\left(\frac{u+b}{\beta}\right) \times$$

$$= \delta(a-b), \qquad \text{if } \beta = \alpha,$$
$$= \frac{1}{|\beta^3-\alpha^3|^{\frac{1}{3}}} \text{Ai}\left(\frac{b-a}{|\beta^3-\alpha^3|^{\frac{1}{3}}}\right), \quad \text{if } \beta > \alpha. \qquad (60)$$

Hence, we have a relation between the inner product in $\mathcal{L}^2(v,u)$ and the inner product in $\mathcal{L}^2(k)$:

$$\int_{\mathbb{R}^2} \frac{du\,dv}{2\pi}\, \psi_E^*(v,u)\, \psi_{E'}(v,u) = \gamma^{-2}\delta(E-E') \int_\mathbb{R} dk\, \widetilde{\psi}_E^*(k)\, \widetilde{\psi}_{E'}(k). \qquad (61)$$

It expresses the completeness of the eigenfunctions $\psi_E(v,u)$, provided that:

$$\int_\mathbb{R} dk\, \widetilde{\psi}_E^*(k)\, \widetilde{\psi}_E(k) < \infty. \qquad (62)$$

Hence, $\widetilde{\psi}_E(k) \in \mathcal{L}^2(k)$. □

Theorem 2. *The integral mapping $\psi_E(v,u) \in \mathcal{L}^2(v,u) \to \widetilde{\psi}_E(k) \in \mathcal{L}^2(k)$ is invertible, i.e.,*

$$\widetilde{\psi}_{E'}(k') = \int_\mathbb{R} dE \int_{\mathbb{R}^2} du\,dv\, \psi_E(u,v)\, e^{-ik'v}\, \text{Ai}\left(\gamma u + \frac{k'^2}{\gamma^2} - \gamma E'\right). \qquad (63)$$

Proof. By integrating both sides of Equation (58) with the following equation:

$$\int_{\mathbb{R}^2} \frac{du\,dv}{2\pi}\, e^{-ik'v}\, \text{Ai}\left(\gamma u + \frac{k'^2}{\gamma^2} - \gamma E'\right), \qquad (64)$$

we get the following integral on the right-hand side:

$$\int_{\mathbb{R}^2} \frac{du\,dv}{2\pi}\, e^{-ik'v}\, \text{Ai}\left(\gamma u + \frac{k'^2}{\gamma^2} - \gamma E'\right) \int_\mathbb{R} dk\, e^{ikv}\, \widetilde{\psi}_E(k)\, \text{Ai}\left(\gamma u + \frac{k^2}{\gamma^2} - \gamma E\right). \qquad (65)$$

Assuming Fubini's theorem hypothesis, we can exchange integrations and the $\frac{dv}{2\pi}$-integration would yield $\delta(k-k')$. Then, after integration on dk, the right-hand side becomes:

$$\widetilde{\psi}_E(k') \int_\mathbb{R} \frac{du}{2\pi}\, \text{Ai}\left(\gamma u + \frac{k^2}{\gamma^2} - \gamma E\right) \text{Ai}\left(\gamma u + \frac{k'^2}{\gamma^2} - \gamma E'\right), \qquad (66)$$

which, upon application of the integration formula (60), yields the following:

$$\int_{\mathbb{R}^2} \frac{du\,dv}{2\pi}\, \psi_E(u,v)\, e^{-ik'v}\, \text{Ai}\left(\gamma u + \frac{k'^2}{\gamma^2} - \gamma E'\right) = \widetilde{\psi}_E(k')\, \gamma^{-2}\delta(E-E'). \qquad (67)$$

A last integration on dE on both sides of this equation gives the final result:

$$\widetilde{\psi}_{E'}(k') = \gamma^2 \int_\mathbb{R} dE \int_{\mathbb{R}^2} du\,dv\, \psi_E(v,u)\, e^{-ik'v}\, \text{Ai}\left(\gamma u + \frac{k'^2}{\gamma^2} - \gamma E'\right). \qquad (68)$$

Hence, the integral mapping $\psi_E(v,u) \to \widetilde{\psi}_E(k)$ is *invertible*, provided that a summation on E is performed or uses complete data. In this respect, it may be called the

Fourier-Airy Transform. These two amplitudes describe the same physics in two different contexts. □

3.3. Representation of the Dynamical Symmetry Algebra $\widehat{\mathfrak{A}}_{FF}$ by Klink's Algebra

Proposition 12. *The action of the $\widehat{\mathfrak{A}}_{FF}$ generators on the wave function $\psi_E(v, u)$ is easily transferred to the wave function $\widetilde{\psi}_E(k)$ according to the following:*

$$\widehat{H}\psi_E(v,u) = E\int_\mathbb{R} dk\, \widetilde{\psi}_E(k)\, e^{ikv}\, \text{Ai}\left(\gamma u + \frac{k^2}{\gamma^2} - \gamma E\right),$$

$$\widehat{F}_\rho \psi_E(v,u) = \int_\mathbb{R} dk\, (k\widetilde{\psi}_E(k))\, e^{ikv}\, \text{Ai}\left(\gamma u + \frac{k^2}{\gamma^2} - \gamma E\right),$$

$$\widehat{F}_\sigma \psi_E(v,u) = \int_\mathbb{R} dk\, \left(i\frac{d}{dk}\right)\widetilde{\psi}_E(k)\, e^{ikv}\, \text{Ai}\left(\gamma u + \frac{k^2}{\gamma^2} - \gamma E\right),$$

$$\widehat{F}_\tau \psi_E(v,u) = \int_\mathbb{R} dk\, (k^2 - E)\widetilde{\psi}_E(k)\, e^{ikv}\, \text{Ai}\left(\gamma u + \frac{k^2}{\gamma^2} - \gamma E\right). \quad (69)$$

Proof. The action of $\widehat{H}$ is E since $\psi_E(v, u)$ is an eigenfunction of $\widehat{H}$.

As $\widehat{F}_\rho$ and $\widehat{F}_\tau$ are represented in $\mathcal{L}^2(v, u)$ by v-derivatives acting under the integral sign on e^{ikv}, we successively get k and k^2 acting on $\widetilde{\psi}_E(k)$.

For the action of $\widehat{F}_\sigma = \left(-\frac{\partial^2}{\partial u\, \partial v} + v\right)$, we observe that:

$$v\, \psi_E(v,u) = \int_\mathbb{R} dk\, \widetilde{\psi}_E(k) \left(-i\frac{\partial}{\partial k} e^{ikv}\right) \text{Ai}\left(\gamma u + \frac{k^2}{\gamma^2} - \gamma E\right). \quad (70)$$

Since the Airy function vanishes for $k = \pm\infty$, we perform a partial integration in k to get the following:

$$v\, \psi_E(v,u) =$$
$$\int_\mathbb{R} dk\, \left(i\frac{\partial}{\partial k}\widetilde{\psi}_E(k)\right) e^{ikv}\, \text{Ai}\left(\gamma u + \frac{k^2}{\gamma^2} - \gamma E\right) + \int_\mathbb{R} dk\, \widetilde{\psi}_E(k)\, e^{ikv}\, \text{Ai}'\left(\gamma u + \frac{k^2}{\gamma^2} - \gamma E\right) i\gamma k. \quad (71)$$

where $\text{Ai}'(x)$ is the derivative of $\text{Ai}(x)$. We then observe that the second integral is as follows:

$$\frac{\partial^2}{\partial u\, \partial v}\psi_E(v,u). \quad (72)$$

Hence, the action of $\widehat{F}_\sigma$ on $\psi_E(v, u)$ is replaced by $i\frac{\partial}{\partial k}\widetilde{\psi}_E(k)$ in $\mathcal{L}^2(k)$. □

Corollary 3. *As a consequence, the action of $\widehat{F}^2$ on $\psi_E(v, u)$ is translated into action on $\widetilde{\psi}_E(k)$ as the action of the one-dimensional confining quartic anharmonic oscillator Hamiltonian in the k variable on the wave function $\widetilde{\psi}_E(k)$:*

$$\widehat{F}^2 \psi_E(v,u) = -\int_\mathbb{R} dk\, \left\{\left(-\frac{1}{2}\frac{d^2}{dk^2} - Ek^2 + \frac{1}{2}k^4\right)\widetilde{\psi}_E(k)\right\} e^{ikv}\, \text{Ai}\left(\gamma u + \frac{k^2}{\gamma^2} - \gamma E\right) \quad (73)$$

Proof. Use the expression of $\widehat{F}^2$ in terms of the following dynamical symmetry algebra generators $(\widehat{F}_\rho, \widehat{F}_\sigma, \widehat{F}_\tau)$:

$$\widehat{F}^2 = (\widehat{H}\widehat{F}_\rho^2 - \frac{1}{2}\widehat{F}_\sigma^2 - \frac{1}{2}(\widehat{F}_\tau + \widehat{H})^2), \quad (74)$$

and the previous proposition. □

Corollary 4. *The mapping $\psi_E(v, u) \to \widetilde{\psi}_E(k)$ induces an isomorphism between the two-variable algebra generated by $(\widehat{F}_\rho, \widehat{F}_\sigma, \widehat{F}_\tau)$, and the one-variable algebra generated by $(k, i\frac{d}{dk}, k^2)$, which is known as Klink's algebra for quartic anharmonic oscillator and for which the quadratic integral of motion $\widehat{F}^2$ is the Schrödinger Hamiltonian of this quartic anharmonic oscillator* [15].

Proof. The commutators of $(\widehat{F}_\rho, \widehat{F}_\sigma, \widehat{F}_\tau)$ are isomorphic to those of $(k, i\frac{d}{dk}, k^2)$ by computational checking. □

This relation shows how the one-dimensional confining quartic anharmonic oscillator is linked to the two-dimensional free fall. Note that this relation cannot be established if one had started with the one-dimensional free fall problem. In classical physics, the presence of quartic anharmonic oscillators in the free fall problem arises only when parabolic coordinates are introduced. This is not the case here. There is still another curious relation between the two systems discovered by Voros in [16].

3.4. Finite Quantum Canonical Transforms

Proposition 13. *The finite quantum unitary transforms generated by $(\widehat{F}_\rho, \widehat{F}_\sigma, \widehat{F}_\tau)$ are expressed as follows:*

$$\widehat{U}_\rho = \exp -i\rho\,\widehat{F}_\rho, \quad \widehat{U}_\sigma = \exp -i\sigma\,\widehat{F}_\sigma, \quad \widehat{U}_\tau = \exp -i\tau\,\widehat{F}_\tau. \tag{75}$$

They produce the quantum version of the classical Iwai–Rew transform (see Equation (8)).

Proof. The proof is trivial and involves the use of the Baker–Hausdorff–Campbell formula:

$$e^{\widehat{Y}} \widehat{X} e^{\widehat{Y}} = \widehat{X} + [\widehat{Y}, \widehat{X}] + \frac{1}{2!}[\widehat{Y}, +[\widehat{Y}, \widehat{X}]] + \dots$$

where $(\widehat{X}, \widehat{Y})$ are operators. Because of the commutation relations of $\widehat{\mathfrak{A}}_{FF}$, the computation of this Baker–Hausdorff–Campbell on $(\widehat{Q}_v, \widehat{P}_v, \widehat{Q}_u, \widehat{P}_u)$ yields only a few terms, whose coefficients are precisely those in Equation (8). □

Proposition 14. *The action of $(\widehat{U}_\rho, \widehat{U}_\sigma, \widehat{U}_\tau)$ on the wave function $\psi_E(v, u)$ may be represented by the following integral transforms:*

$$
\begin{aligned}
\widehat{U}_\rho \psi_E(v, u) &= \int_\mathbb{R} dk \left(e^{-ik\rho} \widetilde{\psi}_E(k)\right) e^{ikv} \operatorname{Ai}\left(\gamma(u - E) + \frac{k^2}{\gamma^2}\right), \\
\widehat{U}_\sigma \psi_E(v, u) &= \int_\mathbb{R} dk\, \widetilde{\psi}_E(k + \sigma)\, e^{ikv} \operatorname{Ai}\left(\gamma(u - E) + \frac{k^2}{\gamma^2}\right), \\
\widehat{U}_\tau \psi_E(v, u) &= \int_\mathbb{R} dk \left(e^{-i\tau(k^2 - E)} \widetilde{\psi}_E(k)\right) e^{ikv} \operatorname{Ai}\left(\gamma(u - E) + \frac{k^2}{\gamma^2}\right).
\end{aligned} \tag{76}
$$

Proof. The proof is straightforward. It uses the action of each generator given by Equation (69) and then exponentiates it as action on $\widetilde{\psi}_E(k)$. This leads to unitary factors for $\widehat{U}_\rho$ and $\widehat{U}_\tau$ as well as a shift in the argument of $\widetilde{\psi}_E(k)$ for $\widehat{U}_\sigma$. This is an alternative form to the one obtained by Iwai–Rew [4]. □

3.5. Quantum Integrals of Motion and Consequences on the Schrödinger Wave Functions

In this subsection, we study the nature of the quantum integrals of motion. In particular, when a manifest separation of variables occurs in an operator $\widehat{O}(v, u) = \widehat{O}(v) + \widehat{O}(u)$, the operator $\widehat{F}(v, u) = \widehat{O}(v) - \widehat{O}(u)$ automatically commutes with $\widehat{O}(v, u)$.

(a1) $\widehat{F}_\rho$ exists since $[\widehat{H}, \widehat{F}_\rho] = 0$ because ρ is a cyclic variable, as in the classical case.

(a2) $\widehat{F}_\tau$ is due to the separation in the (v, u) Cartesian variables. Hence,

$$\widehat{H} = \left(-\frac{1}{2}\frac{\partial^2}{\partial v^2}\right) + \left(-\frac{1}{2}\frac{\partial^2}{\partial u^2} + u\right), \text{ and } \widehat{F}_\tau = \left(-\frac{1}{2}\frac{\partial^2}{\partial v^2}\right) - \left(-\frac{1}{2}\frac{\partial^2}{\partial u^2} + u\right) \tag{77}$$

commute.

(b1) $\widehat{F}_\sigma$ is due to the separation in $\pi/4$ rotated Cartesian variables (ξ, ζ). With the following change in variables:

$$(v, u) = \frac{1}{2}((\xi + \zeta), (\xi - \zeta)), \quad (\xi, \zeta) = ((u + v), (u - v)), \tag{78}$$

$(\widehat{H}, \widehat{F}_\sigma)$ becomes $(\widehat{H}(\xi,\zeta), \widehat{F}_\sigma(\xi,\zeta))$ given by the following equations:

$$\left(\left(-\frac{\partial^2}{\partial \xi^2}+\frac{1}{2}\xi\right)+\left(-\frac{\partial^2}{\partial \zeta^2}+\frac{1}{2}\zeta\right), \left(-\frac{\partial^2}{\partial \xi^2}+\frac{1}{2}\xi\right)-\left(-\frac{\partial^2}{\partial \zeta^2}+\frac{1}{2}\zeta\right)\right) \tag{79}$$

which commute.

(b2) However, there was an unproven claim of separation of variables in displaced parabolic coordinates [10] as a justification for the existence of $(\widehat{F}_\sigma)$. Hereafter, we provide an argument that may explain this claim. Consider the intermediate space (k, u), obtained after a partial v-Fourier transform from the (v, u) space. The Schrödinger equation in the (k, u) space for stationary states of the eigenvalue E admits solutions in the form of:

$$\text{Ai}\left(\gamma u + \frac{k^2}{\gamma^2} - \gamma E\right). \tag{80}$$

This solution has a constant value C on the parabolas of equation $\left(u + \frac{1}{2}k^2 - E\right) = C$, where C is a shift of the parabolas. Consequently, the tangential derivative of any function of (k, u) along these parabolas is zero. This tangential derivative is just $\mathbf{t} \cdot \nabla$, where $\nabla = \left(\frac{\partial}{\partial k}, \frac{\partial}{\partial u}\right)$ is the gradient operator and $\mathbf{t} = (1, -k)$ is the tangent vector to the parabola. Thus, we have the following:

$$\left(-k\frac{\partial}{\partial u}+\frac{\partial}{\partial k}\right)\text{Ai}\left(\gamma u + \frac{k^2}{\gamma^2} - \gamma E\right) = 0. \tag{81}$$

Going back to the (v, u) space by the k-Fourier inverse transform, this tangential derivative reappears as a (v, u)−partial differential operator $\left(\frac{\partial^2}{\partial u \partial v} - v\right)$, which is just $-\widehat{F}_\sigma$. Therefore, the claim of reference [13] is only valid in the (k, u)-space.

(c) $\widehat{F}^2$ is due to a special form of separation of variables (Stäckel separation of variables), when one changes from Cartesian (v, u) to parabolic (x, y) coordinates by the following formulas: $(u, v) = \left(\frac{x^2-y^2}{2}, xy\right)$. After working out the expressions of the (x, y) partial derivatives, we end up with new expressions in (x, y) for $(\widehat{H}, \widehat{F}^2)$:

$$\begin{aligned}\widehat{H}(x,y) &= \frac{1}{x^2+y^2}\left\{\left(-\frac{1}{2}\frac{\partial^2}{\partial x^2}+\frac{1}{2}x^4\right)+\left(-\frac{1}{2}\frac{\partial^2}{\partial y^2}-\frac{1}{2}y^4\right)\right\},\\ \widehat{F}^2(x,y) &= \frac{(-1)}{x^2+y^2}\left\{y^2\left(-\frac{1}{2}\frac{\partial^2}{\partial x^2}+\frac{1}{2}x^4\right)-x^2\left(-\frac{1}{2}\frac{\partial^2}{\partial y^2}-\frac{1}{2}y^4\right)\right\}.\end{aligned} \tag{82}$$

These expressions can also be obtained from the quantization of classical expressions by replacing (p_x, p_y) with $\left(-i\frac{\partial}{\partial x}, -i\frac{\partial}{\partial y}\right)$. The meaning of this Stäckel separation of variables is given in the following proposition.

Proposition 15. *As the total energy is conserved as $\widehat{H} = E$, the eigenfunctions of $\widehat{H}$ in parabolic coordinates are products of Schrödinger eigenfunctions of confining and non-confining quartic oscillators in the form of $\psi_E(v,u) \sim \overline{\psi}_E(x,y) = \overline{\psi}_E^{(+)}(x)\overline{\psi}_E^{(-)}(y)$. Then, $\widehat{F}^2\overline{\psi}_E^{(+)}(x)\overline{\psi}_E^{(-)}(y) = -E\overline{\psi}_E^{(+)}(x)\overline{\psi}_E^{(-)}(y)$.*

Proof. Using the expression of $\widehat{H}$ in parabolic coordinates given by Equation (82) and calling the corresponding wave function $\overline{\psi}_E(x,y)$, we can transform the stationary Schrödinger equation $\widehat{H}\overline{\psi}_E(x,y) = E\overline{\psi}_E(x,y)$ for all $(x,y) \neq (0,0)$ into the following condition:

$$\left\{\left(-\frac{1}{2}\frac{\partial^2}{\partial x^2}-Ex^2+\frac{1}{2}x^4\right)+\left(-\frac{1}{2}\frac{\partial^2}{\partial y^2}-Ey^2-\frac{1}{2}y^4\right)\right\}\overline{\psi}_E(x,y) = 0. \tag{83}$$

Since this condition is to be satisfied for all $(x,y) \neq (0,0)$, $\overline{\psi}_E(x,y)$ should be a product of eigenfunctions of the two separate Schrödinger operators in x and in y, with opposite eigenvalues. For the confining quartic potential in x, it is known that this Schrödinger operator has discrete non-degenerate point spectrum with the eigenvalues $E = \epsilon_n$, with $n \in \mathbb{N}$, which are bounded below. On the other hand, the Schrödinger operator with non-confining quartic potential in y has a continuous, real, non-degenerate, scattering-type spectrum. Thus, $\psi_E(v,u) \sim \overline{\psi}_E(x,y)$ must be of the product form (up to a multiplicative constant):

$$\overline{\psi}_E(x,y) = \overline{\psi}_E^{(+)}(x) \overline{\psi}_E^{(-)}(x),$$

verifying the stationary Schrödinger equations:

$$\begin{aligned}
\left(-\frac{1}{2}\frac{\partial^2}{\partial x^2} - Ex^2 + \frac{1}{2}x^4\right)\overline{\psi}_E^{(+)}(x) &= \mathcal{H}_{osc}^{(+)}(x)\overline{\psi}_E^{(+)}(x) = +E\,\overline{\psi}_E^{(+)}(x), \\
\left(-\frac{1}{2}\frac{\partial^2}{\partial y^2} - Ey^2 - \frac{1}{2}y^4\right)\overline{\psi}_E^{(-)}(y) &= \mathcal{H}_{osc}^{(-)}(y)\overline{\psi}_E^{(-)}(y) = -E\,\overline{\psi}_E^{(-)}(y).
\end{aligned} \quad (84)$$

Then, after rewriting the expression of $\hat{F}^2$ under the following form:

$$\hat{F}^2 = \frac{(-1)}{x^2+y^2}\left\{y^2\left(-\frac{1}{2}\frac{\partial^2}{\partial x^2} - Ex^2 + \frac{1}{2}x^4\right) - x^2\left(-\frac{1}{2}\frac{\partial^2}{\partial y^2} - Ey^2 - \frac{1}{2}y^4\right)\right\}, \quad (85)$$

we see that

$$\hat{F}^2 \overline{\psi}_E(x,y) = \hat{F}^2 \overline{\psi}_E^{(+)}(x)\,\overline{\psi}_E^{(-)}(y) = -E\,\overline{\psi}_E^{(+)}(x)\,\overline{\psi}_E^{(-)}(y). \quad (86)$$

Hence, for $\hat{H} = E$, we have necessarily $-E$ as an eigenvalue of $\hat{F}^2$, where $E = \epsilon_n$ is any eigenvalue of the confining quartic potential $+\frac{1}{2}x^4$. □

3.6. Integral Relation for Schrödinger Eigenfunctions of Quartic Oscillators

The passage to parabolic coordinates has a remarkable consequence on the eigenfunctions of quantum quartic oscillators, as stated in the following theorem.

Theorem 3. *The eigenfunctions of quantum confining and non-confining quartic oscillators verify the following integral relation:*

$$\overline{\psi}_E^{(+)}(x)\,\overline{\psi}_E^{(-)}(y) = \mu_E \int_{\mathbb{R}} dk\, \overline{\psi}_E^{(+)}(k)\, e^{ikxy}\, \mathrm{Ai}\!\left(\frac{x^2 - y^2 + k^2 - 2E}{\gamma^2}\right). \quad (87)$$

Proof. Recast the expression of $\psi_E(v,u)$ in parabolic coordinates to get the following:

$$\overline{\psi}_E^{(+)}(x)\,\overline{\psi}_E^{(-)}(y) = \mu_E \int_{\mathbb{R}} dk\, \tilde{\psi}_E(k)\, e^{ikxy}\, \mathrm{Ai}\!\left(\frac{x^2 - y^2 + k^2 - 2E}{\gamma^2}\right), \quad (88)$$

where μ_E takes care of the fact that $\psi_E(v,u) \sim \overline{\psi}_E(x,y)$ up to a multiplicative constant. Let $\hat{F}^2$ operate on both sides of this equation. On the left-hand side, we get $-E\,\overline{\psi}_E^{(+)}(x)\,\overline{\psi}_E^{(-)}(y)$ according to Equation (86). On the right-hand side, we obtain:

$$-\mu_E \int_{\mathbb{R}} dk \left(-\frac{1}{2}\frac{d^2}{dk^2} - Ek^2 + \frac{1}{2}k^4\right)\tilde{\psi}_E(k)\, e^{ikxy}\, \mathrm{Ai}\!\left(\frac{x^2-y^2+k^2-2E}{\gamma^2}\right), \quad (89)$$

according to Equation (73). Hence, to have consistency between the two sides, one must require that $\tilde{\psi}_E(k) = \overline{\psi}_E^{(+)}(k)$. Therefore, it follows that the eigenfunctions of the confining and non-confining quartic oscillators should verify the integral relation above. □

Since the confining quartic oscillator is of dominant physical interest as a non-trivial field theory model in zero space dimension, an integral identity for its eigenfunctions can be derived from the previous identity via a "Wick-rotation" [17].

Corollary 5. *The eigenfunctions $\overline{\psi}_E^{(+)}(x)$ of the confining quartic oscillator fulfil the following integral identity:*

$$\overline{\psi}_E^{(+)}(x)\,\overline{\psi}_E^{(+)}(x') = \mu_E \int_{\mathbb{R}} dk\, \overline{\psi}_E^{(+)}(k) \begin{pmatrix} \cosh xx'k \\ \sinh xx'k \end{pmatrix} \mathrm{Ai}\!\left(\frac{x^2 + x'^2 + k^2 - 2E}{\gamma^2}\right). \tag{90}$$

Proof. Since

$$\mathcal{H}_{osc}^{(-)}(ix') = -\mathcal{H}_{osc}^{(+)}(x'), \quad \text{and} \quad \overline{\psi}_E^{(-)}(ix') = \overline{\psi}_E^{(+)}(x'), \tag{91}$$

replace y with $y = ix'$ in Equation (88). Then, as parity is a good quantum number for $\mathcal{H}_{osc}^{(+)}$, the factor e^{-kxy} in the integrand should be replaced by either $\sinh xx'k$ or $\cosh xx'k$, according to the parity of the state of energy $E = \epsilon_n$ for $n \in \mathbb{N}$. This last integral identity was discovered long ago via the Weyl quantization of anharmonic oscillators [18] and recently rediscovered by [19]. □

4. An Application to Wave Propagation in Duct of Varying Section

In 2003, B J Forbes et al. [20] observed that the Webster equation describing the excess pressure $p(u,t)$ in a fluid flowing in a duct with a slowly varying circular cross-sectional area $S(u)$:

$$\frac{1}{c^2}\frac{\partial^2 p(u,t)}{\partial t^2} = \frac{1}{S(u)}\frac{\partial}{\partial u}\!\left(S(u)\frac{\partial p(u,t)}{\partial u}\right), \tag{92}$$

where u is the space coordinate along the axis of the duct, t the time, and c is the constant wave speed in the fluid, which can be turned into a Klein–Gordon equation with a potential $V(u)$ for a wave function $\psi(u,t)$:

$$\left(\frac{1}{c^2}\frac{\partial^2}{\partial t^2} - \frac{\partial^2}{\partial u^2} + V(u)\right)\psi(u,t) = 0, \tag{93}$$

if $\psi(u,t) = p(u,t)\sqrt{S(u)}$ and

$$V(u) = \frac{1}{\sqrt{S(u)}}\frac{d^2 \sqrt{S(u)}}{du^2}. \tag{94}$$

Hence, if we require that the potential to be linear $V(u) = (u - d)$, where d is an arbitrary distance, Equation (94) shows that the duct section should vary as $S(u) = \mathrm{Ai}(u - d)$, and consequently, Equation (93) is obtained from the two-dimensional free fall Schrödinger equation by "Wick-rotation" $v = ict$ [17] (without a factor $1/2$ in the partial derivatives). This Klein–Gordon Equation (93) admits a dynamical symmetry algebra which is the Wick rotated $\widehat{\mathfrak{A}}_{FF}$. Note that the free fall $\widehat{F}_\tau$ integral of motion now becomes the Klein–Gordon operator, and conversely, the free fall Hamiltonian operator is now an integral of the motion for the Klein–Gordon operator. The quadratic integral operator of the Klein–Gordon problem is now the Wick rotated free fall $\widehat{F}^2$ operator. In this case, one directly gets the integral identity for the Schrödinger eigenfunctions of the confining quartic anharmonic oscillator; see Equation (90).

The general solution of this Klein–Gordon wave equation $\phi_d(t,u)$ is formally analogous to $\psi_E(v,u)$; it is given by the ω-integral for $t > 0$:

$$\phi_d(t,u) = \int_{\mathbb{R}} d\omega\, f(\omega)\, \mathrm{Ai}\!\left(\gamma(u-d) + \frac{\omega^2}{c^2\gamma^2}\right), \tag{95}$$

with the initial conditions $\phi_d(0,u)$ and $\dot{\phi}_d(0,u) = \frac{d}{dt}\phi_d(t,u)|_{t=0}$.

Theorem 4. *The angular frequency distribution $f(\omega)$ of $\phi_d(t,u)$ is fully determined by the initial conditions.*

Proof. The proof uses the integral given by Equation (60) and also by correctly integrating the delta function $\delta(\omega^2 - \omega'^2)$:

$$\frac{1}{2}(f(\omega') + f(-\omega')) = \frac{|\omega'|}{c^2} \int_{\mathbb{R}} du\, \phi_d(0,u)\, \text{Ai}\left((u-d) + \frac{\omega'^2}{c^2}\right),$$

$$\frac{1}{2}(-f(\omega') + f(-\omega')) = \frac{\text{sgn}\omega'}{c^2} \int_{\mathbb{R}} du\, \phi_d(0,u)\, \text{Ai}\left((u-d) + \frac{\omega'^2}{c^2}\right) \quad (96)$$

where $\text{sgn}\omega = \frac{\omega}{|\omega|}$ is the sign of ω. This fully determines $f(\omega)$. □

5. Conclusions and Perspectives

In this paper, a complete account of the dynamical symmetries of the two-dimensional free fall is provided both classically and quantum mechanically. The results may open the way towards the construction of the representation of its dynamical algebra. This is a challenging task which makes it necessary to understand the *zonal* character of the integral relation fulfilled by the eigenfunctions of the confining quartic oscillator [21]. An extension to dimensions higher than two, which may reveal new unexpected features of this simple physical problem as it did in two dimensions, is foreseen as future work.

Funding: This research received no external funding.

Institutional Review Board Statement: Not applicable.

Informed Consent Statement: Not applicable.

Conflicts of Interest: The author declares no conflict of interest.

References

1. Halliday, D.; Resnick, R.; Walker, J. *Fundamentals of Physics*; John Wiley and Sons, Inc.: New York, NY, USA, 2005; pp. 24–26.
2. Wheeler, N. Classical/Quantum Dynamics in a Uniform Gravitational Field: An Unobstructed Free Fall. Available online: https://www.reed.edu/physics/faculty/wheeler/documents/Quantum%20Mechanics/Miscellaneous%20Essays/Quantum%20Bouncer/E4.%20Free%20Fall.pdf (accessed on 27 October 2021).
3. Engelhardt, W. Free fall in gravitational theory. *Phys. Essays* **2017**, *30*, 294. [CrossRef]
4. Iwai, T.; Rew, S.G. Classical and quantum symmetry groups of a free-fall particle. *J. Math. Phys.* **1985**, *26*, 55–61. [CrossRef]
5. Barut, A.O.; Kleinert, H. Transition probabilities of the Hydrogen atom from noncompact dynamical groups. *Phys. Rev.* **1967**, *156*, 1541–1545. [CrossRef]
6. Alliluev, S.P. On the relation between "accidental" degeneracy and "hidden" symmetry of a system. *Sov. Phys. JETP* **1958**, *6*, 156–159.
7. Whittaker, E.T. *A Treatise on the Analytical Dynamics of Particles and Rigid Bodies*, 4th ed.; Cambridge University Press: Cambridge, UK, 1959.
8. Fordy, A.P. Quantum super-integrable systems as exactly solvable systems. *Symmetry Integr. Geom. Methods Appl.* **2007**, *3*, 025.
9. Jauch, J.M.; Hill, E.L. On the problem of degeneracy in Quantum Mechanics. *Phys. Rev.* **1940**, *57*, 641–645. [CrossRef]
10. Miller, W.; Kalnins, E.G.; Kress, J.M.; Pogosyan, G.S. Infinite-order Symmetries for Quantum Separable Systems. *Phys. At. Nucl.* **2005**, *68*, 1756–1763. [CrossRef]
11. Englefield, M.J. *Group Theory and the Coulomb Problem*; Wiley-Interscience: New York, NY, USA, 1972.
12. Wheeler, N. *Classical/Quantum Theory of the 2-Dimensional Hydrogen*; Lecture Notes Reed College: Portland Oregon, OR, USA, 1999. Available online: www.reed.edu/physics/faculty/wheeler/documents/QuantumMechanics/Miscellaneous (accessed on 26 October 2021).
13. Popper, I.; Post, S.; Winternitz, P. Third order super-integrable systems separable in parabolic coordinates. *J. Math. Phys.* **2012**, *53*, 062105. [CrossRef]
14. Vallée, O.; Soares, M. *Airy Functions and Applications to Physics*; World Scientific: London, UK, 2004.
15. Klink, W.H. Nilpotent groups and anharmonic oscillators. In *Noncompact Lie Groups and Some of Their Applications*; Tanner, E.A., Wilson, R., Eds.; Kluwer Academic Publishers: Norwell, MA, USA, 1994; pp. 301–313.
16. Voros, A. Airy function—Exact WBK results for potential of odd degree. *J. Phys. A Math. Gen.* **1999**, *32*, 1301–1311. [CrossRef]
17. Wick, G.C. Properties of Bethe-Salpeter Wave Functions. *Phys. Rev.* **1954**, *96*, 1124–1134. [CrossRef]
18. Truong, T.T. Weyl quantization of anharmonic oscillators. *J. Math. Phys.* **1975**, *16*, 1034–1043. [CrossRef]
19. Hallnas, M.; Langmann, E. A product formula for the eigenfunctions of a quartic oscillator. *J. Math. Anal. Appl.* **2015**, *426*, 1012–1025. [CrossRef]

20. Forbes, B.J.; Pike, E.R.; Sharp, D.B. The acoustical Klein-Gordon equation: The wave-mechanical step and barrier potential functions. *J. Acoust. Soc. Am.* **2003**, *114*, 1291–1302. [CrossRef] [PubMed]
21. Berezin, F.A.; Gelfand, I.M. Some remarks on the theory of spherical functions. *Tr. Mosk. Mat. Obs.* **1956**, *5*, 311–351.

Article

The Surjective Mapping Conjecture and the Measurement Problem in Quantum Mechanics

Fritz Wilhelm Bopp

Department Physik, University Siegen, 57076 Siegen, Germany; bopp@physik.uni-siegen.de

Abstract: Accepting a time-symmetric quantum dynamical world with ontological wave functions or fields, we follow arguments that naturally lead to a two-boundary interpretation of quantum mechanics. The usual two boundary picture is a valid superdeterministic interpretation. It has, however, one unsatisfactory feature. The random selection of a chosen measurement path of the universe is far too complicated. To avoid it, we propose an alternate two-boundary concept called surjective mapping conjecture. It takes as fundamental a quantum-time running forward like the usual time on the wave-function side and backward on the complex conjugate side. Unrelated fixed arbitrary boundary conditions at the initial and the final quantum times then determine the measurement path of the expanding and contracting quantum-time universe in the required way.

Keywords: time symmetric quantum dynamics; two-boundary interpretation of quantum mechanics; the resurrection of macroscopic causality; cosmological epochs without macroscopic descriptions

Citation: Bopp, F.W. The Surjective Mapping Conjecture and the Measurement Problem in Quantum Mechanics. *Symmetry* **2021**, *13*, 2155. https://doi.org/10.3390/sym13112155

Academic Editors: Tuong Trong Truong and Young S. Kim

Received: 10 September 2021
Accepted: 1 November 2021
Published: 11 November 2021

Publisher's Note: MDPI stays neutral with regard to jurisdictional claims in published maps and institutional affiliations.

Copyright: © 2021 by the author. Licensee MDPI, Basel, Switzerland. This article is an open access article distributed under the terms and conditions of the Creative Commons Attribution (CC BY) license (https://creativecommons.org/licenses/by/4.0/).

1. Introduction

The wave function and the fields are the draft horses of quantum mechanics and quantum field theory. We here take a realist view about their ontological existence. Measurement decisions involving collapses require the elimination of wave function components all the way back. So, the realist view requires backward causation. The wave function and its conjugate follow the Schrödinger equation, which is entirely symmetric in time. For field theory, with its Hamilton function, the situation is analogous.

To accept time symmetry is a severe step. It involves some non-locality. Interpretations used different strategies to get around it. The Copenhagen interpretation tries to avoid backward causation by denying wave function ontological existence. In "An intricate quantum statistical effect and the foundation of quantum mechanics" [1], we argued that it is not successful in a broader quantum statistical domain and that backward causation is unavoidable. The basic argument is: If identical particles are produced with a certain probability at the time t_1 one can decide at a later time t_2 to allow them to mingle or to keep them separate. This decision at the time t_2 introduces or prevents interference terms which enhance or deplete the production probability a the time t_1. As particle production is ontological, it means backward causation. In some way, the acceptance of time symmetry on a quantum level is a paradigm change. As we argued in "Time Symmetric Quantum Mechanics and Causal Classical Physics" [2], there is nothing wrong with time-symmetric quantum mechanics as long in our rapidly expanding universe an approximate classical time directed physics can be obtained.

The term "conjecture" instead of "interpretation" was used as there might be unthought-of problems. The most dangerous aspects are infinities which often lead to problems if involved limits are not considered carefully. We, therefore, decided to take the universe as finite with a vast lifetime from a big bang to a highly extended final state. As the universe is almost empty and sparsely interacting at its end, the limit $t_f \to \infty$ presumably exists but is not taken.

As Sakurai [3] pointed out, most of the spectacular successes of Quantum Mechanics (QM) lie in the domain of Quantum Dynamics (QD), meaning QM without the Measurement Process involving collapses. Many applications of QM involve just static situations, and for most processes, one just needs to know the given initial and the possible final states.

2. The Components of Measurements

Nobody doubts QD; differences in interpretations concern the Theory of Measurements. There is a lot of splitting and merging in QD in which no components are eliminated. Such eliminations are an essential element of Measurement Processes. Empirically one knows that such measurement processes in which components disappear require witnesses. Usually, witnesses are observers who remember the choices taken. Here they can be just objects whose existence documents past selections.

Some basic facts about witnesses seem not fully realized. The fastest witness production involves phase changes [4]. For traceable witnesses, the by far dominant part is low-energy photons. Any drift chamber, any circuit board, and, if he visits, Wigner's friend's brain all emit measurable radiofrequency electromagnetic waves.

Individual photons with a radio wavelength between 0.03 and 20 m carry unmeasurable energies of 10^{-23} and 10^{-26} Joule. There have to be a lot of them, and there are known to be penetrating. A substantial fraction of them will escape the apparatus, the laboratory, and there is a special window for these frequencies in the ionosphere. In our expanding universe, it means a large number of them will enter a part of the sky where they will never encounter any interaction. In this way, the information of the measurement will be stored in the final state. The final hypersphere ($\propto c \cdot t_f^3$) is enormous and even a single photon will somehow reversed point to the precise emission region reflecting the measurement decision. The stored information will not be disturbed by other particles. The same applies to the other measurements. As emission takes a specific time, the number of measurements in a finite universe is a finite huge number, say, ζ for "zillion". In an expanding universe ζ grows faster than its lifetime.

A measurement result will be definite, such as up or down. It means the same choice on the wavefunction side and the complex conjugated one as described by the following projection:

$$P_{\text{no interference}}(t) = |\uparrow>^{\text{wave}} <\uparrow|^{\text{wave}} \otimes |\uparrow>^{\text{conjugate}} <\uparrow|^{\text{conjugate}} \\ + |\downarrow>^{\text{wave}} <\downarrow|^{\text{wave}} \otimes |\downarrow>^{\text{conjugate}} <\downarrow|^{\text{conjugate}} \quad (1)$$

To just add such an operator is not acceptable. It eliminates contributions that can only be done in processes involving witnesses.

However, in a finite universe, no problem arises. The final state on the wave-function side has to agree with that on the complex conjugated one. So, no mixed terms can contribute, e.g., with $\uparrow$ witnesses on the wave-function side and with $\downarrow$ witnesses on the other side. Formally for an unconstrained universe, one can replace the sum over all final states by a unit operator connecting the wavefunction and the complex conjugate one as done in Equation (2), depicting the evolution of the wave function and its complex conjugate part.

$$\sum_{j,j',j'',j'''} <i|U_j(t-t_i) \quad O \quad U_{j'}(t_f-t)\,\mathbf{1}\,U_{j''}(t-t_f) \quad O \quad U_{j'''}(t_i-t)|i>$$
$$\searrow \to \text{up} \to 1 \to \text{up} \to \nearrow$$

or

$$\sum_{j,j',j'',j'''} <i|U_j(t-t_i) \quad O \quad U_{j'}(t_f-t)\,\mathbf{1}\,U_{j''}(t-t_f) \quad O \quad U_{j'''}(t_i-t)|i> \quad (2)$$
$$\searrow \to \text{down} \to 1 \to \text{down} \to \nearrow$$

That the witnesses on both sides have to agree to contribute is now apparent. I kept the unit matrix to remember the separation. The argument stayed entirely in the domain of QD, and no addition was needed.

Next, the measurement will have to determine an actual choice, like a projection operator which takes $\uparrow$ and eliminates $\downarrow$:

$$P_{\text{jump}}(t) = |\uparrow><\uparrow|. \quad (3)$$

At what time should the operator act? Clearly, this should occur outside of the quantum domain where all components are needed. This means it should occur behind a good part of the witness production, but it can be later. As there is backward causation, the only requirement is that states coming from the chosen component can be distinguished from the unchosen one. As some witnesses stay forever, it is possible to move the projection to the end, i.e., just before t_f.

With everything except these measurement decisions integrated out, the number of measurement-decision paths in the universe will again be a finite number like 2^ζ. Of course, it is somewhat lower as there are correlations in measurement results like that of Alice and Bob and non-binary choices?

3. Two Boundary Quantum Dynamics

If we put all measurement projections to the end, we have QD everywhere except at the boundaries. One can define a final density operator containing all those projections and formally obtain the two-boundary interpretation [5–8] of QM. It is a serious, valid, super-deterministic interpretation of QM [9].

An unpleasant feature with these and similar theories is the required random selection of measurement path determined by the final boundary. For each measurement path i, it has to obtain the absolute square $|\text{amplitude}_i|^2$. Witnesses prevent interference contributions. It has then to choose a random number $rand \in \left[0, \sum_{i=1}^{\zeta!} |\text{amplitude}_i|^2\right]$ and determine the chosen J so that $rand \sim \sum_{i=1}^{J} |\text{amplitude}_i|^2$. In our opinion, it is far too complicated to be attractive.

There is a modified two-boundary interpretation, the surjective mapping conjecture, which avoids this unpleasant random choice. No measurement choice projections are needed. It takes the quantum time $\tau \in [0, 2t_f]$ as fundamental and replaces the usual expression for the density.

$$\psi \cdot \psi^*|_{(t)}$$

by

$$\psi(\tau_1 = t) \cdot \psi(\tau_2 = 2 \cdot t_f - t)$$

For simplicity we inserted the definitions of the arguments. The mapping $(\tau_1, \tau_2) \to t$ involved is in the set theory called surjection. With the condition:

$$\psi(\tau_2) = \psi(\tau_1 = 2t_f - \tau_2)^* \tag{4}$$

the usual description would be obtained.

Equation (4) trivially holds around the unit operator **1** for $t = t_f \pm \epsilon$. The highly extended state at t_f with all its witnesses strongly influences the smaller universe around it to keep Equation (4) sufficiently accurate for a large part of the universe, including our epoch. Where approximately valid, Equation (4) allows for a macroscopic description.

The formalism removes the constrain of this Equation (4). It allows for unequal boundaries at the initial and final quantum time:

$$\psi(2t_f) \neq \psi(0)^*$$

They are result from a missing theory of the total quantum universe depicted in Figure 1.

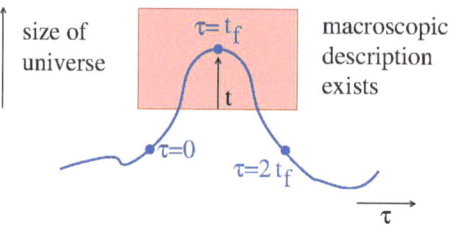

Figure 1. The unknown quantum universe.

One can evolve and revolve both boundary states to the state of maximum extend t_f:

$$< \psi(0) | U(t_f, 0) \quad \text{and} \quad < \psi(2t_f) | U(t_f, 2t_f)$$

In the conventional quantum theory, they would be equal, and their product would be one. Now both sides are unrelated. Each occupies a tiny fraction of the vast phase space of the universe at t_f, presumably of the order of t_f^{-3}. No overlap is expected except for an extremely tiny region:

$$< \psi(0) | U(t_f, 0) \, 1 \, U(t_f, 2t_f)^* | \psi(2t_f) > = \text{extremely tiny}$$

There is no reason for the product to vanish.

It changes the picture of measurements. Consider a generic Stern-Gerlach experiment shown in Figure 2 separately showing the wave function and the complex conjugate side. They are split in an inhomogeneous magnetic field and then enter drift chambers, where witnesses connected to the border state at t_f fix a choice observable by charge coupled electronics. The setup of the initial particle to a spin $\otimes$ is common to both outcomes, $\uparrow$ and $\downarrow$, and can be ignored on both sides. The spin $\uparrow$-component and its future on both sides, the wave function and the conjugate one, following the dotted red line is then given in Equation (5):

$$< \psi(\otimes) | \psi(\uparrow) >< \psi(t, \uparrow) | U | \psi(t_f, \uparrow) >_{\psi(0)} \cdot$$
$$< \psi(t_f, \uparrow) | U | \psi(2t_f - t, \uparrow) >_{\psi(2t_f)} < \psi(\uparrow) | \psi(\otimes) > \quad (5)$$

An analogue expression holds for the spin $\downarrow$-component. The limited overlap at t_f makes both terms tiny.

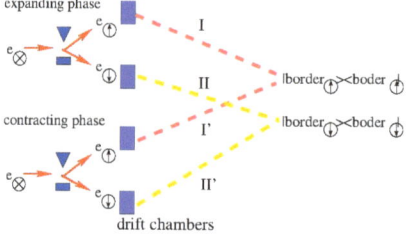

Figure 2. Generic Stern-Gerlach experiment.

For statistical reasons, two independent, very tiny quantities are extremely unlikely of the same magnitude. So one will be dominant and the other irrelevant. (Exceptional situations—we ignore—might lead to multi-world choice between coexisting observers.) It looks like a random process choosing one and collapsing the other, but it reflects the influence of the available future paths.

This situation is not unimportant. A discontinuous dynamical evolution of the universe, i.e., one with jumps, is considered unacceptable on philosophical grounds. Furthermore, Einstein and other important physicists could not accept QM as a complete theory as it involved random dynamics, i.e., as "der Alte würfelt nicht". The Surjective Mapping Conjecture avoids random dynamics and offers the envisioned completion.

The boundaries at the initial and final quantum-times are fixed. We can still average over many settings with different positions in the universe. The first and last factor in Equation (5) is known to be independent of settings and can be pulled out. As the process is basically symmetric one has for the average of the remaining factors:

$$\begin{aligned}&\left[<\psi(t,\uparrow)|U|\psi(t_f,\uparrow)>_{\psi(0)}<\psi(t_f,\uparrow)|U|\psi(2t_f-t,\uparrow)>_{\psi(2t_f)}\right]\\&=\left[<\psi(t,\downarrow)|U|\psi(t_f,\downarrow)>_{\psi(0)}<\psi(t_f,\downarrow)|U|\psi(2t_f-t,\downarrow)>_{\psi(2t_f)}\right]\end{aligned}. \quad (6)$$

It means the relative probability is given just by the first and last factors:

$$|<\psi(\otimes)|\psi(\uparrow)>|^2 \quad \text{and} \quad |<\psi(\otimes)|\psi(\downarrow)>|^2 \quad (7)$$

as required by the Born rule. It comes out automatically, no special choice had to be made.

It reflects a general statistical property of QM (in Figure 3). For any setup with a measurement defined by witnesses reaching t_f these witnesses, in turn, allow the matching at t_f to determine individual measurements unpredictably:

```
setups
QM calculations
measurements  ←——  ←—— environment
```

Figure 3. General situation of a measurement.

No predictions are possible. Averaging, however, eliminates this seemingly random influence of the environment, and predictive calculations are possible.

It is in some way a hidden variable theory. However, the hidden variable

$$\psi(t=0)^* - \psi(t=2t_f)$$

does not sit on individual particles where they would create problems [10]. It produces a narrow overlap region at t_f. which determines all measurements, like in the usual two-boundary picture.

It is intrinsically somewhat less deterministic than the usual two boundary interpretations [11]. What happens at a time $\tau_1 = t$ and $\tau_2 = 2t_f - t$ affects the evolution in between, including the "final" t_f state. So, if one likes, one can add or imagine to have added something like an outside, free-willed manipulation affecting this "final" t_f state.

4. Conclusions

To conclude, as a multi-particle physicist analysing Bose–Einstein correlations, I was forced to accept backward causation. Backward causation and, consequently, non-locality changes paradigms in the interpretation of QM, suggesting a two-boundary interpretation. This note showed how a particular choice of boundaries could eliminate an un-pleasant feature of this interpretation.

Funding: This research received no external funding.

Institutional Review Board Statement: Not applicable.

Informed Consent Statement: Not applicable.

Data Availability Statement: Not applicable.

Acknowledgments: I have to thank many colleagues for fruitful discussions.

Conflicts of Interest: The authors declare no conflict of interest.

References

1. Bopp, F.W. An intricate quantum statistical effect and the foundation of quantum mechanics. *arXiv* **2019**, arXiv:1909.01391.
2. Bopp, F.W. Time Symmetric Quantum Mechanics and Causal Classical Physics. *Found. Phys.* **2017**, *47*, 490–504. [CrossRef]
3. Sakurai, J.J.; Napolitano, J.J. *Modern Quantum Mechanics*; Pearson Higher Ed; Cambridge University Press, Cambridge, UK, 2017.
4. Zurek, W.H. Decoherence, einselection, and the quantum origins of the classical. *Rev. Mod. Phys.* **2003**, *75*, 715. [CrossRef]
5. Hartle, J.B. Arrows of Time and Initial and Final Conditions in the Quantum Mechanics of Closed Systems Like the Universe. *arXiv* **2020**, arXiv:2002.07093.
6. Wharton, K. Time-symmetric boundary conditions and quantum foundations. *Symmetry* **2010**, *2*, 272–283. [CrossRef]
7. Wharton, K. Quantum Theory without Quantization. *arXiv* **2011**, arXiv:1106.1254.
8. Aharonov, Y.; Cohen, E.; Landsberger, T. The Two-Time Interpretation and Macroscopic Time-Reversibility. *Entropy* **2017**, *19*, 111. [CrossRef]
9. Hossenfelder, S.; Palmer, T. Rethinking superdeterminism. *Front. Phys.* **2020**, *8*, 139. [CrossRef]
10. Cabello, A.; Gu, M.; Gühne, O.; Larsson, J.Å.; Wiesner, K. Thermodynamical cost of some interpretations of quantum theory. *Phys. Rev. A* **2016**, *94*, 052127. [CrossRef]
11. Bopp, F.W. How to Avoid Absolute Determinism in Two Boundary Quantum Dynamics. *Quantum Rep.* **2020**, *2*, 442–449. [CrossRef]

Article

An Application of the Madelung Formalism for Dissipating and Decaying Systems

Maedeh Mollai and Seyed Majid Saberi Fathi *

Department of Physics, Faculty of Science, Ferdowsi University of Mashhad, Azadi Square, Mashhad 9177948974, Iran; maedeh.mollai@mail.um.ac.ir
* Correspondence: saberifathi@um.ac.ir

Abstract: This paper is concerned with the modeling and analysis of quantum dissipation and diffusion phenomena in the Schrödinger picture. We derive and investigate in detail the Schrödinger-type equations accounting for dissipation and diffusion effects. From a mathematical viewpoint, this equation allows one to achieve and analyze all aspects of the quantum dissipative systems, regarding the wave equation, Hamilton–Jacobi and continuity equations. This simplification requires the performance of "the Madelung decomposition" of "the wave function", which is rigorously attained under the general Lagrangian justification for this modification of quantum mechanics. It is proved that most of the important equations of dissipative quantum physics, such as convection-diffusion, Fokker–Planck and quantum Boltzmann, have a common origin and can be unified in one equation.

Keywords: decaying systems; Schrödinger equation; Madelung formulation

Citation: Mollai, M.; Saberi Fathi, S.M. An Application of the Madelung Formalism for Dissipating and Decaying Systems. *Symmetry* **2021**, *13*, 812. https://doi.org/10.3390/sym13050812

Academic Editor: Tuong Trong Truong

Received: 23 February 2021
Accepted: 27 March 2021
Published: 6 May 2021

Publisher's Note: MDPI stays neutral with regard to jurisdictional claims in published maps and institutional affiliations.

Copyright: © 2021 by the authors. Licensee MDPI, Basel, Switzerland. This article is an open access article distributed under the terms and conditions of the Creative Commons Attribution (CC BY) license (https:// creativecommons.org/licenses/by/ 4.0/).

1. Introduction

Since the early 20th century, the challenging problem of dissipation and diffusion modeling has been widely studied in quantum theory because a comprehensive understanding of dissipation in quantum mechanics is fundamental to the foundations of this theory [1,2]. The quantum–mechanical treatment of dissipative processes and other nonequilibrium phenomena has been the subject of much attention due to its applicability in various fields such as solid-state and statistical physics, incoherent solitons, photochemistry, Brownian dynamics, heavy-ion scattering, quantum gravity theories, dynamical modes of plasma physics, propagation of optical pulses and damping effects in nonlinear media [3–5].

In 1926 (the same year Schrödinger published his celebrated articles), Madelung reformulated the Schrödinger equation into a set of real, non-linear partial differential equations comparable with the Euler equations which were used in hydrodynamics. Madelung showed that the two equations were mathematically equivalent [6,7], and if one writes the wave function in the form of e^{R+iS}, the Schrödinger equation implies that, first, S is governed by a classical Hamilton–Jacobi-like equation, or alternatively that $\vec{v} = \nabla S$ is formulated by a Newton-like equation; second, ρ (which is defined as $\rho(x,t) = |\psi|^2 = R(x,t)^2$) is governed by a classical continuity equation [8]. The only formal difference between these equations and their purely classical counterparts is the existence of an additional "quantum" potential. Since that time these equations have provided the basis for numerous classical interpretations of quantum mechanics, including the hydrodynamic interpretation first proposed by Madelung [6], the theory of stochastic mechanics due to Nelson and others [3,8–12], the hidden-variable and double-solution theories of Bohm and de Broglie [13–15] and quite possibly other interpretations as well [8,16,17]. In some of these theories, such as the hydrodynamic interpretation and stochastic mechanics, the Madelung equations are taken as fundamental, and the Schrödinger equation is viewed as a mathematical consequence [8].

In the Schrödinger picture, quantum diffusion and dissipation effects have been effectively modeled by nonlinear terms of the type $\lambda \Delta_\psi \psi$, which was first formulated by Kostin [18,19] to describe nonlinear Schrödinger–Langevin dynamics, where λ is a friction constant and $\Delta_\psi = -i \log \frac{\psi}{|\psi|}$ is the argument of the complex wave function $\psi(x,t)$, and also later by logarithmic nonlinearities with the form $\log(|\psi|^2)\psi$, which was studied by Bialynicki-Birula and Mycielski [20,21]. They suggested that this logarithmic form maintains the lack of correlation between noninteracting particles. In a system under observation, there are many degrees of freedom such that information would be lost in the coupling process, which leads to dissipation [22]. Caldeira and Legget showed by using the influence-functional method [23] that dissipation tends to destroy quantum interference in a time scale shorter than the relaxation time of the system [22]. This result has given justification for the use of logarithmic nonlinear wave equations [12,18,22,24] to describe quantum dissipation. These equations are acceptable as a proper, practical bath functional in time-dependent density functional theory for open quantum systems [22,24].

In this paper the main purpose is to obtain a deeper understanding of some equations of quantum dissipation and their interrelations in a more satisfactory way based on the influence-functional method. We start from the fact that the Lagrangian density equation can be generalized to accommodate arbitrary wave functions. This is done using a substitution prescription for the principal function by applying a general complex function of $e^{\beta(x,t)}$ to introduce a wide-ranging Schrödinger-type wave equation and their corresponding Hamilton–Jacobi and continuity equations to access desirable results. It is shown that this approach provides a unified framework to aggregate and reproduce the wide class of Schrödinger-type equations compatible with the convection–diffusion, quantum Boltzmann and Fokker–Planck continuity equations; this leads to a deeper understanding of the nature of dissipative systems. We use Madelung's fluid dynamical formulation of the Schrödinger equation, in order to quantize the generalized Hamilton–Jacobi equation [25], which is intimately related to Hamilton–Jacobi theory. In this work, it is shown that the spatial dependence of the real part of $e^{\beta(x,t)}$ leads to the dissipative form of quantum potential, which is first introduced here.

This paper includes two main parts. First, the Schrödinger-type equation for dissipative systems via Lagrangian density and Euler–Lagrange equations is derived, and the related Madelung decomposition is discussed. In the next section it is shown that by assigning appropriate functions for $\beta(x,t)$, a wide range of important equations in nonequilibrium quantum physics, such as convection–diffusion, Fokker–Planck and quantum Boltzmann, are obtained.

2. Schrödinger-Type Equations of Dissipative Systems

2.1. Derivation

The Lagrangian approach based on the Principle of Least Action has been a unifying principle in almost all areas of physics to obtain dynamical equations. The Lagrangian L is a functional of field amplitude $\psi(x,t)$. It can usually be expressed as the integral overall space of a Lagrangian density $\mathcal{L}(\psi)$. If the field Lagrangian density $\mathcal{L}(\psi)$ is given, we can obtain the dynamical field equation from the Euler–Lagrange equations [26–29],

$$\frac{\partial \mathcal{L}}{\partial \psi} - \frac{\partial}{\partial t}\left(\frac{\partial \mathcal{L}}{\partial \dot{\psi}}\right) - \sum_{i=1}^{3} \frac{\partial}{\partial x_i}\left(\frac{\partial \mathcal{L}}{\partial(\partial \psi/\partial x_i)}\right) = 0, \qquad (1)$$

where $\frac{\partial}{\partial \psi}$ is a partial derivative.

Based on the Lagrangian approach, one can obtain the Schrödinger equation by its corresponding Lagrangian density [26,27], which is

$$\mathcal{L}_{sh}(\psi) = i\hbar \psi^* \frac{\partial \psi}{\partial t} - \frac{\hbar^2}{2m} \sum_{i=1}^{3} \left(\frac{\partial \psi^*}{\partial x_i}\right)\left(\frac{\partial \psi}{\partial x_i}\right),$$

If we define $\dot{\psi} = \frac{\partial \psi}{\partial t}$ and the gradient as $\vec{\nabla} \equiv \left(\frac{\partial}{\partial x}, \frac{\partial}{\partial y}, \frac{\partial}{\partial z}\right)$, the above equation can be written as

$$\mathcal{L}_{sh}(\psi) = i\hbar \psi^* \dot{\psi} - \frac{\hbar^2}{2m} \vec{\nabla} \psi^* \cdot \vec{\nabla} \psi, \tag{2}$$

Here $\psi(x,t)$ is a complex function, and we can treat ψ and ψ^* as independent fields. Now we can put the Lagrangian density (2) in the Euler–Lagrange Equation (1) for the field ψ^* to obtain the Schrödinger equation (or ψ for its complex conjugate), and end up with the Schrödinger equation for free particles [26,27,30]:

$$-i\hbar \frac{\partial \psi}{\partial t} = \frac{\hbar^2}{2m} \nabla^2 \psi \tag{3}$$

where $\nabla^2 \equiv \left(\frac{\partial^2}{\partial x^2}, \frac{\partial^2}{\partial y^2}, \frac{\partial^2}{\partial z^2}\right)$ is the Laplacian.

Theorem 1. *If $\psi(x,t)$ satisfied Equation (3), and it can be written in the form $\psi(x,t) = e^{-\beta(x,t)} \psi_D(x,t)$ where $\beta(x,t) \epsilon \mathbb{C}$ and $x \in \mathbb{R}^3$, $t \in \mathbb{R}$, then $\psi(x,t)$ is also the solution of the following differential equation,*

$$-i\hbar \frac{\partial \psi}{\partial t} = \frac{\hbar^2}{2m} \left(\vec{\nabla} + \vec{\nabla} \beta\right)^2 \psi. \tag{4}$$

Proof. By putting $\psi_D = e^{\beta(x,t)} \psi(x,t)$ in $\mathcal{L}_{sh}$ we have

$$\begin{aligned}\mathcal{L}_{sh}(\psi_D) &= i\hbar (e^\beta \psi)^* \frac{\partial (e^\beta \psi)}{\partial t} - \frac{\hbar^2}{2m} \vec{\nabla} (e^\beta \psi)^* \cdot \vec{\nabla} (e^\beta \psi) \\ &= i\hbar e^{\beta^*} e^\beta \psi^* \left(\frac{\partial \beta}{\partial t} \psi + \frac{\partial \psi}{\partial t}\right) - \frac{\hbar^2}{2m} e^{\beta^*} e^\beta \left(\psi^* \vec{\nabla} \beta^* + \vec{\nabla} \psi^*\right) \cdot \left(\psi \vec{\nabla} \beta + \vec{\nabla} \psi\right)\end{aligned}$$

Now $e^{\beta^*} e^\beta = e^{(\beta_r - i\beta_i) + (\beta_r + i\beta_i)} = e^{2\beta_r}$, where $\beta_r(x,t)$ and $\beta_i(x,t)$ are the real and imaginary parts of $\beta(x,t)$. By defining $\mathcal{L}_D(\psi) = \mathcal{L}_{sh}(\psi_D)$, the above equation gives

$$\mathcal{L}_D(\psi) = i\hbar e^{2\beta_r} \psi^* \left(\frac{\partial \beta}{\partial t} \psi + \frac{\partial \psi}{\partial t}\right) - \frac{\hbar^2}{2m} e^{2\beta_r} \left(\vec{\nabla} \beta^* \cdot \vec{\nabla} \beta \psi^* \psi + \vec{\nabla} \beta \cdot \vec{\nabla} \psi^* \psi + \vec{\nabla} \beta^* \cdot \vec{\nabla} \psi \psi^* + \vec{\nabla} \psi^* \cdot \vec{\nabla} \psi\right). \tag{5}$$

This is the Lagrangian density of dissipation wave function. Now to obtain the dynamical equation of dissipation wave function, we substitute $\mathcal{L}_D$ in the Euler–Lagrange Equation (1) to obtain

$$\begin{aligned}&i\hbar \frac{\partial \beta}{\partial t} \psi + i\hbar \frac{\partial \psi}{\partial t} - \frac{\hbar^2}{2m} (\vec{\nabla} \beta^* \cdot \vec{\nabla} \beta) \psi - \frac{\hbar^2}{2m} \vec{\nabla} \beta^* \cdot \vec{\nabla} \psi + \frac{\hbar^2}{2m} 2(\vec{\nabla} \beta_r \cdot \vec{\nabla} \beta) \psi \\ &+ \frac{\hbar^2}{2m} (\nabla^2 \beta) \psi + \frac{\hbar^2}{2m} \vec{\nabla} \beta \cdot \vec{\nabla} \psi + \frac{\hbar^2}{2m} 2 \vec{\nabla} \beta_r \cdot \vec{\nabla} \psi + \frac{\hbar^2}{2m} \nabla^2 \psi = 0.\end{aligned} \tag{6}$$

To simplify the above equation we replace $\beta = (\beta_r + i\beta_i)$ and $\beta^* = (\beta_r - i\beta_i)$ in some terms in Equation (6), then we have

$$\begin{aligned}\left\{i\hbar \frac{\partial \beta}{\partial t} + \frac{\hbar^2}{2m} \nabla^2 \beta\right\} \psi &+ \left\{i\hbar \frac{\partial \psi}{\partial t} + \frac{\hbar^2}{2m} \nabla^2 \psi\right\} - \frac{\hbar^2}{2m} (\vec{\nabla} \beta_i)^2 \psi + \frac{\hbar^2}{2m} (\vec{\nabla} \beta_r)^2 \psi + i\frac{\hbar^2}{2m} 2\vec{\nabla} \beta_r \cdot \vec{\nabla} \beta_i \psi + \frac{\hbar^2}{2m} \vec{\nabla} \beta_r \cdot \vec{\nabla} \psi \\ &+ i\frac{\hbar^2}{2m} \vec{\nabla} \beta_i \cdot \vec{\nabla} \psi = 0.\end{aligned} \tag{7}$$

By using the two following equations

$$\left(\vec{\nabla} \beta\right)^2 \psi = \left(\vec{\nabla} \beta_r + i\vec{\nabla} \beta_i\right)^2 \psi = \left(\left(\vec{\nabla} \beta_r\right)^2 - \left(\vec{\nabla} \beta_i\right)^2 + i2\vec{\nabla} \beta_r \cdot \vec{\nabla} \beta_i\right) \psi \tag{8}$$

and

$$\vec{\nabla} \beta \cdot \vec{\nabla} \psi = (\vec{\nabla} \beta_r + i\vec{\nabla} \beta_i) \cdot \vec{\nabla} \psi \tag{9}$$

Equation (7) is rephrased in terms of β as

$$\left\{i\hbar\frac{\partial \beta}{\partial t}\psi + i\hbar\frac{\partial \psi}{\partial t}\right\} + \frac{\hbar^2}{2m}\nabla^2\psi + \frac{\hbar^2}{2m}\nabla^2\beta\psi + \frac{\hbar^2}{2m}(\vec{\nabla}\beta)^2\psi + \frac{\hbar^2}{2m}2\vec{\nabla}\beta\cdot\vec{\nabla}\psi = 0.$$

Finally, to shorten Equation (9) and have an optimal form we use

$$(\vec{\nabla} + \vec{\nabla}\beta)^2\psi = \nabla^2\psi + \psi\nabla^2\beta + (\vec{\nabla}\beta)^2\psi + 2\vec{\nabla}\beta\cdot\vec{\nabla}\psi$$

Therefore, we conclude

$$-i\hbar\frac{\partial\psi}{\partial t} = \frac{\hbar^2}{2m}(\vec{\nabla} + \vec{\nabla}\beta)^2\psi + i\hbar\frac{\partial\beta}{\partial t}\psi \tag{10}$$

Equation (10) is a dissipative Schrödinger-like equation (DSE). □

Theorem 1, to some extent, is comparable to the Stoker [25] method, which has a variable function from which different equations can be obtained. Stocker's method was rudimentary and was not covered thoroughly. In the quantum hydrodynamical framework, Nassar [22,31] proposed a generalized nonlinear equation covering some of the famous equations due to Kostin [18], Süssmann and Hasse [32], Bialynicki-Birula–Mycielski [20], Stocker–Albrecht [25] and Schuch–Chung–Hartmann [33]. His equation had a variable parameter to produce different equations. Zander, Plastino and Díaz-Alonso [34] have investigated the nonlinear equation proposed by Nassar [22], and in its corresponding Hamilton–Jacobi and continuity parts some terms are left undefined. Recently, Gonçalves and Olavo [35] have derived, from first principles, a generalized Schrödinger equation that encompasses dissipative phenomena. Their results are not applicable to different equations of quantum dissipation systems.

Our approach is based on a variable complex valued function of $e^{\beta(x,t)}$, which can produce Fokker–Plank, convection–diffusion and quantum Boltzmann equations. These equations are only some examples of the applicability of this approach.

In this work, to the best of our knowledge, for the first time we present:

(A) a formalism which aggregates most of the important equations of dissipative quantum systems in a comprehensive manner; this leads to a deeper understanding of the nature of dissipative systems, see Section 3;

(B) a general form for quantum potential that would appear when the dissipation parameter depends on space, such as plasma currents where the quantum Boltzmann equation is used [36], see Sections 2.2 and 3.4.

Corollary 1. *if* $\beta = -\gamma t$ *($\gamma > 0$), we have $e^{-\gamma t}\psi$ which describes a decaying wave function, then its relevant wave equation is*

$$-i\hbar\frac{\partial\psi}{\partial t} = \frac{\hbar^2}{2m}\nabla^2\psi - i\hbar|\gamma|\psi. \tag{11}$$

Corollary 2. *if* $\beta = i\frac{q\varphi}{\hbar c}$ *(where $\frac{q\varphi}{\hbar c}$ is the Aharonov–Bohm parameter [27]), we have $e^{i\frac{q\varphi}{\hbar c}}\psi$ which describes an Aharonov–Bohm effect, and its relevant wave equation is the same as the standard Schrödinger equation.*

2.2. Madelung Decomposition of DSE

The equations of quantum hydrodynamics (Madelung equations) are Madelung's corresponding alternative formulation of the Schrödinger equation [6], which is written in terms of hydrodynamical variables. The derivation of Madelung equations is similar to the de Broglie–Bohm formulation, which represents the Schrödinger equation as a quantum Hamilton–Jacobi equation [37]. The Madelung equations, by their virtue of being formulated in the language of Newtonian mechanics, make it possible to construct a wide class of quantum theories by making the same coordinate-independent modifications found in Newtonian mechanics, without any need to construct a quantization algorithm [38].

Consider the one-dimensional time-dependent Schrödinger Equation (3) and write the polar form of wave function in terms of amplitude $R(x,t)$ and phase function (or action function) $S(x,t)$, i.e.,

$$\psi(x,t) = R(x,t)e^{\frac{iS(x,t)}{\hbar}}, \quad (12)$$

in which both are real valued functions. The probability density associated with this wave function is $\rho(x,t) = R(x,t)^2$.

By substituting this wave function into the Schrödinger Equation (3), one obtains a system of two coupled partial differential equations, which are the Madelung hydrodynamical formulation of wave mechanics [7,15,25,39]. Now by equating the real and imaginary parts, first we find the continuity equation for the imaginary part,

$$\frac{\partial \rho}{\partial t} + \vec{\nabla} \cdot \vec{\mathcal{J}} = 0, \quad (13)$$

in which the probability flux is $\vec{\mathcal{J}}$, defined as $\vec{\mathcal{J}} = \rho \frac{\vec{\nabla}S}{m}$, and $\vec{v} = \frac{\vec{\nabla}S}{m}$ is the flow velocity (or drift velocity) of the probability current.

The second equation or the real part is the quantum Hamilton–Jacobi equation (for $V = 0$), given by

$$-\frac{\partial S}{\partial t} = \frac{1}{2m}(\vec{\nabla}S)^2 - \frac{\hbar^2}{2m}\frac{\nabla^2 R}{R}; \quad (14)$$

in which the total energy is equal to the kinetic energy plus a quantum potential Q ($Q = -\frac{\hbar^2}{2m}\frac{\nabla^2 R}{R}$). Because of the explicit dependence of the quantum potential on $\hbar$, it brings all quantum effects into hydrodynamic formulation [39].

Now by applying Madelung decomposition on DSE (10), and after some manipulations, with the use of the identity

$$(\vec{\nabla} + \vec{\nabla}\beta_r)^2 R = \nabla^2 R + R\nabla^2\beta_r + 2\vec{\nabla}\beta_r \cdot \vec{\nabla}R + (\vec{\nabla}\beta_r)^2 R \quad (15)$$

the general quantum Hamilton–Jacobi and the general continuity equations, respectively, become

$$-\frac{\partial}{\partial t}(S + \hbar\beta_i) = \frac{1}{2m}(\vec{\nabla}S + \hbar\vec{\nabla}\beta_i)^2 - \frac{\hbar^2}{2m}\frac{(\vec{\nabla} + \vec{\nabla}\beta_r)^2 R}{R} \quad (16)$$

and

$$\left(\frac{\partial}{\partial t} + 2\frac{\partial \beta_r}{\partial t}\right)\rho + (\vec{\nabla} + 2\vec{\nabla}\beta_r) \cdot \left(\rho \frac{\vec{\nabla}S + \hbar\vec{\nabla}\beta_i}{m}\right) = 0 \quad (17)$$

Equation (16) is the Hamilton–Jacobi equation, in which on the left-hand side is the total energy and the right-hand side includes a dissipative form of the kinetic energy plus the general form of quantum potential. It is noteworthy to emphasize, when $\vec{\nabla}\beta_r \neq 0$, the quantum potential equation appears as

$$Q_D = -\frac{\hbar^2}{2m}\frac{(\vec{\nabla} + \vec{\nabla}\beta_r)^2 R}{R}. \quad (18)$$

This means that the spatial part of β_r changes the quantum properties and modifies the effect of the quantum potential. Since β_r is responsible for dissipative phenomena, one could refer (18) as a dissipative quantum potential (DQP).

It can be shown that the mean value of the DQP is proportional to the Fisher information probability density about the observable $\hat{x}$. By definition of Fisher information

$$\mathcal{F}(\theta) := E\left[-\frac{\partial^2}{\partial \theta^2}\log f(X|\theta)\right] = -\int f\nabla^2(\ln f)d^3x \tag{19}$$

and since

$$\nabla^2(\ln f) = \frac{\nabla^2 f}{f} - \left(\frac{\nabla f}{f}\right)^2 \tag{20}$$

it can be proved that, similar to quantum potential [40–42], the mean value of the DQP is proportional to Fisher information

$$\langle Q_D \rangle = \int \psi_D^* Q_D \psi_D dr = \frac{\hbar^2}{8m}\mathcal{F} \tag{21}$$

and this is important because quantum fluctuations and quantum geometry are related to the quantum potential via Fisher information [43,44].

Temporal and spatial dependence of the imaginary part β (which adds to the phase) changes the total energy and the kinetic energy, respectively. Additionally, the spatial dependence of the real part β changes the well-known quantum potential, but its temporal part has no effect on the quantum Hamilton–Jacobi Equation (16).

Equation (17) is the origin of several important equations in open quantum systems. In the following sections, Equation (17) will be further discussed.

3. Applications

In this section, we will provide further discussion of Equations (16) and (17). We show that by selecting an appropriate function for $\beta(x,t)$, a wide range of important equations in quantum physics are obtained. Moreover, with enough knowledge about the wave function, it would be possible to predict the nature of the Hamilton–Jacobi, continuity and more importantly wave equations and vice versa.

For example, here is a brief description of the Berry's phase. Berry's phase [43] is a quantum phase effect arising in systems that undergo a slow, cyclic evolution. In an adiabatic evolution of the Hamiltonian, a quantum system in an nth eigenstate, remains in this n^{th} eigenstate of the Hamiltonian, while picking up a phase factor. Under adiabatic approximation, the coefficient of the nth eigenstate is given by

$$C_n(t) = C_n(0)\exp\left[-\int_0^t \langle \psi_n(t\prime)|\dot{\psi}_n(t\prime)\rangle dt\prime\right] = C_n(0)e^{i\gamma_n(t)} \tag{22}$$

where $\gamma_n(t)$ is the Berry's phase with respect of parameter t. According to (22) and comparison with (16) and (17) one concludes that the adiabatic evolution has no effect on the continuity part, but the Hamilton-Jacobi equation takes the form of

$$-\frac{\partial S}{\partial t} = \frac{1}{2m}(\vec{\nabla}S)^2 - \frac{\hbar^2}{2m}\frac{\vec{\nabla}^2 R}{R} + \hbar\frac{\partial \gamma(t)}{\partial t} \tag{23}$$

and the corresponding wave equation is obtained as

$$i\hbar\frac{\partial \psi}{\partial t} = -\frac{\hbar^2}{2m}\nabla^2\psi - \hbar\frac{\partial \gamma(t)}{\partial t}\psi \tag{24}$$

In the following, for better clarification, we will apply this approach to obtain the continuity equation, the Fokker–Planck equation, the convection–diffusion equation and the quantum Boltzmann equation. The important point is that they are all derived from a single equation produced by the "Lagrangian density of the Srodinger equation". Therefore,

Equation (10) can be considered as a consistent and constructive generalization for the Schrödinger equation.

3.1. Continuity Equation with Source or Sink

In the simplest example we take

$$\beta_i = 0 \text{ and } \beta_r = \gamma t, \gamma \in \mathbb{R} \tag{25}$$

so, the wave function is $\psi_D = e^{\gamma t}\psi$, in which γ is a real number. Therefore, the corresponding wave equation takes the form

$$i\hbar \frac{\partial \psi}{\partial t} = -\frac{\hbar^2}{2m}\nabla^2 \psi + i\gamma\hbar\psi. \tag{26}$$

In this case, the Hamilton–Jacobi equation remains without change

$$-\frac{\partial S}{\partial t} = \frac{1}{2m}(\vec{\nabla}S)^2 - \frac{\hbar^2}{2m}\frac{\nabla^2 R}{R}, \tag{27}$$

and, as expected, the probability current is no longer conserved

$$\frac{\partial R^2}{\partial t} + \vec{\nabla} \cdot \vec{\mathcal{J}} = -2\gamma\rho \tag{28}$$

where $2\gamma\rho$ represents source or sink for the current [17].

3.2. The Fokker–Plank Equation

As an important example, we take

$$\beta_r = 0 \text{ and } \beta_i = \ln R(x,t) \tag{29}$$

where $R(x,t)$ is the real valued amplitude of the wave function, and upon substituting in DSE (10), its corresponding wave equation takes the form

$$\begin{aligned}i\hbar \frac{\partial \psi}{\partial t} = &-\frac{\hbar^2}{2m}\nabla^2\psi - i\frac{\hbar^2}{2m}\nabla^2(\ln|\psi|)\psi - i\frac{\hbar^2}{m}\vec{\nabla}\ln|\psi| \cdot \vec{\nabla}\psi \\ &+ \frac{\hbar^2}{2m}(\vec{\nabla}\ln|\psi|)^2\psi - \hbar\frac{\partial \ln|\psi|}{\partial t}\psi.\end{aligned} \tag{30}$$

Next, by setting (29) in Equations (16) and (17), we obtain its relevant continuity and the Hamilton–Jacobi equations, respectively

$$-\frac{\partial}{\partial t}(S + \hbar \ln R) = \frac{1}{2m}(\vec{\nabla}S + \hbar\vec{\nabla}\ln R)^2 - \frac{\hbar^2}{2m}\frac{\nabla^2 R}{R} \tag{31}$$

$$\frac{\partial \rho}{\partial t} + \vec{\nabla} \cdot \vec{\mathcal{J}} = -\frac{\hbar}{2m}\nabla^2\rho. \tag{32}$$

Equation (31) is the Hamilton–Jacobi equation of a diffusive system, whose total energy is on the left-hand side and total kinetic energy plus quantum potential is on the right-hand side. Total kinetic energy includes kinetic energy corresponding to drift velocity in which there is an added non-classical, stochastic diffusion velocity (either of Markovian or of non-Markovian type) [1].

Equation (32) is the well-known Fokker–Planck equation. Some fundamental considerations of quantum theory suggest a general, complex nonlinear Schrödinger equation (outside the classes most often studied), which follows from admitting quantum diffusion currents, so that its probability density satisfies the Fokker–Planck equation [17,45–47].

It is worth mentioning that the diffusion current, and thus the Fokker–Planck equation, are due to the presence of $\ln R(x,y)$ in the phase of the wave function

$$\psi \to \psi_D = e^{(\beta_r + i\beta_i)}\psi$$
$$\psi_D = Re^{i\kappa \ln R}e^{i\frac{S}{\hbar}} = Re^{\frac{i(S+\hbar\kappa\ln R)}{\hbar}}$$

where κ is a real number.

Doebner and Goldin in [45] propose a group-theoretical justification for a nonlinear modification of quantum mechanics, as the most general class of Schrödinger-type equations compatible with the Fokker–Planck continuity equation; they start with

$$i\hbar\frac{\partial \psi}{\partial t} = H_0\psi + \mathcal{R}[\psi]\psi + i\mathcal{I}[\psi]\psi \text{ with } \mathcal{I}[\psi] = \frac{\hbar D}{2}\left(\frac{\nabla^2(\overline{\psi}\psi)}{\overline{\psi}\psi}\right) \tag{33}$$

where $\mathcal{R}[\psi]$ and $\mathcal{I}[\psi]$ are the real and imaginary parts of the nonlinear functional multiplying ψ; and $\mathcal{I}[\psi]$ is the key phrase to create the Fokker–Planck equation. Then, they declare that the theory at this point gives no further information about $\mathcal{R}[\psi]$, but it is reasonable to assume that $\mathcal{R}[\psi]$ is of a form similar to $\mathcal{I}[\psi]$.

If Equation (30) is rewritten as (34), its compliance with the Doebner–Goldin Equation (33) is determined [35,45,46,48]

$$i\hbar\frac{\partial \psi}{\partial t} = -\frac{\hbar^2}{2m}\nabla^2\psi + \left\{\frac{\hbar}{m}\frac{\vec{\nabla}|\psi|}{|\psi|}\vec{\nabla}S + \frac{\hbar^2}{2m}\left(\frac{\vec{\nabla}|\psi|}{|\psi|}\right)^2 + \hbar\frac{\partial \ln|\psi|}{\partial t}\right\}\psi + i\left\{\frac{-\hbar^2}{2m}\left(\frac{\nabla^2\psi^2}{2\psi^2}\right)\right\}\psi \tag{34}$$

The general Lagrangian justification for this nonlinear modification of quantum mechanics seems to suggest it as a minimal nonlinear generalization of the Schrödinger equation; it contains the least terms and still satisfies the Fokker–Planck equation. In addition, it satisfies the general requirements of a nonlinear Schrödinger equation: (a) the probability is conserved, (b) the equation is homogeneous, (c) non-interacting particle subsystems remain uncorrelated, (d) plane waves are solutions for the free equation and (e) the free equation is also time and space translation invariant [45,48].

Logarithmic terms which appear in (34) have a great importance in quantum friction and diffusion effects. Dissipation tends to destroy quantum interference in a time scale shorter than the relaxation time of the system [21]. This result has given justification for the use of logarithmic nonlinear wave equations [12,18,22,24] to describe quantum dissipation. Additionally, the logarithmic form guarantees non-interacting particle subsystems remain uncorrelated [20].

3.3. Convection–Diffusion Equation

If we combine the diffusion (the Fokker–Planck equation) (32) with source or sink (28), the result would be a convection–diffusion equation. Thus, based on previous sections we pick $\psi_D = e^{(\beta_r + i\beta_i)}\psi = e^{\frac{\gamma}{2}t + i\ln R}\psi$ ($\gamma \in \mathbb{R}$), and substituting in (10), (16) and (17), respectively, the wave equation, the Hamilton–Jacobi equation and continuity equations are obtained as

$$i\hbar\frac{\partial \psi}{\partial t} = -\frac{\hbar^2}{2m}(\vec{\nabla} + i\vec{\nabla}\ln|\psi|)^2\psi + \hbar\frac{\partial}{\partial t}\ln|\psi|\,\psi + i\hbar\frac{\gamma}{2}\psi \tag{35}$$

$$-\frac{\partial}{\partial t}(S + \hbar\ln R) = \frac{1}{2m}(\vec{\nabla}S - \hbar\vec{\nabla}\ln R)^2 - \frac{\hbar^2}{2m}\frac{\nabla^2 R}{R} \tag{36}$$

$$\frac{\partial \rho}{\partial t} + \vec{\nabla}\cdot\vec{\mathcal{J}} = -\frac{\hbar}{2m}\nabla^2\rho - 2\gamma\rho. \tag{37}$$

Equation (37) is the convection–diffusion equation [49], and $2\gamma\rho$ is the source or sink of the probability current.

3.4. Quantum Boltzmann Equation

As a last example, we take $\Gamma(x,t)$ as a general real function and again the diffusion factor of $i \ln R(x,t)$ we have

$$\beta = \frac{-\Gamma(x,t)}{2} + i \ln R(x,t), \tag{38}$$

Therefore, wave Equation (10) takes the form of

$$i\hbar \frac{\partial \psi}{\partial t} = -\frac{\hbar^2}{2m}\left(\vec{\nabla} - \vec{\nabla}\Gamma + i\vec{\nabla}\ln|\psi|\right)^2 \psi + \hbar \frac{\partial \ln|\psi|}{\partial t}\psi + \frac{i\hbar}{2}\frac{\partial \Gamma}{\partial t}\psi \tag{39}$$

and corresponding Hamilton–Jacobi and continuity equations become

$$-\frac{\partial}{\partial t}(S + \hbar \ln R) = \frac{1}{2m}\left(\vec{\nabla}S + \hbar\vec{\nabla}\ln R\right)^2 - \frac{\hbar^2}{2m}\frac{\left(\vec{\nabla} - \frac{\vec{\nabla}\Gamma}{2}\right)^2 R}{R} \tag{40}$$

$$\left(\frac{\partial}{\partial t} - \frac{\partial \Gamma}{\partial t}\right)\rho + \left(\vec{\nabla} - \vec{\nabla}\Gamma\right)\cdot\left(\vec{J} + \frac{\hbar}{2m}\rho\vec{\nabla}\ln\rho\right) = 0. \tag{41}$$

We see in Equation (40) that DQP is revealed. It means that if $\vec{\nabla}\Gamma(x,t) \neq 0$, the well-known quantum potential will take a new form of DQP.

Equation (41) is the continuity equation, and with some manipulations it takes a familiar form

$$\frac{\partial \rho}{\partial t} + \vec{\nabla}\cdot(\rho\vec{v}) - \frac{\hbar}{2m}\nabla^2\rho = \rho\left(\frac{\partial}{\partial t} + \vec{v}\cdot\vec{\nabla} + \frac{\hbar}{2m}\vec{\nabla}\ln\rho\cdot\vec{\nabla}\right)\Gamma. \tag{42}$$

where $\vec{v} = \frac{\vec{\nabla}S}{m}$ is the drift velocity. The left-hand side of the above equation is the Fokker–Plank equation and the right-hand side resembles the quantum Boltzmann equation

$$\left[\frac{\partial}{\partial t} + \vec{v}\cdot\vec{\nabla} + \vec{F}\cdot\vec{\nabla}_p\right]f(\vec{x},\vec{p},t) = \mathcal{Q}[f](\vec{x},\vec{p}). \tag{43}$$

The quantum Boltzmann equation gives the non-equilibrium time evolution of a gas of quantum mechanically interacting particles. In Equation (43), $f(\vec{x},\vec{p},t)$ is a general distribution function, $\vec{F}$ is external applied force and $\mathcal{Q}$ is quantum collision operator, accounting for the interactions between the gas particles; if it is zero then the particles do not collide. Thus, the Fokker–Planck term at the left-hand side of (42) represents the effects of particle collisions [50].

Go back to the right-hand side of Equation (42). Since we can write

$$\vec{\nabla}\Gamma = \frac{\partial \Gamma}{\partial x}\frac{\partial \vec{\nabla}S}{\partial \vec{\nabla}S} = \nabla^2 S \frac{\partial \Gamma}{\partial \vec{\nabla}S}, \tag{44}$$

then by substituting (44) and (43) into the right-hand side of (42), we have

$$\left(\vec{F}\cdot\frac{\partial}{\partial \vec{p}}\right)f = \left(\frac{\hbar}{2m^2}\nabla^2 S \vec{\nabla}\ln\rho\cdot\frac{\partial}{\partial \vec{\nabla}S}\right)\Gamma. \tag{45}$$

Thus, we can write Equation (42) as

$$\frac{\partial \rho}{\partial t} + \vec{\nabla}\cdot(\rho\vec{v}) - \frac{\hbar}{2m}\nabla^2\rho = \rho\left(\frac{\partial}{\partial t} + \vec{v}\cdot\vec{\nabla} + \frac{\hbar}{2m}(\vec{\nabla}\cdot\vec{v})\vec{\nabla}\ln\rho\cdot\vec{\nabla}_p\right)\Gamma. \tag{46}$$

So, the external applied force is defined as $\vec{F} = \frac{\hbar}{2m}(\vec{\nabla} \cdot \vec{v})\vec{\nabla} \ln \rho$, which is related to both drift and diffusion velocities.

The Fokker–Planck equation, which was firstly derived to treat the Brownian motion of molecules, has been extensively used to evaluate the collision term of the Boltzmann equation for describing small-angle binary collisions of the inverse-square type of force [36]. In stellar dynamics, Chandrasekhar first discussed this theory for stochastic effects of gravity [36,51]. The applications of this equation to classical plasma physics were first treated by Landau, Spitzer, as well as Cohen, Spitzer and Routly, and an elegant mathematical treatment was completed by Rosenbluth, MacDonald and Judd [36,52].

4. Summery and Conclusions

In this approach based on the influence-functional method we gain a better understanding on quantum friction and diffusion effects in a more satisfactory way. We generalize the Lagrangian density equation to accommodate arbitrary wave functions, to provide an approach to general quantum dissipation and diffusion modeling by introducing a dissipative Schrödinger-type equation (DSE). The key point of this approach is $\beta(x,t)$, as a complex function, which is responsible for taking the problem from equilibrium phenomena to nonequilibrium ones. The approach provides the achievable analysis, for all aspects of the phenomena, concerning the wave equation, the Hamilton–Jacobi equation and the continuity equation. It is shown that by applying a general complex function of $e^{\beta(x,t)}$ one can produce a wide-ranging Schrödinger-type wave equation and their corresponding Hamilton–Jacobi and continuity equations to access desirable results. To show the widespread applications of this approach, we provided some examples of the Berry phase, the continuity equation, the Fokker–Planck equation, the convection–diffusion equation and the quantum Boltzmann equation. All of these equations are obtained by step-by-step generalizations of the beta function, see Table 1.

Table 1. Dissipation quantum equations and their related β functions.

Quantum Equations	β_r	β_i
Continuity equation with Source or Sink	$\beta_r(x,t) = \gamma t$	$\beta_i(x,t) = 0$
Fokker–Planck equation	$\beta_r(x,t) = 0$	$\beta_i(x,t) = \ln R(x,t)$
Convection–diffusion equation	$\beta_r(x,t) = \frac{\gamma t}{2}$	$\beta_i(x,t) = \ln R(x,t)$
Quantum Boltzmann equation	$\beta_r(x,t) = \frac{-\Gamma(x,t)}{2}$	$\beta_i(x,t) = \ln R(x,t)$

As a first step, we chose $\beta_r(x,t) = \gamma t$ with no imaginary part, then we would have growing/shrinking amplitude; it means the probability current is no longer conserved, and there would be source/sink in the continuity equation.

Some fundamental considerations of quantum theory suggest a general, complex nonlinear Schrödinger equation, which follows from admitting quantum diffusion currents and must be such that its probability density satisfies the Fokker–Planck equation. To describe the diffusion current there is no need for source or sink; so the real part which affects the amplitude is zero, and the imaginary part which controls the phase gets the $\beta_i = \ln R(x,t)$ value. From this choice the Fokker–Planck equation results in the continuity part, but it also changes the usual appearance of the Hamilton–Jacobi (31). As expected, by quantum friction and diffusion effects, logarithmic nonlinearities appear in the corresponding wave Equation (34).

Now if we have a combination of growing/shrinking amplitude with diffusion current, $\beta_r(x,t) = \frac{\gamma t}{2}$ and $\beta_i(x,t) = \ln R(x,t)$, the result is the convection–diffusion Equation (37).

Finally, choosing for general real part $\Gamma(x,t)$, which means $\beta_r(x,t) = \frac{-\Gamma(x,t)}{2}$ and $\beta_i(x,t) = \ln R(x,t)$, this leads to the quantum Boltzmann equation. Because of the spatial dependence of $\beta_r(x,t)$, the new appearance for quantum potential (DQP) has been revealed (18). Since the quantum potential depends on the amplitude of a wave function, it is deformed as DQP for a dissipative system. This form of quantum potential is first presented in this work.

All of above equations are derived from a single equation produced by the Dissipative Schrödinger Equation (10). This unified framework provides a common ground for a better understanding of quantum friction and diffusion effects and allows a deeper understanding of the nature of the dissipative systems. The present approach is useful to the study of non-equilibrium quantum mechanical systems, such as mesoscopic systems.

Author Contributions: All authors have contributed equally. All authors have read and agreed to the published version of the manuscript.

Funding: Research for this project has been supported by grant number 3/49685 from Ferdowsi University of Mashhad.

Data Availability Statement: Not applicable.

Conflicts of Interest: The authors declare no conflict of interest.

References

1. Salesi, G. Spin and Madelung Fluid. *Mod. Phys. Lett. A* **1996**, *11*, 1815–1823. [CrossRef]
2. Hirschfelder, J.; Tang, K. Quantum mechanical streamlines. III. Idealized reactive atom-diatomic molecule collision. *Diatomic Mol. Collis.* **1976**, *64*, 760. [CrossRef]
3. Nagasawa, M. *Schrödinger Equations and Diffusion Theory*; Springel Basel: Boston, MA, USA, 2012.
4. Weiss, U. *Quantum Dissipative Systems*; Singapore World Sientific: Singapore, 2012.
5. Celeghini, E.; Rasetti, M.; Vitiello, G. Quantum dissipation. *Ann. Phys.* **1992**, *215*, 156–170. [CrossRef]
6. Madelung, E. Eine anschauliche Deutung der Gleichung von Schrödinger. *Naturwissenschaften* **1926**, *14*, 1004. [CrossRef]
7. Madelung, E. Quantentheorie in hydrodynamischer Form. *Eur. Phys. J. A* **1927**, *40*, 322–326. [CrossRef]
8. Wallstrom, T.C. Inequivalence between the Schrödinger equation and the Madelung hydrodynamic equations. *Phys. Rev. A* **1994**, *49*, 1613–1617. [CrossRef]
9. Nelson, E. *Quantum Fluctuations*; Princeton University: Princeton, NY, USA, 1985.
10. Wang, M. Derivation of Feynman's path integral theory based on stochastic mechanics. *Phys. Lett. A* **1989**, *137*, 437–439. [CrossRef]
11. Köppe, J.; Patzold, M.; Grecksch, W.; Paul, W. Quantum Hamilton equations of motion for bound states of one-dimensional quantum systems. *J. Math. Phys.* **2018**, *59*, 062102. [CrossRef]
12. Yasue, K. Stochastic calculus of variations. *J. Funct. Anal.* **1981**, *41*, 327–340. [CrossRef]
13. Bohm, D. A Suggested Interpretation of the Quantum Theory in Terms of "Hidden" Variables. II. *Phys. Rev.* **1952**, *85*, 180–193. [CrossRef]
14. de Broglie, L. *Current Interpretation of Wave Mechanics, A CriticaL Study*; Elsevier: Amsterdam, The Netherlands, 1964.
15. Holland, P.R. *The Quantum Theory of Motion*; Cambridge University Press: Cambridge, UK, 1993.
16. Jammer, M. *The Philosophy of Quantum Mechanics*; Wiley: NewYork, NY, USA, 1974.
17. Razavy, M. *Classical and Quantum Dissipative Systems*; Imperial College Press: London, UK, 2005.
18. Kostin, M. On the Schrödinger-Langevin equation. *J. Chem. Phys.* **1972**, *57*, 3589–3591. [CrossRef]
19. Kostin, M.D. Friction and dissipative phenomena in quantum mechanics. *J. Stat. Phys.* **1975**, *12*, 145–151. [CrossRef]
20. Bialynicki-Birula, I.; Mycielski, J. Nonlinear wave mechanics. *Ann. Phys.* **1976**, *100*, 62–93. [CrossRef]
21. Guerrero, P.; López, J.L.; Montejo-Gámez, J.; Nieto, J. Wellposedness of a Nonlinear, Logarithmic Schrödinger Equation of Doebner–Goldin Type Modeling Quantum Dissipation. *J. Nonlinear Sci.* **2012**, *22*, 631–663. [CrossRef]
22. Nassar, A.; Miret-Arte, S. Dividing Line between Quantum and Classical Trajectories in a Measurement Problem:Bohmain Time Constant. *Phys. Rev. Lett.* **2013**, *111*, 150401. [CrossRef] [PubMed]
23. Caldeira, A.; Leggett, A. Influence of damping on quantum interference: An exactly soluble model. *Phys. Rev. A* **1985**, *31*, 1059. [CrossRef] [PubMed]
24. Yuen-Zhou, J.; Tempel, D.; Rodríguez-Rosario, C.; Aspuru-Guzik, A. Time-Dependent Density Functional Theory for Open Quantum Systems with Unitary Propagation. *Phys. Rev. Lett.* **2010**, *104*, 043001. [CrossRef] [PubMed]
25. Stoker, W.; Albrecht, K. A formalism for the construction of quantum friction equations. *Ann. Physic* **1979**, *117*, 436–446. [CrossRef]
26. Xiang-Yao, W.; Bai-Jun, Z.; Xiao-Jing, L.; Xiao, L.; Yi-Heng, W.; Yan, W.; Qing-Cai, W.; Cheng, S. Derivation of Nonlinear Schrödinger Equation. *Int. J. Theor. Phys.* **2010**, *49*, 2437–2445.
27. Greiner, W.; Reinhardt, J. *Field Quantization*; Springer: Berlin/Heidelberg, Germany, 1996.
28. Yourgrau, W.; Mandelstam, S. *Variational Principles in Dynamics and Quantum Theory*; Dover Publications: Philadelphia, PA, USA, 1968.
29. Cassel, K. *Variational Methods with Applications in Science and Engineering*; Cambridge University Press: Cambridge, MA, USA, 2013.
30. Fujita, T. *Symmetry and Its Breaking in Quantum Field Theory*; New York Nova Science Publishers: New York, NY, USA, 2006.

31. Nassar, A.B. Time-dependent invariant associated to nonlinear Schrödinger-Langevin equation. *J. Math. Phys.* **1986**, *27*, 2949–2952. [CrossRef]
32. Hasse, R.W. On the quantum mechanical treatment of dissipative systems. *J. Math. Phys.* **1975**, *16*, 2005–2011. [CrossRef]
33. Schuch, D.; Chung, K.; Hartmann, H. Nonlinear Schrödinger-type field equation for the description of dissipative systems. I. Derivation of the nonlinear field equation and one-dimensional example. *J. Math. Phys.* **1983**, *24*, 1652–1660. [CrossRef]
34. Zander, C.; Plastino, A.; Diaz-Alonso, J. Wave packet dynamics for a non-linear Schrödinger equation describing continuous position measurements. *Ann. Phys.* **2015**, *362*, 36–56. [CrossRef]
35. Gonçalves, L.; Olavo, L.S.F. Schrödinger equation for general linear velocity-dependent forces. *Phys. Rev. A* **2018**, *97*, 022102. [CrossRef]
36. Petrasso, R.D.; Chi-Kang, L. Fokker-Planck equation for moderately coupled plasmas. *Phys. Rev. Lett.* **1993**, *70*, 3063–3066.
37. Sanz, A.S.; Miret-Artes, S. A Trajectory Description of Quantum Processes. In *Fundamentals: A Bohmian 1prespective*; Springer: Berlin/Heidelberg, Germany, 2012.
38. Reddiger, M. The Madelung Picture as a Foundation of Geometric Quantum Theory. *Found. Phys.* **2017**, *47*, 1317–1367. [CrossRef]
39. Wyatt, R.E. *Quantum Dynamics with Trajectories, Introduction to Quantum Hydrodynamics*; Springer: Berlin/Heidelberg, Germany, 2005.
40. Reginatto, M. Derivation of the equations of nonrelativistic quantum mechanics using the principle of minimum Fisher information. *Phys. Rev. A* **1998**, *58*, 1775–1778. [CrossRef]
41. Tsekov, R. Towards nonlinear quantum Fokker-Planck equation. *Int. J. Theor. Phys.* **2009**, *48*, 1431–1435. [CrossRef]
42. Carroll, R.W. *Fluctuations, Information, Gravity and the Quantum Potential*; Springer: Berlin/Heidelberg, Germany, 2006.
43. Carollo, A.; Fuentes-Guridi, I.; Santos, M.F.; Vedral, V. Geometric phase in open systems. *Phys. Rev. Lett.* **2003**, *90*, 160402. [CrossRef]
44. Licata, I.; Fiscaletti, D. *Quantum Potential: Physics, Geometry And Algebra*; Springer: Berlin/Heidelberg, Germany, 2014.
45. Doebner, H.D.; A Goldin, G. Properties of nonlinear Schrodinger equations associated with diffeomorphism group representations. *J. Phys. A Math. Gen.* **1994**, *27*, 1771–1780. [CrossRef]
46. Doebner, H.D.; Goldin, A.G. On a general nonlinear Schrodinger equation admitting diffusion currents. *Phys. Lett. A* **1992**, *162*, 397–401. [CrossRef]
47. Risken, H. *The Fokker-Planck Equation: Methods of Solution and Applications*; Springer: Berlin/Heidelberg, Germany, 1989.
48. Nattermann, P.; Scherer, W.; Ushveridze, A. Exact solutions of the general Doebner-Goldin equation. *Phys. Lett. A* **1994**, *184*, 234–240. [CrossRef]
49. Atangana, A. *Fractional Operators with Constant and Variable Order with Application to Geo-Hydrology*; Academic Press: Cambridge, MA, USA, 2018; pp. 49–72.
50. Castella, F. On the derivation of a quantum Boltzmann equation from the periodic Von-Neumann equation. *ESAIM Math. Model. Numer. Anal.* **1999**, *33*, 329–349. [CrossRef]
51. Chandrasekhar, S. *Principle of Stellar Dynamics*; Dover Publications: New York, NY, USA, 1943.
52. Rosenbluth, M.N.; Macdonald, W.M.; Judd, D.L. Fokker-Planck Equation for an Inverse-Square Force. *Phys. Rev.* **1957**, *107*, 1–6. [CrossRef]

Review

Quantum Spin-Wave Theory for Non-Collinear Spin Structures, a Review

Hung T. Diep [†]

Laboratoire de Physique Théorique et Modélisation, CY Cergy Paris Université, CNRS, UMR 8089, 2 Avenue Adolphe Chauvin, 95302 Cergy-Pontoise, France; diep@cyu.fr

† CY Cergy Paris Université (Former Name, University of Cergy-Pontoise).

Abstract: In this review, we trace the evolution of the quantum spin-wave theory treating non-collinear spin configurations. Non-collinear spin configurations are consequences of the frustration created by competing interactions. They include simple chiral magnets due to competing nearest-neighbor (NN) and next-NN interactions and systems with geometry frustration such as the triangular antiferromagnet and the Kagomé lattice. We review here spin-wave results of such systems and also systems with the Dzyaloshinskii–Moriya interaction. Accent is put on these non-collinear ground states which have to be calculated before applying any spin-wave theory to determine the spectrum of the elementary excitations from the ground states. We mostly show results obtained by the use of a Green's function method. These results include the spin-wave dispersion relation and the magnetizations, layer by layer, as functions of T in 2D, 3D and thin films. Some new unpublished results are also included. Technical details and discussion on the method are shown and discussed.

Keywords: quantum spin-wave theory; Green's function theory; frustrated spin systems; non-collinear spin configurations; Dzyaloshinskii–Moriya interaction; phase transition; Monte Carlo simulation

Citation: Diep, H.T. Quantum Spin-Wave Theory for Non-Collinear Spin Structures, a Review. *Symmetry* **2022**, *14*, 1716. https://doi.org/10.3390/sym14081716

Academic Editors: Sergey Troshin and Sergei D. Odintsov

Received: 14 July 2022
Accepted: 15 August 2022
Published: 17 August 2022

Publisher's Note: MDPI stays neutral with regard to jurisdictional claims in published maps and institutional affiliations.

Copyright: © 2022 by the author. Licensee MDPI, Basel, Switzerland. This article is an open access article distributed under the terms and conditions of the Creative Commons Attribution (CC BY) license (https://creativecommons.org/licenses/by/4.0/).

1. Introduction

In a solid the interactions between its constituent atoms or molecules gives rise to elementary excitations from its ground state (GS) when the temperature increases from zero. One has examples of elementary excitations due to atom-atom interactions, known as phonons, or due to spin-spin interactions, known as magnons. Note that magnons are spin waves (SW) when they are quantized. Elementary excitations are defined also for interactions between charge densities in plasma, or for electric dipole-dipole interactions in ferroelectrics, among others. Elementary excitations are thus collective motions which dominate the low-temperature behaviors of solids in general.

For a given system, there are several ways to calculate the energy of elementary excitations from classical treatments to quantum ones. Since those collective motions are waves, their energy depends on the wave vector **k**. The **k**-dependent energy is often called the SW spectrum for spin systems. Note that though the calculation of the SW spectrum is often for periodic crystalline structures, it can also be performed for symmetry-reduced systems such as in thin films or in semi-infinite solids in which the translation symmetry is broken by the presence of a surface.

In this review we focus on the SW excitations in magnetically ordered systems. The history began with ferromagnets and antiferromagnets with collinear spin GSs, parallel or antiparallel configurations in the early 1950s. Most of the works on the SW used either the classical method or the quantum Holstein–Primakoff transformation. The Green's function (GF) technique has also been introduced in a pioneering paper of Zubarev [1]. The first application of this method to thin films has been done [2]. Note that unlike the SW theory, the GF can treat the SW up to higher temperatures. We will come back to this point later.

Let us recall some important breakthroughs in the study of non-collinear spin configurations. The first discovery of the helical spin configuration has been published in 1959 [3,4]. Some attempts to treat this non-collinear case have been done in the 1970 and

1980. Let us cite two noticeable works on this subject in Refs. [5,6]. In these works, a local system of spin coordinates have been introduced in the way that each spin lies on its quantization axis. One can therefore use the commutation relations between spin deviation operators. These works took into account magnon-magnon interactions by expanding the Hamiltonian up to three-operator terms at temperature $T = 0$ [5] or up to four-operator terms at low T [6]. Nevertheless, since these works used the Holstein–Primakoff method, the case of higher T cannot be dealt with. In Ref. [7], the GF method has been employed for the first time to calculate the SW spectrum in a frustrated system where the GS spin configuration is non-collinear. Using the SW spectrum, the local order parameters and the specific heat were calculated. Since this work, we have applied the GF method to a variety of systems where the GS is non-collinear. In this review, we will recall results of some of these published works.

Let us comment on the frustration which is the origin of the non-collinear GS. The frustration is caused by either the competing interactions in the system or a geometry frustration as in the triangular lattice with only the antiferromagnetic interaction between the nearest neighbors (NN) (see Ref. [8]). The frustration causes high GS degeneracy, and for the vector spins (XY and Heisenberg cases) the spin configurations are non-collinear, making the calculation of the SW spectrum harder. A number of examples will be shown in this review paper.

In addition to competing interactions, the Dzyaloshinskii–Moriya (DM) interaction [9,10] is also the origin of non-collinear spin configurations in spin systems. While the Heisenberg model between two spins is written as $-J_{ij}\mathbf{S}_i \cdot \mathbf{S}_j$ giving rise to two collinear spins in the GS, the DM interaction is written as $\mathbf{D}_{ij} \cdot \mathbf{S}_i \times \mathbf{S}_j$ giving rise to two perpendicular spins. The DM model was historically proposed to explain the phenomenon of weak ferromagnetism observed in Mn compounds [11]. However, the DM interaction is at present known in various materials, in particular at the interface of a multilayer [12–16]. Although in this review we do not show the effect of the DM interaction in a magnetic field which gives rise to topological spin swirls known as skyrmions, we should mention a few of the important works given in Refs. [17–21]. Skyrmions are among the most studied subjects at the time being due to their potential applications in spin electronics [22]. We refer the reader to the rich biography given in our recent papers in Refs. [23,24].

Since this paper is a review on the method and the results of published works on SW in non-collinear GS spin configurations, it is important to recall the method and show main results of some typical cases. We would like to emphasize that, on the GF technique, to our knowledge there are no authors other than us working with this method. Therefore, the works mentioned in the references of this paper are our works published over the last 25 years. The aim of this review is two-fold. First we show technical details of the GF method by selecting a number of subjects which are of current interest in research: helimagnets, systems including a DM interaction, and surface effects in thin films. Second, we show that these systems possess many striking features due to the frustration.

This paper is organized as follows. In Section 2, we express the Hamiltonian in a general non-collinear GS and define the local system of spin coordinates. Here, we also present the calculation of the GS and the foundation of the self-consistent GF technique and the calculation of the SW dispersion relation and layer magnetizations at arbitrary temperature (T). We show in Section 3 the numerical results obtained from the GF. Section 4 shows interesting examples using various kinds of interaction including the DM interaction in a variety of systems from two dimensions, to thin films and superlattices. Section 5 treats a case where the DM interaction competes with the antiferromagnetic interaction in the frustrated antiferromagnetic triangular lattice. Section 6 presents the surface effect in a thin film where its surface is frustrated. Concluding remarks are given in Section 7.

2. Hamiltonian of a Chiral Magnet—Local Coordinates

Chiral order in helimagnets has been subject of recent extensive investigations. In Ref. [25], the surface structure of thin helimagnetic films has been studied. In Ref. [26] exotic spin configurations in ultrathin helimagnetic holmium films have been investigated. In Refs. [27,28]

chiral structure and spin reorientations in MnSi thin films have been theoretically studied. In these works, the chiral structures have been considered at $T = 0$, but not the SW even at $T = 0$. The main difficulty was due to the non-collinear, non-uniform spin configurations. We have shown that this was possible using the GFs generalized for such spin configurations given in Ref. [7].

To demonstrate the method, let us follow Ref. [29]: we consider the body-centered tetragonal (bct) lattice with Heisenberg spins. Each spin interacts with its nearest neighbors (NN) via the exchange constant J_1 and with its next NN (NNN) on the c-direction via the exchange J_2 (see Figure 1).

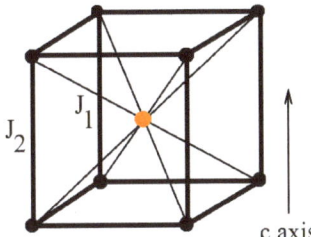

Figure 1. Interactions J_1 (thin solid lines) between nearest neighbors and J_2 between next nearest neighbors along the c axis in a bct lattice.

We consider the simplest model of a helimagnet, given by the following Hamiltonian

$$\mathcal{H} = -J_1 \sum_{i,j} \mathbf{S}_i \cdot \mathbf{S}_j - J_2 \sum_{i,k} \mathbf{S}_i \cdot \mathbf{S}_k \qquad (1)$$

where $\mathbf{S}_i$ is a quantum spin of magnitude 1/2, the first sum is performed over all NN pairs, and the second sum over pairs on the c-axis (cf. Figure 1).

In the case of an infinite crystal, the chiral state occurs when J_1 is ferromagnetic and J_2 is antiferromagnetic and $|J_2|/J_1$ is larger than a critical value, as will be shown below.

Let us suppose that the energy of a spin E_C in a chiral configuration when the angle between two NN spin in the neighboring planes is θ, one has (omitting the factor S^2)

$$E = -8J_1 \cos\theta - 2J_2 \cos(2\theta) \qquad (2)$$

The lowest-energy state corresponds to

$$\begin{aligned}\frac{dE}{d\theta} &= 0 \\ \to 8J_1 \sin\theta + 4 \sin(2\theta) &= 0 \\ 8J_1 \sin\theta(1 + \frac{J_2}{J_1}\cos\theta) &= 0\end{aligned} \qquad (3)$$

There are two solutions, $\sin\theta = 0$ and $\cos\theta = -\frac{J_1}{J_2}$. The first solution corresponds to the ferromagnetic state, and the second solution exists if $-\frac{J_1}{J_2} \leq 1$ which corresponds to the chiral state.

For a thin helimagnetic film, the angle between spins in adjacent layers varies due to the surface. We can use the method of energy minimization for each layer, then we have a set of coupled equations to solve (see Ref. [29]). Figure 2 displays an example of the angle distribution across the film thickness N_z.

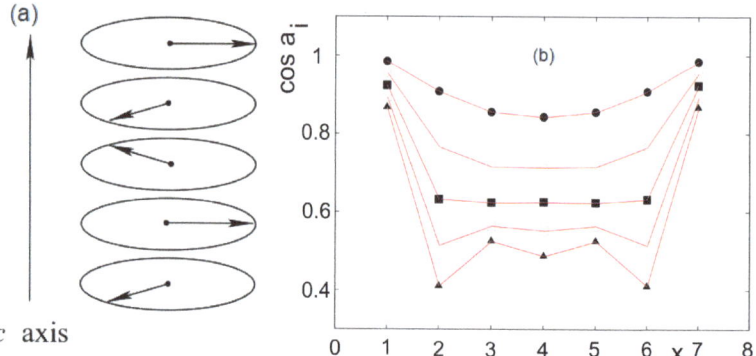

Figure 2. (a) Chiral structure along the c-axis for an infinite crystal, in the case $\theta = 2\pi/3$, namely $J_2/J_1 = -2$; (b) Cosine of $\alpha_1 = \theta_1 - \theta_2, \ldots, \alpha_7 = \theta_7 - \theta_8$ across the film for several values $J_2/J_1 = -1.2, -1.4, -1.6, -1.8, -2$ (from top) with $N_z = 8$: a_i stands for $\theta_i - \theta_{i+1}$ and x indicates the film layer i where the angle a_i with the layer $(i+1)$ is shown. See text for comments.

In order to calculate the SW spectrum for systems of non-collinear spin configurations, let us emphasize that the commutation relations between spin operators are established when the spin lies on its quantization z. In the non-collinear cases, each spin has its own quantization axis. It is therefore important to choose a quantization axis for each spin. We have to use the system of local coordinates defined as follows. In the Hamiltonian, the spins are coupled two by two. Consider a pair $\mathbf{S}_i$ and $\mathbf{S}_j$. As seen above, in the general case these spins make an angle $\theta_{i,j} = \theta_j - \theta_i$ determined by the competing interactions in the systems. For quantum spins, in the course of calculation we need to use the commutation relations between the spin operators S^z, S^+, S^-. As said above, these commutation relations are derived from the assumption that the spin lies on its quantization axis z. We show in Figure 3 the local coordinates assigned to spin $\mathbf{S}_i$ and $\mathbf{S}_j$. We write

$$\mathbf{S}_i = S_i^x \hat{\xi}_i + S_i^y \hat{\eta}_i + S_i^z \hat{\zeta}_i \tag{4}$$

$$\mathbf{S}_j = S_j^x \hat{\xi}_j + S_j^y \hat{\eta}_j + S_j^z \hat{\zeta}_j \tag{5}$$

Expressing the axes of $\mathbf{S}_j$ in the frame of $\mathbf{S}_i$ one has

$$\hat{\zeta}_j = \cos\theta_{i,j} \hat{\zeta}_i + \sin\theta_{i,j} \hat{\xi}_i \tag{6}$$

$$\hat{\xi}_j = -\sin\theta_{i,j} \hat{\zeta}_i + \cos\theta_{i,j} \hat{\xi}_i \tag{7}$$

$$\hat{\eta}_j = \hat{\eta}_i \tag{8}$$

so that

$$\mathbf{S}_j = S_j^x(-\sin\theta_{i,j} \hat{\zeta}_i + \cos\theta_{i,j} \hat{\xi}_i)$$
$$+ S_j^y \hat{\eta}_i + S_j^z(\cos\theta_{i,j} \hat{\zeta}_i + \sin\theta_{i,j} \hat{\xi}_i) \tag{9}$$

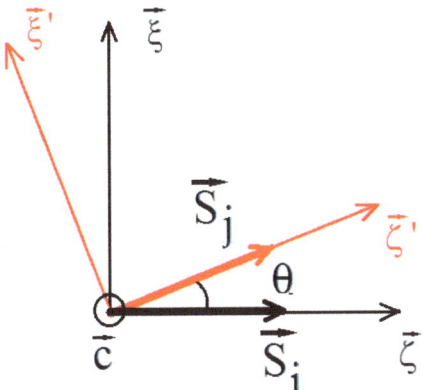

Figure 3. Spin $\mathbf{S}_i$ lies along the $\vec{\zeta}$ axis (its quantization axis), while spin $\mathbf{S}_j$ lies along its quantization axis $\vec{\zeta}'$ which makes an angle θ with the $\vec{\zeta}$ axis. The axes $\vec{\xi}$ and $\vec{\xi}'$ are perpendicular respectively to $\vec{\zeta}$ and $\vec{\zeta}'$. The perpendicular axes $\hat{\eta}_i$ and $\hat{\eta}_j$ coincide with the $\vec{c}$ axis, perpendicular to the basal plane of the bct lattice.

Using Equation (9) to express $\mathbf{S}_j$ in the $(\hat{\xi}_i, \hat{\eta}_i, \hat{\zeta}_i)$ coordinates, we calculate $\mathbf{S}_i \cdot \mathbf{S}_j$, we get the following Hamiltonian from (28):

$$\begin{aligned}
\mathcal{H}_e &= -\sum_{<i,j>} J_{i,j} \Big\{ \frac{1}{4}(\cos\theta_{i,j} - 1)\left(S_i^+ S_j^+ + S_i^- S_j^-\right) \\
&\quad + \frac{1}{4}(\cos\theta_{i,j} + 1)\left(S_i^+ S_j^- + S_i^- S_j^+\right) \\
&\quad + \frac{1}{2}\sin\theta_{i,j}(S_i^+ + S_i^-)S_j^z - \frac{1}{2}\sin\theta_{i,j} S_i^z \left(S_j^+ + S_j^-\right) \\
&\quad + \cos\theta_{i,j} S_i^z S_j^z \Big\}
\end{aligned} \quad (10)$$

This explicit Hamiltonian in terms of the angle between two NN spins is common for a non-collinear spin configuration due to exchange interactions $J_{i,j}$. For other types of interactions such as the DM interaction, the explicit Hamiltonian in terms of the angle will be different as shown in Section 4.

We define the following GFs for the above Hamiltonian:

$$\begin{aligned}
G_{i,j}(t,t') &= <<S_i^+(t); S_j^-(t')>> \\
&= -i\theta(t-t')<\left[S_i^+(t), S_j^-(t')\right]>
\end{aligned} \quad (11)$$

$$\begin{aligned}
F_{i,j}(t,t') &= <<S_i^-(t); S_j^-(t')>> \\
&= -i\theta(t-t')<\left[S_i^-(t), S_j^-(t')\right]>
\end{aligned} \quad (12)$$

Writing their equations of motion we have

$$\begin{aligned}
i\hbar \frac{d}{dt} G_{i,j}(t,t') &= \left\langle \left[S_i^+(t), S_j^-(t')\right] \right\rangle \delta(t-t') \\
&\quad - \left\langle\left\langle [\mathcal{H}, S_i^+(t)]; S_j^-(t')\right\rangle\right\rangle,
\end{aligned} \quad (13)$$

$$\begin{aligned}
i\hbar \frac{d}{dt} F_{i,j}(t,t') &= \left\langle \left[S_i^-(t), S_j^-(t')\right] \right\rangle \delta(t-t') \\
&\quad - \left\langle\left\langle [\mathcal{H}, S_i^-(t)]; S_j^-(t')\right\rangle\right\rangle,
\end{aligned} \quad (14)$$

where

$$S_j^\pm = S_j^x \hat{\xi}_j \pm i S_j^y \hat{\eta}_j$$
$$\left[S_j^+, S_l^-\right] = 2 S_j^z \delta_{j,l}$$
$$\left[S_j^z, S_l^\pm\right] = \pm S_j^\pm \delta_{j,l}$$

Note that the equation of motion of the G Green's function generates the F Green's functions, and vice-versa. Performing the commutators in Equations (13) and (14), and using the Tyablikov approximation [30] for higher-order GFs, for instance $<< S_{i'}^z S_i^+(t); S_j^-(t') >> \simeq < S_{i'}^z ><< S_i^+(t); S_j^-(t') >>$ etc., we obtain

$$\begin{aligned} i\hbar \frac{dG_{i,j}(t,t')}{dt} =\ & 2<S_i^z>\delta_{i,j}\delta(t-t') \\ & - \sum_{i'} J_{i,i'}[<S_i^z>(\cos\theta_{i,i'}-1) \times \\ & \times\ F_{i',j}(t,t') \\ & +\ <S_i^z>(\cos\theta_{i,i'}+1)G_{i',j}(t,t') \\ & -\ 2<S_{i'}^z>\cos\theta_{i,i'}G_{i,j}(t,t')] \\ & +\ 2\sum_{i'} I_{i,i'}<S_{i'}^z>\cos\theta_{i,i'}G_{i,j}(t,t') \end{aligned} \quad (15)$$

$$\begin{aligned} i\hbar \frac{dF_{i,j}(t,t')}{dt} =\ & \sum_{i'} J_{i,i'}[<S_i^z>(\cos\theta_{i,i'}-1) \times \\ & \times\ G_{i',j}(t,t') \\ & +\ <S_i^z>(\cos\theta_{i,i'}+1)F_{i',j}(t,t') \\ & -\ 2<S_{i'}^z>\cos\theta_{i,i'}F_{i,j}(t,t')] \\ & -\ 2\sum_{i'} I_{i,i'}<S_{i'}^z>\cos\theta_{i,i'}F_{i,j}(t,t') \end{aligned} \quad (16)$$

Note that the Tyablikov decoupling scheme is equivalent to the so-called "random-phase-approximation" (RPA).

For the sake of clarity, we write separately the NN and NNN sums, we have

$$\begin{aligned} i\hbar \frac{dG_{i,j}(t,t')}{dt} =\ & 2<S_i^z>\delta_{i,j}\delta(t-t') \\ & - \sum_{k'\in NN} J_{i,k'}[<S_i^z>(\cos\theta_{i,k'}-1) \times \\ & \times\ F_{k',j}(t,t') \\ & +\ <S_i^z>(\cos\theta_{i,k'}+1)G_{k',j}(t,t') \\ & -\ 2<S_{k'}^z>\cos\theta_{i,k'}G_{i,j}(t,t')] \\ & +\ 2\sum_{k'\in NN} I_{i,k'}<S_{k'}^z>\cos\theta_{i,k'}G_{i,j}(t,t') \\ & - \sum_{i'\in NNN} J_{i,i'}[<S_i^z>(\cos\theta_{i,i'}-1) \times \\ & \times\ F_{i',j}(t,t') \\ & +\ <S_i^z>(\cos\theta_{i,i'}+1)G_{i',j}(t,t') \\ & -\ 2<S_{i'}^z>\cos\theta_{i,i'}G_{i,j}(t,t')] \end{aligned} \quad (17)$$

$$\begin{aligned}
i\hbar \frac{dF_{k,j}(t,t')}{dt} &= \sum_{i'\in NN} J_{k,i'}[<S_k^z>(\cos\theta_{k,i'}-1)\times \\
&\quad \times G_{i',j}(t,t') \\
&\quad + <S_k^z>(\cos\theta_{k,i'}+1)F_{i',j}(t,t') \\
&\quad - 2<S_{i'}^z>\cos\theta_{k,i'}F_{k,j}(t,t')] \\
&\quad - 2\sum_{i'\in NN} I_{k,i'}<S_{i'}^z>\cos\theta_{k,i'}F_{k,j}(t,t') \\
&\quad + \sum_{k'\in NNN} J_{k,k'}[<S_k^z>(\cos\theta_{k,k'}-1)\times \\
&\quad \times G_{k',j}(t,t') \\
&\quad + <S_k^z>(\cos\theta_{k,k'}+1)F_{k',j}(t,t') \\
&\quad - 2<S_{k'}^z>\cos\theta_{k,k'}F_{k,j}(t,t')]
\end{aligned} \qquad (18)$$

For simplicity, we suppose in the following $J_{k,k'}$ are all equal to J_1 for NN interactions and to J_2 for NNN interactions. $I_{k,k'}$ is taken to be I_1 for NN pairs. In addition, in the film coordinates defined above, we denote the Cartesian components of the spin position $\mathbf{R}_i$ by three indices (ℓ_i, m_i, n_i) in three directions x, y and z.

Since there is the translation invariance in the xy plane, the in-plane Fourier transforms of the above equations in the xy plane are

$$\begin{aligned}
G_{i,j}(t,t') &= \frac{1}{\Delta}\int\int_{BZ} d\mathbf{k}_{xy}\frac{1}{2\pi}\int_{-\infty}^{+\infty} d\omega e^{-i\omega(t-t')} \\
&\quad \times g_{n_i,n_j}(\omega,\mathbf{k}_{xy})e^{i\mathbf{k}_{xy}\cdot(\mathbf{R}_i-\mathbf{R}_j)},
\end{aligned} \qquad (19)$$

$$\begin{aligned}
F_{k,j}(t,t') &= \frac{1}{\Delta}\int\int_{BZ} d\mathbf{k}_{xy}\frac{1}{2\pi}\int_{-\infty}^{+\infty} d\omega e^{-i\omega(t-t')} \\
&\quad \times f_{n_k,n_j}(\omega,\mathbf{k}_{xy})e^{i\mathbf{k}_{xy}\cdot(\mathbf{R}_k-\mathbf{R}_j)},
\end{aligned} \qquad (20)$$

where ω is the SW frequency, $\mathbf{k}_{xy}$ the wave-vector parallel to xy planes and $\mathbf{R}_i$ the position of $\mathbf{S}_i$. n_i, n_j and n_k denote the z-components of the sites $\mathbf{R}_i$, $\mathbf{R}_j$ and $\mathbf{R}_k$. The integral over $\mathbf{k}_{xy}$ is performed in the first Brillouin zone (BZ) whose surface is Δ in the xy reciprocal plane. $n_i = 1$ denotes the surface layer, $n_i = 2$ the second layer etc.

In the 3D case, the Fourier transformation of Equations (17) and (18) in the three (x, y, z) directions yields the SW spectrum in the absence of anisotropy:

$$\hbar\omega = \pm\sqrt{A^2 - B^2} \qquad (21)$$

where

$$\begin{aligned}
A &= J_1\langle S^z\rangle[\cos\theta + 1]Z\gamma + 2ZJ_1\langle S^z\rangle\cos\theta \\
&\quad + J_2\langle S^z\rangle[\cos(2\theta) + 1]Z_c\cos(k_z a) \\
&\quad + 2Z_c J_2\langle S^z\rangle\cos(2\theta) \\
B &= J_1\langle S^z\rangle(\cos\theta - 1)Z\gamma \\
&\quad + J_2\langle S^z\rangle[\cos(2\theta) - 1]Z_c\cos(k_z a)
\end{aligned}$$

where $Z = 8$ is the NN coordination number, $Z_c = 2$ the NNN number on the c-axis and $\gamma = \cos(k_x a/2)\cos(k_y a/2)\cos(k_z a/2)$ where a is the lattice constant taken the same in three directions. Note that $\hbar\omega$ is zero when $A = \pm B$. This is realized at two points as expected in helimagnets: $k_x = k_y = k_z = 0$ ($\gamma = 1$) and $k_z = 2\theta$ along the helical axis. It is interesting to note that we recover the SW dispersion relation of ferromagnets (antiferromagnets) [2] with NN interaction only by putting $\cos\theta = 1$ (-1) in the above coefficients.

In the case of a thin film, the in-plane Fourier transformation yields the following matrix equation
$$\mathbf{M}(\omega)\mathbf{h} = \mathbf{u}, \tag{22}$$
where $\mathbf{h}$ and $\mathbf{u}$ are given by
$$\mathbf{h} = \begin{pmatrix} g_{1,n'} \\ f_{1,n'} \\ \vdots \\ g_{n,n'} \\ f_{n,n'} \\ \vdots \\ g_{N_z,n'} \\ f_{N_z,n'} \end{pmatrix}, \mathbf{u} = \begin{pmatrix} 2\langle S_1^z \rangle \delta_{1,n'} \\ 0 \\ \vdots \\ 2\langle S_{N_z}^z \rangle \delta_{N_z,n'} \\ 0 \end{pmatrix}, \tag{23}$$

We take $\hbar = 1$ hereafter. Note that $\mathbf{M}(\omega)$ is a $(2N_z \times 2N_z)$ matrix given by Equation (24) where

$$\mathbf{M}(\omega) = \begin{pmatrix} \omega + A_1 & 0 & B_1^+ & C_1^+ & D_1^+ & E_1^+ & 0 & 0 & 0 & 0 & 0 & 0 \\ 0 & \omega - A_1 & -C_1^+ & -B_1^+ & -E_1^+ & -D_1^+ & 0 & 0 & 0 & 0 & 0 & 0 \\ \cdots & \cdots & \cdots & \cdots & \cdots & \cdots & \cdots & \cdots & \cdots & \cdots & \cdots & \cdots \\ \cdots & \cdots & D_n^- & E_n^- & B_n^- & C_n^- & \omega + A_n & 0 & B_n^+ & C_n^+ & D_n^+ & E_n^+ & \cdots \\ \cdots & \cdots & -E_n^- & -D_n^- & -C_n^- & -B_n^- & 0 & \omega - A_n & -C_n^+ & -B_n^+ & -E_n^+ & -D_n^+ & \cdots \\ \cdots & \cdots & \cdots & \cdots & \cdots & \cdots & \cdots & \cdots & \cdots & \cdots & \cdots & \cdots \\ 0 & 0 & 0 & 0 & 0 & 0 & D_{N_z}^- & E_{N_z}^- & B_{N_z}^- & C_{N_z}^- & \omega + A_{N_z} & 0 \\ 0 & 0 & 0 & 0 & 0 & 0 & -E_{N_z}^- & -D_{N_z}^- & -C_{N_z}^- & -B_{N_z}^- & 0 & \omega - A_{N_z} \end{pmatrix} \tag{24}$$

$$\begin{aligned}
A_n &= -8J_1(1+d)\left[\langle S_{n+1}^z \rangle \cos\theta_{n,n+1} \right. \\
&\quad \left. + \langle S_{n-1}^z \rangle \cos\theta_{n,n-1}\right] \\
&\quad - 2J_2\left[\langle S_{n+2}^z \rangle \cos\theta_{n,n+2} \right. \\
&\quad \left. + \langle S_{n-2}^z \rangle \cos\theta_{n,n-2}\right] \\
B_n^\pm &= 4J_1\langle S_n^z \rangle(\cos\theta_{n,n\pm 1} + 1)\gamma \\
C_n^\pm &= 4J_1\langle S_n^z \rangle(\cos\theta_{n,n\pm 1} - 1)\gamma \\
E_n^\pm &= J_2\langle S_n^z \rangle(\cos\theta_{n,n\pm 2} - 1) \\
D_n^\pm &= J_2\langle S_n^z \rangle(\cos\theta_{n,n\pm 2} + 1)
\end{aligned}$$

where we recall that n denotes the layer number, namely $1, 2, \ldots, N_z$ and $d = I_1/J_1$. Note that $\theta_{n,n\pm 1}$ denotes the angle between a spin in the layer n and its NN spins in adjacent layers $n \pm 1$ etc. and $\gamma = \cos\left(\frac{k_x a}{2}\right)\cos\left(\frac{k_y a}{2}\right)$.

In order to obtain the SW frequency ω, we solve the secular equation $\det|\mathbf{M}| = 0$ for each given (k_x, k_y). Since the linear dimension of the square matrix is $2N_z$, we obtain $2N_z$ eigen-values of ω, half positive and half negative, corresponding to two opposite spin precessions as in antiferromagnets. These values depend on the input values $<S_n^z>$ ($n = 1, \ldots, N_z$). Thus, we have to solve the secular equation by iteration until the convergence of input and output values. Note that, even at $T = 0$, $<S_n^z>$ are not equal to $1/2$ due to the zero-point spin contraction [31]. In addition, because of the film surfaces, the spin contractions are not uniform.

The solution for $g_{n,n}$ can be calculated (see Ref. [29]). The spectral theorem [1] can be used to obtain, after a somewhat lengthy algebra (see [29]),:

$$\langle S_n^z \rangle = \frac{1}{2} - \frac{1}{\Delta} \int \int dk_x dk_y \sum_{i=1}^{2N_z} \frac{D_{2n-1}(\omega_i)}{e^{\beta \omega_i} - 1} \quad (25)$$

where $n = 1, \ldots, N_z$, and

$$D_{2n-1}(\omega_i(\mathbf{k}_{xy})) = \frac{|\mathbf{M}|_{2n-1}(\omega_i(\mathbf{k}_{xy}))}{\prod_{j \neq i}[\omega_j(\mathbf{k}_{xy}) - \omega_i(\mathbf{k}_{xy})]}. \quad (26)$$

As $<S_n^z>$ depend each other in $\omega_i (i = 1, \ldots, 2N_z)$, their solutions should be obtained by iteration at a given temperature T. In the particular case where $T = 0$ one has

$$\langle S_n^z \rangle (T = 0) = \frac{1}{2} + \frac{1}{\Delta} \int \int dk_x dk_y \sum_{i=1}^{N_z} D_{2n-1}(\omega_i(\mathbf{k}_{xy})) \quad (27)$$

Note that the sum is performed over N_z negative ω_i since positive ω_i yields the zero Bose–Einstein factor at $T = 0$).

The transition temperature T_c can be calculated self-consistently when all $< S_n^z >$ tend to zero.

We show in the following section, the numerical results using the above formulas.

3. Results for Helimagnets Obtained from the Green's Function Technique

We use the ferromagnetic interaction between NN as unit, namely $J_1 = 1$. Take the helimagnetic case where J_2 is negative with $|J_2| > J_1$. We have determined above the spin configuration across the film for several values of $p = J_2/J_1$. Replacing the angles $\theta_{n,n\pm 1}$ and $\theta_{n,n\pm 2}$ in the matrix elements of $|\mathbf{M}|$, then calculating $\omega_i (i = 1, \ldots, 2N_z)$ for each $\mathbf{k}_{xy}$. For the iterative procedure, the reader is referred to Ref. [29]. The solution $\langle S_n^z \rangle (n = 1, \ldots, N_z)$ is obtained when the input and the output are equal with a desired precision P.

3.1. Spectrum

We calculate the SW spectrum as described above for each a given J_2/J_1. The SW spectrum depends on T. We show in Figure 4 the SW spectrum ω versus $k_x = k_y$ for an eight-layer film with $J_2/J_1 = -1.4$ at $T = 0.1$ and $T = 1.02$ (in units of $J_1/k_B = 1$). We observe that

(i) There are opposite-precession SW modes. Unlike ferromagnets, SW in antiferromagnets and non-collinear spin structures have opposite spin precessions [31]. The negative sign does not mean SW negative energy, but it indicates just the precession contrary to the trigonometric sense,

(ii) There are two degenerate acoustic "surface" branches one on each side. These degenerate "surface" modes stem from the symmetry of the two surfaces. These surface modes propagate parallel to the film surface but are damped when going to the bulk,

(iii) With increasing T, layer magnetizations decrease as seen hereafter, this reduces therefore the SW frequency (see Figure 4b),

(iv) Surface and bulk SW spectra have been observed by inelastic neutron scattering in collinear magnets (ferro- and antiferromagnetic films) [32,33]. However, such experiments have not been reported for helimagnetic thin films.

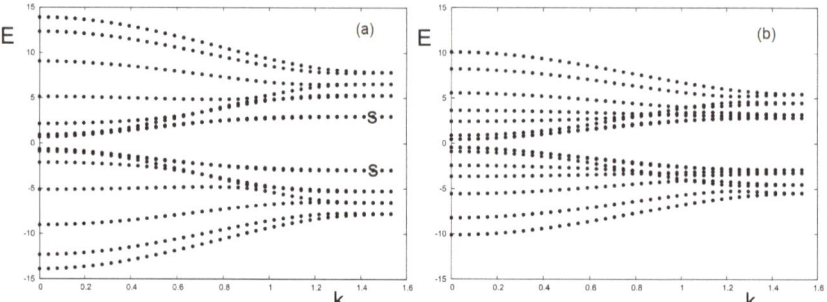

Figure 4. (a) Spectrum $E = \hbar\omega$ versus $k \equiv k_x = k_y$ for $J_2/J_1 = -1.4$ at $T = 0.1$ and (b) $T = 1.02$, for $N_z = 8$ and $d = 0.1$. The surface branches are indicated by s.

3.2. Zero-Point Spin Contraction and Transition Temperature

It is known that, in antiferromagnetic materials, quantum fluctuations cause a contraction of the spin length, namely the spin length is shorter than the spin magnitude, at $T = 0$ [31]. We demonstrate here that a spin with a stronger antiferromagnetic interaction has a stronger contraction: spins in the first and in the second layers have only one antiferromagnetic NNN on the c-axis while interior spins have two NNN. The contraction at a given J_2/J_1 is thus expected to be stronger for interior spins. This is shown in Figure 5: with increasing $|J_2|/J_1$, i.e., the antiferromagnetic interaction becomes stronger, the contraction is stronger. Of course, there is no contraction when the system is ferromagnetic, namely when $J_2 \to -1$.

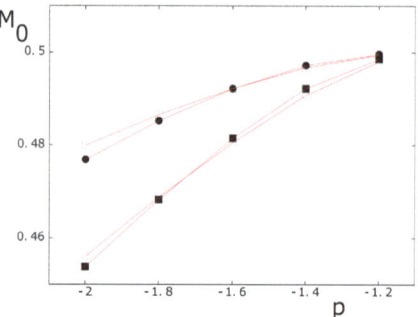

Figure 5. Spin lengths of the first four layers at $T = 0$ for several values of $p = J_2/J_1$ with $d = 0.1$, $N_z = 8$. As seen, all spins are contracted to values smaller than the spin magnitude $1/2$. Black circles, void circles, black squares and void squares are for first, second, third and fourth layers, respectively.

3.3. Layer Magnetizations

We show now the layer ordering in Figures 6 and 7 where $J_2/J_1 = -1.4$ and -2, respectively, in the case of $N_z = 8$. Consider first the case $J_2/J_1 = -1.4$. We note that the surface magnetization, having a large value at $T = 0$ as seen in Figure 5, crosses the interior layer magnetizations at $T \simeq 0.42$ to become much smaller than interior magnetizations at higher temperatures. This crossover phenomenon is due to the competition between quantum fluctuations, which dominate low-T behavior, and the low-lying surface SW modes which reduce the surface magnetization at higher T. Note that the second-layer magnetization makes also a crossover at $T \simeq 1.3$ which is more complicated to analyze. Similar crossovers have been observed in other quantum systems such as antiferromagnetic films [34] and superlattices [35].

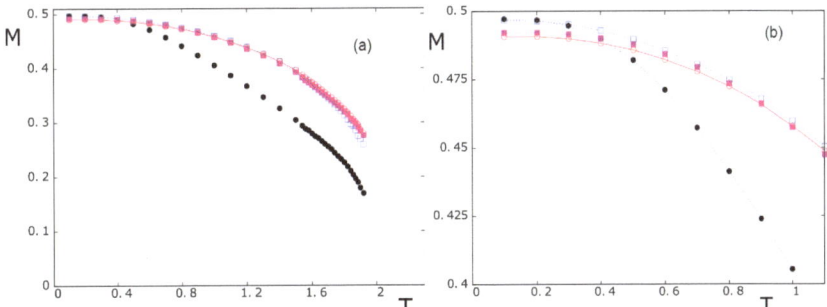

Figure 6. (**a**) Layer magnetizations as functions of T for $J_2/J_1 = -1.4$ with $d = 0.1$, $N_z = 8$, (**b**) Zoom of the region at low T to show crossover. Black circles, blue void squares, magenta squares and red void circles are for first, second, third and fourth layers, respectively. See text.

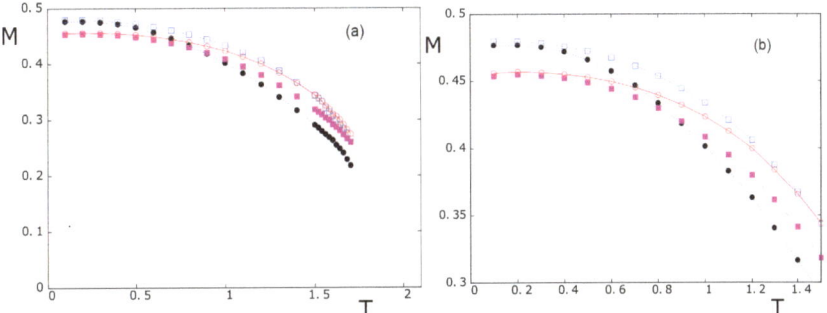

Figure 7. (**a**) Layer magnetizations as functions of T for $J_2/J_1 = -2$ with $d = 0.1$, $N_z = 8$, (**b**) Zoom of the region at low T to show crossover. Black circles, blue void squares, magenta squares and red void circles are for first, second, third and fourth layers, respectively. See text.

Similar remarks are also hold for $J_2/J_1 = -2$ shown in Figure 7.

Note that the results shown above have been calculated with an in-plane anisotropy interaction $d = 0.1$. Larger d yields stronger layer magnetizations and larger T_c.

To close this section on SW in helimagnetic bct thin films, we mention that the same investigation was done in the case of simple-cubic helimagnetic films where the surface spin reconstruction and the surface SW have been shown [36]. We have also studied the frustrated bct Heisenberg helimagnet in which the SW spectrum of the non-collinear spin configuration has been calculated [37].

4. Dzyaloshinskii–Moriya Interaction in Thin Films

Let us consider a thin film made of N square lattices stacked in the y direction perpendicular to the film surface. The results for this system have been published in Ref. [38]. Hereafter, we review some of these important results. The Hamiltonian is given by

$$\mathcal{H} = \mathcal{H}_e + \mathcal{H}_{DM} \tag{28}$$

$$\mathcal{H}_e = -\sum_{\langle i,j \rangle} J_{i,j} \mathbf{S}_i \cdot \mathbf{S}_j \tag{29}$$

$$\mathcal{H}_{DM} = \sum_{\langle i,j \rangle} \mathbf{D}_{i,j} \cdot \mathbf{S}_i \times \mathbf{S}_j \tag{30}$$

where $J_{i,j}$ and $\mathbf{D}_{i,j}$ are the exchange and DM interactions, respectively, between two quantum Heisenberg spins $\mathbf{S}_i$ and $\mathbf{S}_j$ of magnitude $S = 1/2$.

We suppose in this section the in-plane and inter-plane exchange interactions between NN are both ferromagnetic and denoted by J_1 and J_2, respectively. The DM interaction is defined only between NN in the plane for simplicity. The J term favors the collinear spin configuration while the DM term favors the perpendicular one. This will lead to a compromise where S_i makes an angle $\theta_{i,j}$ with its neighbor S_j. It is obvious that the quantization axes of S_i and S_j are different. Therefore, the transformation using the local coordinates, Equations (4)–(9), is necessary. Let us suppose that the vector $D_{i,j}$ is along the y axis, namely the $\hat{\eta}_i$ axis. We write

$$D_{i,j} = D e_{i,j} \hat{y}_i \tag{31}$$

where $e_{i,j} = +1(-1)$ if $j > i$ ($j < i$) for NN j on the $\hat{x}$ or $\hat{z}$ axis. One has by definition $e_{j,i} = -e_{i,j}$.

The easiest way to determine the GS is to minimize the local energy at each spin: taking a spin and calculating the local field acting on it from its neighbors. Then, we align the spin in its local-field direction to minimize its energy. Repeating this procedure for all spins, we say we realize one sweep. We have to make a sufficient number of sweeps to obtain the convergence with a desired precision (see details in Ref. [39]). This local energy minimization is called "the steepest descent method". We show in Figure 8 the configuration obtained for $D = -0.5$ using $J_1 = J_2 = 1$.

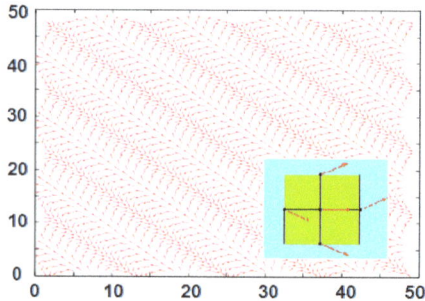

Figure 8. The ground state is a planar configuration on the xz plane. The figure shows the case where $\theta = \pi/6$ ($D = -0.577$), $J_1 = J_\perp = 1$ using the steepest descent method. The inset shows a zoom around a spin with its nearest neighbors.

We see that each spin has the same angle with its four NN in the plane (angle between NN in adjacent planes is zero). We demonstrate now the dependence of θ on J_1: the energy of the spin S_i is written as

$$E_i = -4J_1 S^2 \cos\theta - 2J_2 S^2 + 4DS^2 \sin\theta \tag{32}$$

where $\theta = |\theta_{i,j}|$ minimizing E_i with respect to θ one obtains

$$\frac{dE_i}{d\theta} = 0 \Rightarrow -\frac{D}{J_1} = \tan\theta \Rightarrow \theta = \arctan\left(-\frac{D}{J_1}\right) \tag{33}$$

The result is in agreement with that obtained by the steepest descent method. An example has been shown in Figure 8.

We rewrite the DM term of Equation (30) as

$$\begin{aligned}\mathbf{S}_i \times \mathbf{S}_j &= (-S_i^z S_j^y - S_i^y S_j^x \sin\theta_{i,j} + S_i^y S_j^z \cos\theta_{i,j})\hat{\zeta}_i \\ &+ (S_i^x S_j^x \sin\theta_{i,j} + S_i^z S_j^z \sin\theta_{i,j})\hat{\eta}_i \\ &+ (S_i^x S_j^y - S_i^y S_j^z \sin\theta_{i,j} - S_i^y S_j^x \cos\theta_{i,j})\hat{\zeta}_i\end{aligned} \tag{34}$$

From Equation (31), we obtain

$$\begin{aligned}
\mathcal{H}_{DM} &= \sum_{\langle i,j\rangle} \mathbf{D}_{i,j} \cdot \mathbf{S}_i \times \mathbf{S}_j \\
&= D\sum_{\langle i,j\rangle}(S_i^x S_j^x e_{i,j}\sin\theta_{i,j} + S_i^z S_j^z e_{i,j}\sin\theta_{i,j}) \\
&= \frac{D}{4}\sum_{\langle i,j\rangle}[(S_i^+ + S_i^-)(S_j^+ + S_j^-)e_{i,j}\sin\theta_{i,j} \\
&\quad + 4 S_i^z S_j^z e_{i,j}\sin\theta_{i,j}]
\end{aligned} \qquad (35)$$

where we have replaced S^x by $(S^+ + S^-)/2$. Note that $e_{i,j}\sin\theta_{i,j}$ is always positive since for a NN on the positive axis direction, $e_{i,j} = 1$ and $\sin\theta_{i,j} = \sin\theta$ where θ is positively defined, while for a NN on the negative axis direction, $e_{i,j} = -1$ and $\sin\theta_{i,j} = \sin(-\theta) = -\sin\theta$.

4.1. Formulation of the Green's Function Technique for the Dzyaloshinskii–Moriya System

Using the transformation into the local coordinates, Equations (4)–(9), one has

$$\begin{aligned}
\mathcal{H} = &-\sum_{\langle i,j\rangle} J_{i,j}\Big\{\frac{1}{4}(\cos\theta_{i,j} - 1)\left(S_i^+ S_j^+ + S_i^- S_j^-\right) \\
&+ \frac{1}{4}(\cos\theta_{i,j} + 1)\left(S_i^+ S_j^- + S_i^- S_j^+\right) \\
&+ \frac{1}{2}\sin\theta_{i,j}(S_i^+ + S_i^-)S_j^z - \frac{1}{2}\sin\theta_{i,j}S_i^z\left(S_j^+ + S_j^-\right) \\
&+ \cos\theta_{i,j} S_i^z S_j^z\Big\} \\
&+ \frac{D}{4}\sum_{\langle i,j\rangle}[(S_i^+ + S_i^-)(S_j^+ + S_j^-)e_{i,j}\sin\theta_{i,j} \\
&+ 4 S_i^z S_j^z e_{i,j}\sin\theta_{i,j}]
\end{aligned} \qquad (36)$$

Note that the quantization axes of the spins are in the xz planes as shown in Figure 3.

We emphasize that, while the sine terms of the DM Hamiltonian, Equation (35), remain after summing over the NN, the sine terms of $\mathcal{H}_e$, the 3rd line of Equation (36), are zero after summing over opposite NN because there is no $e_{i,j}$ term.

It is very important to emphasize again that the commutation relations between spin operators S^z and $S^{\pm}$ are valid when the spin lies on its local quantization axis. Therefore, it is necessary ro use the local coordinates for each spin.

In two dimensions (2D) there is no long-range order at non-zero T for isotropic spin models with short-range interaction [40]. Thin films have very small thickness, not far from 2D systems. Thus, in order to stabilize the ordering at very low T, we use a very small anisotropy interaction between between $\mathbf{S}_i$ and $\mathbf{S}_j$ as follows

$$\mathcal{H}_a = -\sum_{\langle i,j\rangle} I_{i,j} S_i^z S_j^z \cos\theta_{i,j} \qquad (37)$$

where $I_{i,j}(>0)$ is positive, small compared to J_1, and limited to NN in the xz plane. For simplicity, we suppose $I_{i,j} = I_1$ for all such NN pairs. As we will see below, the small value of I_1 does stabilize the SW spectrum when D becomes large. The Hamiltonian is finally given by

$$\mathcal{H} = \mathcal{H}_e + \mathcal{H}_{DM} + \mathcal{H}_a \qquad (38)$$

Using the two GF's in the real space given by Equations (11) and (12) and using the same method, we study the effect of the DM interaction. For the DM term, the commutation relations $[\mathcal{H}, S_i^{\pm}]$ lead to:

$$D \sum_l \sin\theta [\mp S_i^z(S_l^+ + S_l^-) + \pm 2 S_i^{\pm} S_l^z] \tag{39}$$

which gives rise, using the Tyablikov decoupling, to the following GF's:

$$\ll S_i^z S_l^{\pm} ; S_j^- \gg \simeq \; <S_i^z> \ll S_l^{\pm} ; S_j^- \gg \tag{40}$$

These functions are in fact the G and F functions. There are thus no new GF's generated by the equations of motion.

As in Section 2, the Fourier transforms in the xz plane $g_{n,n'}$ and $f_{n,n'}$ of the G and F lead to the matrix equation

$$\mathbf{M}(E)\mathbf{h} = \mathbf{u}, \tag{41}$$

$\mathbf{M}(E)$ being given by Equation (42) below

$$\begin{pmatrix} E+A_1 & B_1 & C_1 & 0 & 0 & 0 & 0 & 0 & 0 \\ -B_1 & E-A_1 & 0 & -C_1 & 0 & 0 & 0 & 0 & 0 \\ \cdots & \cdots & \cdots & \cdots & \cdots & \cdots & \cdots & \cdots & \cdots \\ \cdots & 0 & C_n & 0 & E+A_n & B_n & C_n & 0 & 0 \\ \cdots & 0 & 0 & -C_n & -B_n & E-A_n & 0 & -C_n & 0 \\ \cdots & \cdots & \cdots & \cdots & \cdots & \cdots & \cdots & \cdots & \cdots \\ 0 & 0 & 0 & 0 & 0 & C_N & 0 & E+A_N & B_N \\ 0 & 0 & 0 & 0 & 0 & 0 & -C_N & -B_N & E-A_N \end{pmatrix} \tag{42}$$

where $E = \hbar\omega$ is the SW energy and the matrix elements are given by

$$\begin{aligned}
A_n &= -J_1[8 <S_n^z> \cos\theta(1+d_n) \\
&\quad -4 <S_n^z> \gamma(\cos\theta+1)] \\
&\quad -2J_2(<S_{n-1}^z> + <S_{n+1}^z>) \\
&\quad -8D\sin\theta <S_n^z> \gamma \\
&\quad +8D\sin\theta <S_n^z> \\
B_n &= 4J_1 <S_n^z> \gamma(\cos\theta-1) \\
&\quad -8D\sin\theta <S_n^z> \gamma \\
C_n &= 2J_2 <S_n^z>
\end{aligned} \tag{43} \tag{44} \tag{45}$$

where $n = 1, 2, \ldots, N$ denotes the layer numbers, $d_n = I_1/J_1$, $\gamma = (\cos k_x a + \cos k_z a)/2$, k_x and k_z are the wave-vector components in the xz planes, a being the lattice constant. Remarks: (i) if $n = 1$ (surface layer) then there are no $n-1$ terms in the A_n, (ii) if $n = N$ then there are no $n+1$ terms in A_n.

For a thin film, the SW frequencies at a given wave vector $\vec{k} = (k_x, k_z)$ are obtained by diagonalizing (42).

The magnetization of the layer n at finite T is calculated as in the helimagnetic case shown in the previous section. The formula of the zero-point spin contraction is also presented there. The transition temperature T_c can be also calculated by the same method. Let us show in the following the results.

4.2. Results for 2D and 3D Cases

In the 2D case, one has only one layer. The matrix (42) is

$$\begin{aligned}
(E+A_n)g_{n,n'} + B_n f_{n,n'} &= 2<S_n^z>\delta(n,n') \\
-B_n g_{n,n'} + (E-A_n)f_{n,n'} &= 0
\end{aligned} \tag{46}$$

where A_n is given by (43) but without J_2 term for the 2D case. Coefficient B_n is given by (44) and $C_n = 0$. The SW frequencies are determined by the following secular equation

$$(E + A_n)(E - A_n) + B_n^2 = 0$$
$$\rightarrow E^2 - A_n^2 + B_n^2 = 0$$
$$\rightarrow E = \pm\sqrt{(A_n + B_n)(A_n - B_n)} \quad (47)$$

Several remarks are in order:

(i) when $\theta = 0$, the last three terms of A_n and B_n are zero: one recovers the ferromagnetic SW dispersion relation

$$E = 2ZJ_1 <S_n^z> (1 - \gamma) \quad (48)$$

where $Z = 4$ is the coordination number of the square lattice (taking $d_n = 0$),

(ii) when $\theta = \pi$, one has $A_n = 8J_1 <S_n^z>$, $B_n = -8J_1 <S_n^z> \gamma$. One recovers then the antiferromagnetic SW dispersion relation

$$E = 2ZJ_1 <S_n^z> \sqrt{1 - \gamma^2} \quad (49)$$

(iii) when there is a DM interaction, one has $0 < \cos\theta < 1$ ($0 < \theta < \pi/2$). If $d_n = 0$, the quantity in the square root of Equation (47) becomes negative at $\gamma = 1$ when θ is not zero. The SW spectrum is not stable at $k_x = k_y = 0$ because the energy is not real. The anisotropy d_n can remove this instability if it is larger than a threshold value d_c. We solve the equation $(A_n + B_n)(A_n - B_n) = 0$ to find d_c. In Figure 9 we show d_c versus θ. As seen, d_c increases from zero with increasing θ.

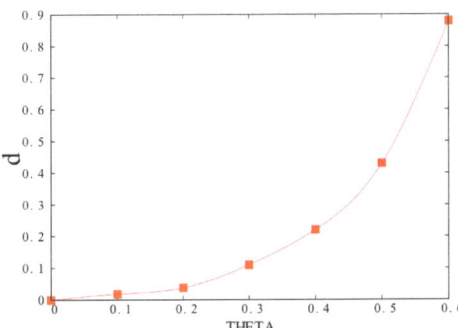

Figure 9. Value d_c at which $E = 0$ at $\gamma = 1$ ($\vec{k} = 0$) vs. θ (in radian). Above this value, E is real. See text for comments.

As we have anticipated, we need to include an anisotropy in order to allow for SW to be excited even at $T = 0$ and for a long-range ordering at non-zero T in 2D as seen below.

We show in Figure 10 the SW dispersion relation calculated from Equation (47) for $\theta = 0.2$ and 0.6 (radian). As seen, the spectrum is symmetric for positive and negative wave vectors. It is also symmetric for left and right precessions. One observes that for small θ, namely small D, $E(k)$ is proportional to k^2 at low k (see Figure 10a). This behavior is that in ferromagnets. For large θ, one observes that $E(k)$ becomes linear in k as seen in Figure 10b. This behavior is similar to that of antiferromagnets. Note that the change of behavior is progressive with increasing θ, we do not observe a sudden transition from k^2 to k behavior. This behavior is also observed in 3D and in thin films as well.

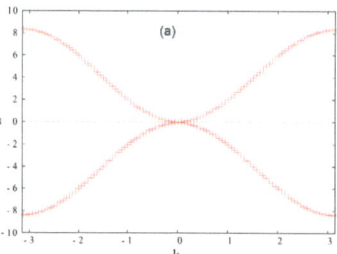 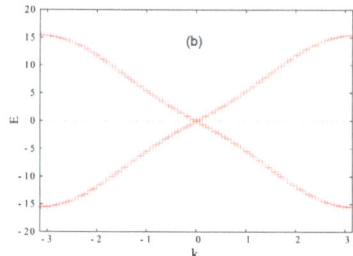

Figure 10. SW frequency $E(k)$ as a function of $k \equiv k_x = k_z$ in the case (**a**) $\theta = 0.2$ and (**b**) $\theta = 0.6$ in 2D. See text for detailed comments.

As said earlier, the inclusion of an anisotropy d permits a long-range ordering at $T \neq 0$ in 2D: Figure 11 displays the magnetization M ($\equiv < S^z >$) calculated by Equation (2) where in each case the limit value d_c has been used. We note that M depends strongly on θ: at high T the larger θ the stronger M. However, at $T = 0$ the spin length is smaller for larger θ due to the zero-point spin contraction [31] calculated by Equation (27). As a consequence there is a cross-over of layer magnetizations at low T as shown in Figure 11b. The spin length at $T = 0$ is shown in Figure 12 for several θ.

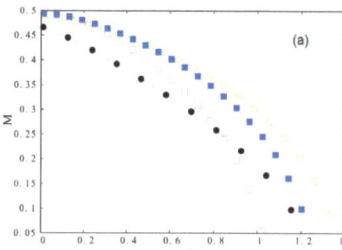 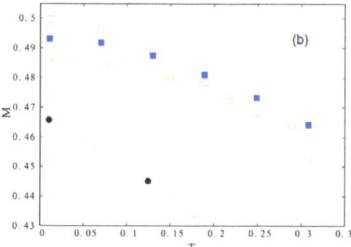

Figure 11. (**a**) Magnetization M as a function of T for the 2D case with $\theta = 0.1, \theta = 0.3, \theta = 0.4, \theta = 0.6$ (void magenta squares, blue filled squares, green void circles and filled black circles, respectively), (**b**) Cross-over of magnetizations is enlarged at low T. See text for comments.

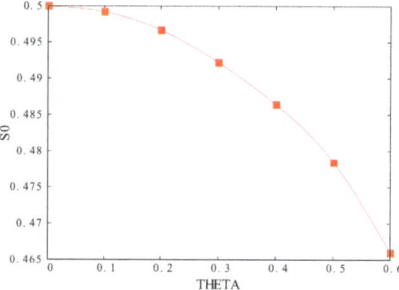

Figure 12. Spin length at $T = 0$ for the 2D case as a function of θ (radian).

We now consider the 3D case. The crystal is infinite in three direction. The Fourier transform in the y direction, namely $g_{n\pm1} = g_n e^{\pm ik_y a}$ and $f_{n\pm1} = f_n e^{\pm ik_y a}$ reduces the matrix (23) to two coupled equations of g and f functions. One has

$$\begin{aligned}(E + A')g + Bf &= 2 <S^z> \\ -Bg + (E - A')f &= 0\end{aligned} \quad (50)$$

where

$$\begin{aligned}A' &= -J_1[8 <S^z> \cos\theta(1+d) \\ &\quad -4 <S^z> \gamma(\cos\theta + 1)] \\ &\quad -4J_2 <S^z> \\ &\quad +4J_2 <S^z> \cos(k_y a) \\ &\quad -8D\sin\theta <S^z> \gamma \\ &\quad +8D\sin\theta <S^z>\end{aligned} \quad (51)$$

$$\begin{aligned}B &= 4J_1 <S^z> \gamma(\cos\theta - 1) \\ &\quad -8D\sin\theta <S^z> \gamma\end{aligned} \quad (52)$$

The spectrum is given by

$$E = \pm\sqrt{(A' + B)(A' - B)} \quad (53)$$

In the ferromagnetic case, $\cos\theta = 1$, thus $B = 0$. Arranging the Fourier transforms in three directions, one gets the 3D ferromagnetic dispersion relation $E = 2Z <S^z> (1 - \gamma^2)$ where $\gamma = [\cos(k_x a) + \cos(k_y a) + \cos(k_z a)]/3$ and $Z = 6$, coordination number of the simple cubic lattice.

As in the 2D case, we find a threshold value d_c for which is the same for a given θ. This is rather obvious because the DM interaction operates in the plane making an angle θ between spins in the plane, therefore its effects act on SW in each plane, not in the y direction perpendicular to the "DM planes". Using Equation (53), we calculate the 3D spectrum displayed in Figure 13 for a small and a large value of θ. As in the 2D case, we observe $E \propto k$ when $k \to 0$ for large θ. The main properties of the system are thus governed by the in-plane DM interaction.

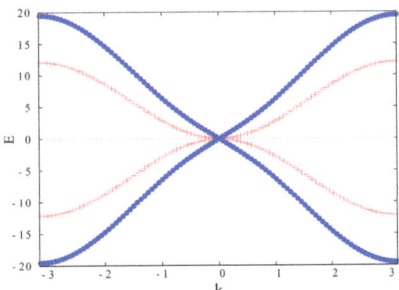

Figure 13. Spin-wave spectrum $E(k)$ versus $k \equiv k_x = k_z$ for $\theta = 0.1$ (red crosses) and $\theta = 0.6$ (blue circles) in three dimensions. Note the linear-k behavior at low k for the large value of θ. See text for comments.

Figure 14 displays the magnetization M versus T for several values of θ. As in the 2D case, when the DM interaction is included, the spins undergo a zero-point contraction which increases with increasing θ. The competition between quantum fluctuations at $T = 0$

and thermal effects at high T gives rise to magnetization cross-over shown in Figure 14b. The spin length at $T = 0$ vs. θ is shown in the inset of Figure 14b. Comparing these results to those of the 2D case, we see that the spin contraction in 2D is stronger than in 3D. This is physically expected because quantum fluctuations are stronger at lower dimensions.

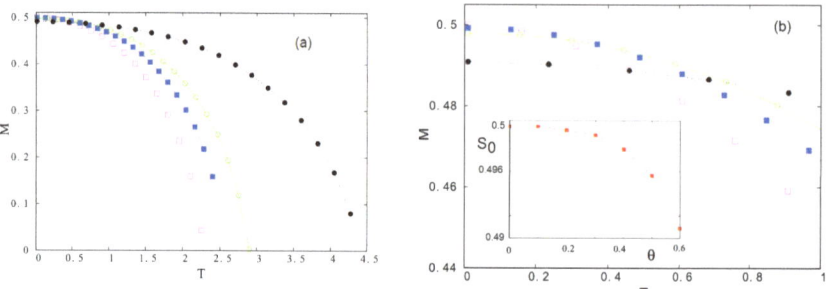

Figure 14. (a) Magnetization M versus temperature T for a 3D crystal $\theta = 0.1$ (radian), $\theta = 0.3$, $\theta = 0.4$, $\theta = 0.6$ (void magenta squares, blue filled squares, green void circles and filled black circles, respectively), (b) Zoom to show the cross-over of magnetizations at low T for different θ, inset shows S_0 versus θ. See text for comments.

4.3. The Case of a Thin Film

As in the 2D and 3D cases, in the case of a thin film it is necessary to use a value for d_n larger or equal to d_c given in Figure 9 to stabilize the SW at long wave-length. Note that for thin films with more than one layer, the value of d_c calculated for the 2D case remains valid.

Figure 15 displays the SW spectrum of a film of eight layers with $J_1 = J_2 = 1$ for a small and a large θ. As in the previous cases, E is proportional to k for large θ (cf. Figure 15b) but only for the first mode. The higher modes are proportional to k^2.

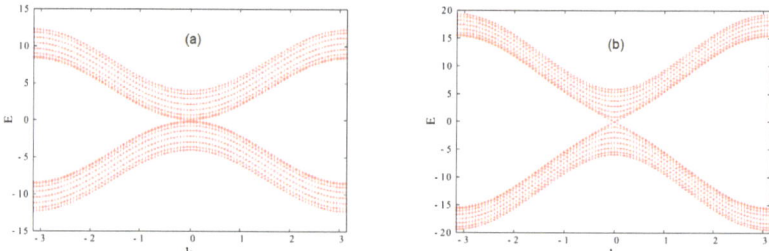

Figure 15. Spin-wave spectrum $E(k)$ versus $k \equiv k_x = k_z$ for a thin film of eight layers: (a) $\theta = 0.2$ (in radian) (b) $\theta = 0.6$, using d_c for each case. Positive and negative branches correspond to right and left precessions. Note the linear-k behavior at low k. See text for comments.

Figure 16 shows the layer magnetizations of the first four layers in a 8-layer film (the other half is symmetric) for two values of θ. One observes that the surface magnetization is smaller than the magnetizations of other interior layers. This is due to the lack of neighbors for surface spins [2].

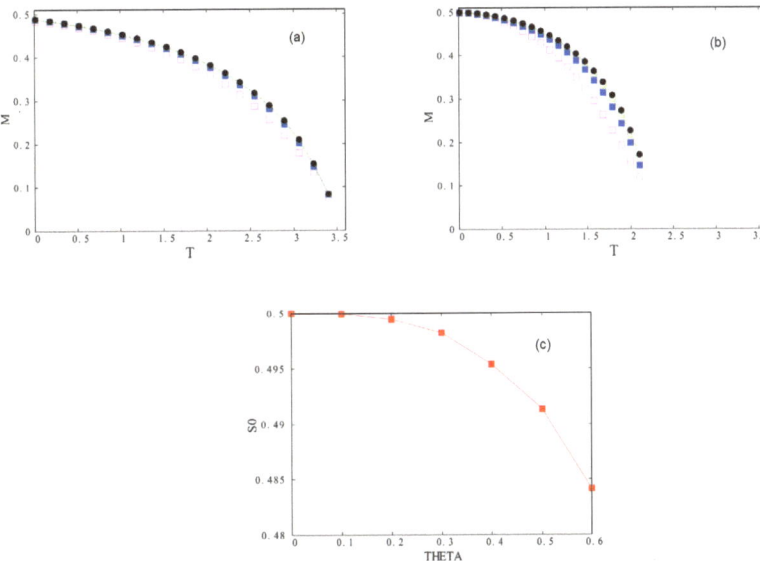

Figure 16. Layer magnetizations M versus temperature T for a film with $N = 8$: (**a**) $\theta = 0.6$ (radian), (**b**) $\theta = 0.2$, (**c**) S_0 versus θ.

The spin contraction at $T = 0$ is displayed Figure 16c.

The effects of the surface exchange and the film thickness have been shown in Ref. [38].

To close this section, let us mention our work [41] on the DM interaction in magneto-ferroelectric superlattices where the SW in the magnetic layer have been calculated. We have also studied the stability of skyrmions at finite T in that work and in Refs. [42,43].

5. Effect of Dzyaloshinskii–Moriya Interaction in a Frustrated Antiferromagnetic Triangular Lattice

The results of this section are not yet published [44]. We will not present this model in details. We show the Hamiltonian, the GS and the SW spectrum.

5.1. Model—Ground State

We consider a triangular lattice occupied by Heisenberg spins of magnitude $1/2$. The DM interaction was introduced historically to explain the weak ferromagnetism in compounds MnO. The superexchange between two Mn atoms is modified with the displacement of the oxygen atom between them. If the displacement of the oxygen is in the xy plane (see Figure 17a), then the DM vector $\mathbf{D}_{i,j}$ is perpendicular to the xy plane and is given by [45,46]

$$\mathbf{D}_{i,j} \propto \mathbf{r}_{iO} \times \mathbf{r}_{Oj} \propto -\mathbf{r}_{ij} \times \mathbf{R} \tag{54}$$

where $\mathbf{r}_{iO} = \mathbf{r}_O - \mathbf{r}_i$ and $\mathbf{r}_{Oj} = \mathbf{r}_j - \mathbf{r}_O$, $\mathbf{r}_{ij} = \mathbf{r}_j - \mathbf{r}_i$. $\mathbf{r}_O$ is the position of non-magnetic ion (oxygen) and $\mathbf{r}_i$ the position of the spin $\mathbf{S}_i$ etc. These vectors are defined in Figure 17a in the particular case where the displacements are in the xy plane. We have therefore $\mathbf{D}_{i,j}$ perpendicular to the xy plane in this case.

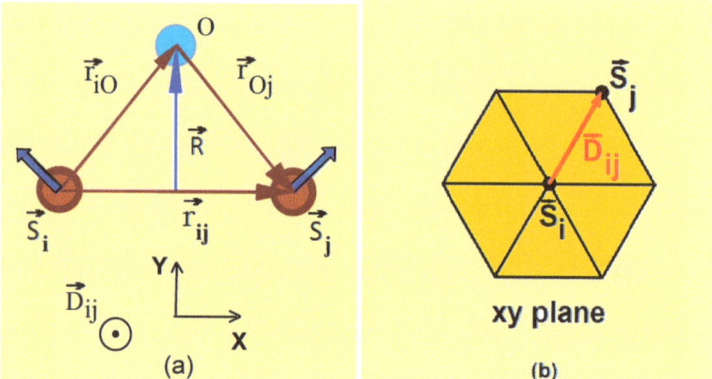

Figure 17. (**a**) D vector along the z direction perpendicular to the xy plane. See the definition of the D vector in the text, (**b**) In-plane $\mathbf{D}_{ij}$ vector chosen along the direction connecting spin $\mathbf{S}_i$ to spin $\mathbf{S}_j$ in the xy plane.

Note, however, that if the atom displacements are in 3D space, $\mathbf{D}_{i,j}$ can be in any direction. In this paper, we consider also the case where $\mathbf{D}_{i,j}$ lies in the xy plane as shown in Figure 17b where $\mathbf{D}_{i,j}$ is taken along the vector connecting spin $\mathbf{S}_i$ to spin $\mathbf{S}_j$.

Note that from Equation (54) one has

$$\mathbf{D}_{j,i} = -\mathbf{D}_{i,j} \tag{55}$$

In the case of perpendicular $\mathbf{D}_{i,j}$, let us define $\mathbf{u}_{i,j}$ as the unit vector on the z axis. From Equations (54) and (55) one writes

$$\mathbf{D}_{i,j} = D\mathbf{u}_{i,j} \tag{56}$$
$$\mathbf{D}_{j,i} = D\mathbf{u}_{j,i} = -D\mathbf{u}_{i,j} \tag{57}$$

where D represents the DM interaction strength. Note however that the DM interaction goes beyond the weak ferromagnetism and may find its origin in various physical mechanisms. So, the form given in (56) is a model, a hypothesis.

In the case of in-plane $\mathbf{D}_{i,j}$, we suppose that $\mathbf{D}_{i,j}$ is given as

$$\mathbf{D}_{i,j} = D(\mathbf{r}_j - \mathbf{r}_i)/|\mathbf{r}_j - \mathbf{r}_i| = D\mathbf{r}_{ij} \tag{58}$$

where D is a constant and $\mathbf{r}_{ij}$ denotes the unit vector along $\mathbf{r}_j - \mathbf{r}_i$. The case of in-plane $\mathbf{D}_{i,j}$ on the frustrated triangular lattice (see Figure 17b) has been recently studied since this case gives rise to a beautiful skyrmion crystal composed of three interpenetrating sublattice skyrmions in a perpendicular applied magnetic field [44,47,48]. A description of this case is however out of the purpose of this review.

5.2. Ground State with a Perpendicular **D** in Zero Field

The Hamiltonian is given by

$$\begin{aligned}\mathcal{H} &= -J\sum_{\langle ij \rangle} \mathbf{S}_i \cdot \mathbf{S}_j - D\sum_{\langle ij \rangle} \mathbf{u}_{i,j} \cdot \mathbf{S}_i \times \mathbf{S}_j \\ &\quad -H\sum_i S_i^z \end{aligned} \tag{59}$$

where $\mathbf{S}_i$ is a classical Heisenberg spin of magnitude 1 occupying the lattice site i. The first sum runs over all spin nearest-neighbor (NN) pairs with an antiferromagnetic exchange interaction J ($J < 0$), while the second sum is performed over all DM interactions between

NN. H is the magnitude of a magnetic field applied along the z direction perpendicular to the lattice xy plane.

In the absence of J, unlike the bipartite square lattice where one can arrange the NN spins to be perpendicular with each order in the xy plane, the triangular lattice cannot fully satisfy the DM interaction for each bond, namely with the perpendicular spins at the ends. For this particular case of interest, we can analytically calculate the GS spin configuration as shown in the following. One considers a triangular plaquette with three spins numbered as 1, 2, and 3 embedded in the lattice. For convenience, in a hexagonal (or triangular) lattice, we define the three sublattices as follows: consider the up-pointing triangles (there are three in a hexagon, see the blue triangles in Figure 18). For the first triangle one numbers in the counter-clockwise sense 1, 2, 3 then one does it for the other two up-pointing triangles of the hexagon, one sees that each lattice site belongs to a sublattice. The DM energy of a plaquette is written as

$$\begin{aligned} H_p &= -2D[\mathbf{u}_{1,2} \cdot \mathbf{S}_1 \times \mathbf{S}_2 + \mathbf{u}_{2,3} \cdot \mathbf{S}_2 \times \mathbf{S}_3 + \mathbf{u}_{3,1} \cdot \mathbf{S}_3 \times \mathbf{S}_1] \\ &= -2D[\sin\theta_{1,2} + \sin\theta_{2,3} + \sin\theta_{3,1}] \end{aligned} \quad (60)$$

where the factor 2 of the D term takes into account the opposite neighbors outside the plaquette, and where $\theta_{1,2} = \theta_2 - \theta_1$ is the oriented angle between $\mathbf{S}_1$ and $\mathbf{S}_2$, etc. Note that the u vectors are in the same direction because we follow the counter-clockwise tour on the plaquette.

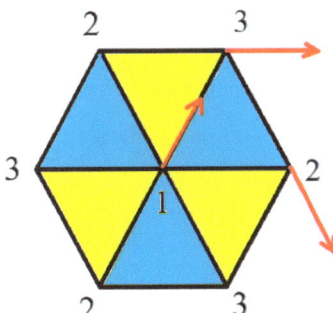

Figure 18. Perpendicular $\mathbf{D}_{i,j}$: Ground-state spin configuration with only Dzyaloshinskii–Moriya interaction on the triangular lattice ($J = 0$) is analytically determined. One angle is 120 degrees and the other two are 60 degrees. Note that the choice of the 120-degree angle in this figure is along the horizontal spin pair. This configuration is one GS and the other two GSs have the 120-degree angles on respectively the two diagonal spin pairs. Note also that the spin configuration is invariant under the global spin rotation in the xy plane. For convenience, the spins are decomposed into three sublattices numbered 1, 2 and 3. See text for explanation.

The minimization of H_p yields

$$\frac{dH_p}{d\theta_1} = 0 = -2D[-\cos(\theta_2 - \theta_1) + \cos(\theta_1 - \theta_3)] \quad (61)$$

$$\frac{dH_p}{d\theta_2} = 0 = -2D[\cos(\theta_2 - \theta_1) - \cos(\theta_3 - \theta_2)] \quad (62)$$

$$\frac{dH_p}{d\theta_3} = 0 = -2D[\cos(\theta_3 - \theta_2) - \cos(\theta_1 - \theta_3)] \quad (63)$$

The solutions for the above equations are

$$\theta_{1,2} = \theta_{3,1} \text{ so that } \theta_{3,2} = \theta_{3,1} + \theta_{1,2} = 2\theta_{1,2} \tag{64}$$
$$\theta_{2,3} = \theta_{1,2} \text{ so that } \theta_{1,3} = \theta_{1,2} + \theta_{2,3} = 2\theta_{2,3} \tag{65}$$
$$\theta_{3,1} = \theta_{2,3} \text{ so that } \theta_{2,1} = \theta_{2,3} + \theta_{3,1} = 2\theta_{3,1} \tag{66}$$

Note that the second and third lines can be obtained by the circular permutation of the indices 1, 2, and 3 using the first line. These three equations, Equations (64)–(66), should be solved. There is more than one solution. We have from Equation (61) $\cos(\theta_{1,2}) = \cos(\theta_{3,1})$. Using Equation (66) one obtains

$$\cos(2\theta_{3,1}) = \cos(\theta_{3,1}) \rightarrow 2\cos^2(\theta_{3,1}) - \cos(\theta_{3,1}) - 1 = 0 \tag{67}$$

This second-degree equation gives $\cos(\theta_{3,1}) = \frac{1\pm\sqrt{1+8}}{4}$. Only the negative solution is acceptable so that $\theta_{3,1} = \theta_{2,3} = \pi/6$. From Equation (66), one has $\theta_{2,1} = \pi/3$. This is one solution given by Equation (68) below. Note that we have taken one of them, Equation (66), to obtain explicit solutions for the three angles given in Equation (68). We can do the same calculation starting with Equations (64) and (65) to get explicit solutions given in Equations (69) and (70). We note that when we make a circular permutation of the indices of Equation (68) we get Equation (69), and a circular permutation of Equation (69) gives Equation (70). One summarizes the three degenerate solutions below

$$\theta_{3,1} = \theta_{2,3} = \pi/6, \ \theta_{2,1} = \pi/3 \tag{68}$$
$$\theta_{1,2} = \theta_{3,1} = \pi/6, \ \theta_{3,2} = \pi/3 \tag{69}$$
$$\theta_{2,3} = \theta_{1,2} = \pi/6, \ \theta_{1,3} = \pi/3 \tag{70}$$

We show in Figure 18 the spin orientations of the solution (68). The GS energy is obtained by replacing the angles into Equation (60). For the three solutions, one gets the energy of the plaquette

$$H_p = -3D\sqrt{3} \tag{71}$$

We have three degenerate GSs.

Note that this solution can be numerically obtained by the steepest descent method described above. The result is shown in Figure 19 for the full lattice. We see in the zoom that the spin configuration on a plaquette is what is obtained analytically, with a global spin rotation as explained in the caption of Figure 18.

As said above, to use the steepest descent method, we consider a triangular lattice of lateral dimension L. The total number of sites N is given by $N = L \times L$. To avoid the finite size effect, we have to find the size limit beyond which the GS does not depend on the lattice size. This is found for $L \geq 100$. Most of calculations have been performed for $L = 100$.

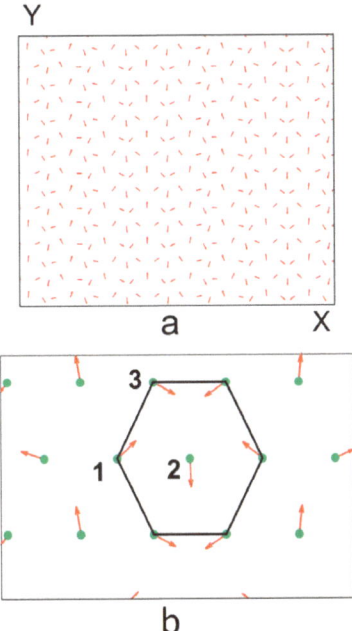

Figure 19. Perpendicular $\mathbf{D}_{i,j}$: (**a**) Ground-state spin configuration with only Dzyaloshinskii–Moriya interaction on the triangular lattice ($J = 0$) obtained numerically by the steepest descent method, (**b**) a zoom on a hexagonal cell, this is exactly what obtained analytically shown in Figure 18 with a global spin rotation in the xy plane: the angle of the horizontal pair (1,2) is 120 degrees, those of (2,3) and (3,1) are equal to 60 degrees.

5.3. Ground State with Both Perpendicular **D** and J in Zero Field- Spin Waves

When both J and perpendicular **D** are present, a compromise is established between these competing interactions. In zero field, the GS shows non-collinear but periodic in-plane spin configurations. The planar spin configuration is easily understood: when **D** is perpendicular and without J, the spins are in the plane. When J is antiferromagnetic without **D**, the spins are also in the plane and form a 120-degree structure. When **D** and J exist together the angles between NN's change but they remain in the plane in order to keep both D and J interactions as low as possible. An example is shown in Figure 20 where one sees that the GS is planar and characterized by two angles $\theta = 102$ degrees and one angle $\beta = 156$ degrees formed by three spins on a triangle plaquette. Note that there are three degenerate states where β is chosen for the pair (1,2) (Figure 20a) or the pair (2,3) or the pair (3,1). Changing the value of D will change the angle values. Changing the sign of D results in a change of the sense of the chirality, but not the angle values.

In the case of perpendicular $\mathbf{D}_{i,j}$ in zero-field, as shown above we find the GS on a hexagon of the lattice is defined by four identical angles β and two angles θ as shown in Figure 20. The values of β and θ depend on the value of D. We take $J = -1$ (antiferromagnetic) hereafter. For $D = 0.5$ we have $\beta = 156$ degrees and $\theta = 102$ degrees. For $D = 0.4$ we obtain $\beta = 108$ degrees and $\theta = 144$ degrees, using $N = 60 \times 60$.

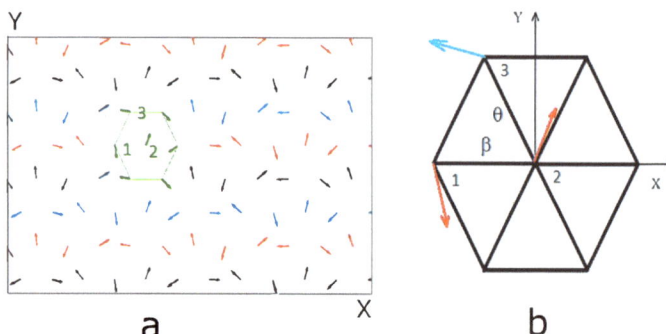

Figure 20. Perpendicular $\mathbf{D}_{i,j}$ with antiferromagnetic J: (**a**) Ground-state spin configuration in zero field for $D = 0.5$, $J = -1$ where the angles in a hexagon are shown in (**b**) with $\beta = 156$ degrees for the pair (1,2) on the horizontal axis and $\theta = 102$ degrees for the pairs (2,3) and (3,1) on the diagonals. Note that there are two other degenerate states where β is chosen for the pair (2,3) or (3,1).

The periodicity of the GS allows us to calculate the SW spectrum in the following.

The model for the calculation of the SW spectrum uses quantum Heisenberg spins of magnitude 1/2, it is given by

$$\mathcal{H} = -J \sum_{\langle i,j \rangle} \mathbf{S}_i \cdot \mathbf{S}_j - D \sum_{\langle i,j \rangle} \mathbf{u}_{i,j} \cdot \mathbf{S}_i \times \mathbf{S}_j - I \sum_{\langle i,j \rangle} S_i^z S_j^z \cos \theta_{ij} \qquad (72)$$

where θ_{ij} is the angle between $\mathbf{S}_i$ and $\mathbf{S}_j$ and the last term is an extremely small anisotropy added to stabilize the SW when the wavelength k tends to zero [31,40]. Note that $\mathbf{u}_{i,j}$ points up and down along the z axis for respective two opposite neighbors.

As before, in order to calculate the SW spectrum for systems of non-collinear spin configurations, we have to use the system of local coordinates. The Hamiltonian becomes

$$\mathcal{H} = -J \sum_{\langle i,j \rangle} \frac{1}{4}(S_i^+ S_j^+ + S_i^- S_j^-)(\cos \theta_{ij} - 1) + \frac{1}{4}(S_i^+ S_j^- + S_i^- S_j^+)(\cos \theta_{ij} + 1)$$

$$+ \frac{1}{2} S_j^z \sin \theta_{ij} (S_i^+ + S_i^-) - \frac{1}{2} \sin \theta_{ij} S_i^z (S_j^+ + S_j^-) + S_i^z S_j^z \cos \theta_{ij}$$

$$- D \sum_{\langle i,j \rangle} S_i^z S_j^z \sin \theta_{i,j} + \frac{1}{4} \sin \theta_{i,j} (S_i^+ S_j^+ + S_i^+ S_j^- + S_i^- S_j^+) + \frac{1}{2} \cos \theta_{i,j} (S_i^z (S_j^+ + S_j^-) - S_j^z (S_i^+ + S_i^-))$$

$$- I \sum_{\langle i,j \rangle} S_i^z S_j^z \cos \theta_{i,j}$$

We define the two GFs by Equations (11) and (12) and use the equations of motion of these functions (13) and (14), we obtain

$$i\hbar \frac{dG_{i,j}(t-t')}{dt} = 2 <S_i^z> \delta_{i,j}\delta(t-t') - J \sum_{\langle l \rangle} <S_i^z> F_{l,j}(t-t')(\cos \theta_{i,l} - 1)$$

$$+ <S_i^z> G_{l,j}(t-t')(\cos \theta_{i,l} + 1) - 2 \cos \theta_{i,l} <S_i^z> G_{i,j}(t-t')$$

$$+ D \sum_{\langle l \rangle} 2 \sin \theta_{i,l} <S_i^z> F_{l,j}(t-t') - \sin \theta_{i,l} <S_i^z> (G_{l,j}(t-t') + F_{l,j}(t-t'))$$

$$- 2I \sum_{\langle l \rangle} \cos \theta_{i,l} <S_i^z> F_{l,j}(t-t')$$

$$i\hbar \frac{dF_{i,j}(t-t')}{dt} = J\sum_{\langle l\rangle} <S_i^z> G_{l,j}(t-t')(\cos\theta_{i,l} - 1)$$
$$+ <S_l^z> F_{l,j}(t-t')(\cos\theta_{i,l} + 1) - 2\cos\theta_{i,l} <S_l^z> F_{i,j}(t-t')$$
$$- D\sum_{\langle l\rangle} 2\sin\theta_{i,l} <S_i^z> G_{l,j}(t-t') - \sin\theta_{i,l} <S_l^z>(G_{l,j}(t-t') + F_{l,j}(t-t'))$$
$$+ 2I\sum_{\langle l\rangle} \cos\theta_{i,l} <S_l^z> G_{l,j}(t-t')$$

Note that $<S_i^z>$ is the average of the spin i on its local quantization axis in the local-coordinates system (see Ref. [38]). We use now the time Fourier transforms of the G and F, we get

$$\hbar\omega g_{i,j} = 2\mu_i \delta_{i,j} - J\sum_{\langle l\rangle} \mu_i f_{lj} e^{-i\mathbf{k}\cdot(\mathbf{R}_i-\mathbf{R}_l)}(\cos\theta_{i,l} - 1)$$
$$+ \mu_i g_{lj} e^{-i\mathbf{k}\cdot(\mathbf{R}_i-\mathbf{R}_l)}(\cos\theta_{i,l} + 1) - 2\mu_l \cos\theta_{i,l} g_{i,j}$$
$$- D\sum_{\langle l\rangle} 2\sin\theta_{i,l} \mu_l g_{i,j} - \sin\theta_{i,l}\mu_i(g_{l,j} e^{-i\mathbf{k}\cdot(\mathbf{R}_i-\mathbf{R}_l)} + f_{l,j} e^{-i\mathbf{k}\cdot(\mathbf{R}_i-\mathbf{R}_l)}) \quad (73)$$
$$+ 2I\sum_{\langle l\rangle} \mu_l \cos\theta_{i,l} g_{i,j}$$

and

$$\hbar\omega f_{i,j} = J\sum_{\langle l\rangle} \mu_i g_{lj} e^{-i\mathbf{k}\cdot(\mathbf{R}_i-\mathbf{R}_l)}(\cos\theta_{i,l} - 1)$$
$$+ \mu_i f_{lj} e^{-i\mathbf{k}\cdot(\mathbf{R}_i-\mathbf{R}_l)}(\cos\theta_{i,l} + 1) - 2\mu_l \cos\theta_{i,l} f_{i,j}$$
$$+ D\sum_{\langle l\rangle} 2\sin\theta_{i,l}\mu_l f_{i,j} - \sin\theta_{i,l}\mu_i(g_{l,j} e^{-i\mathbf{k}\cdot(\mathbf{R}_i-\mathbf{R}_l)} + f_{l,j} e^{-i\mathbf{k}\cdot(\mathbf{R}_i-\mathbf{R}_l)}) \quad (74)$$
$$- 2I\sum_{\langle l\rangle} \mu_l \cos\theta_{i,l} f_{i,j}$$

where $\mu_i \equiv <S_i^z>$, $\mathbf{k}$ is the wave vector in the reciprocal lattice of the triangular lattice, and ω the SW frequency. Note that the index z in S_i^z is not referring to the real space direction z, but to the quantization axis of the spin $\mathbf{S}_i$. At this stage, we have to replace $\theta_{i,j}$ by either β or θ according on the GS spin configuration given above (see Figure 20).

As in the previous sections, writing the above equations under a matrix form, we have

$$\mathbf{M}(\hbar\omega)\mathbf{h} = \mathbf{C}, \quad (75)$$

where $\mathbf{M}(\hbar\omega)$ is a square matrix of dimension 2×2, $\mathbf{h}$ and $\mathbf{C}$ are given by

$$\mathbf{h} = \begin{pmatrix} g_{i,j} \\ f_{i,j} \end{pmatrix}, \quad \mathbf{C} = \begin{pmatrix} 2\langle S_i^z\rangle \delta_{i,j} \\ 0 \end{pmatrix}, \quad (76)$$

and the matrix $\mathbf{M}(\hbar\omega)$ is given by

$$\mathbf{M}(\hbar\omega) = \begin{pmatrix} \hbar\omega + A & B \\ -B & \hbar\omega - A \end{pmatrix}$$

The nontrivial solution of g and f imposes the following secular equation:

$$0 = \begin{pmatrix} \hbar\omega + A & B \\ -B & \hbar\omega - A \end{pmatrix} \quad (77)$$

where

$$A = -J(8\mu_i \cos\beta(1+I) + 4\mu_i \cos\theta(1+I) - 4\mu_i\gamma(\cos\beta+1) - 2\mu_i\alpha(\cos\theta+1))$$
$$- D(4\mu_i \sin\beta\gamma + 2\mu_i \sin\theta\alpha) + D(8\mu_i \sin\beta + 4\mu_i \sin\theta) \tag{78}$$

$$B = J(4\mu_i\gamma(\cos\beta-1) + 2\mu_i\alpha(\cos\theta-1)) - D(4\gamma\mu_i \sin\beta + 2\mu_i\alpha \sin\theta) \tag{79}$$

where the sum on the two NN on the x axis (see Figure 20b) is

$$\sum_l e^{-i\mathbf{k}\cdot(\mathbf{R}_i-\mathbf{R}_l)} = 2\cos(k_x) \equiv 2\alpha \tag{80}$$

and the sum on the four NN on the oblique directions of the hexagon (see Figure 20b) is

$$\sum_l e^{-i\mathbf{k}\cdot(\mathbf{R}_i-\mathbf{R}_l)} = 4\cos(k_x/2)\cos(\sqrt{3}k_y/2) \equiv 4\gamma \tag{81}$$

Solving Equation (77) for each given (k_x, k_y) one obtains the SW frequency $\omega(k_x, k_y)$:

$$(\hbar\omega)^2 = A^2 - B^2 \rightarrow \hbar\omega = \pm\sqrt{A^2 - B^2} \tag{82}$$

Plotting $\omega(k_x, k_y)$ in the space (k_x, k_y) one obtains the full SW spectrum.

The spin length $\langle S_i^z \rangle$ (for all i, by symmetry) is given by (see technical details in Ref. [31]):

$$\langle S^z \rangle \equiv \langle S_i^z \rangle = \frac{1}{2} - \frac{1}{\Delta}\int\int dk_x dk_z \sum_{i=1}^{2} \frac{Q(E_i)}{e^{E_i/k_BT} - 1} \tag{83}$$

where $E_i(i=1,2) = \pm\sqrt{A^2-B^2}$ are the two solutions given above, and $Q(E_i)$ is the determinant (cofactor) obtained by replacing the first column of **M** by **C** at E_i.

The spin length $\langle S^z \rangle$ at a given T is calculated self-consistently by following the method given in Refs. [31,38].

Let us show the SW spectrum ω (taking $\hbar = 1$) for the case of $J = -1$ and $D = 0.5$ in Figure 21 versus k_y with $k_x = 0$ (Figure 21a) and versus k_x for $k_y = 0$ (Figure 21b). In order to see the effect of the DM interaction alone we take the anisotropy $I = 0$. One observes here that for a range of small wave-vectors the SW frequency is imaginary. The SW corresponding to these modes do not propagate in the system. Why do we have this case here? The answer is that when the NN make a large angle (perpendicular NN, for example), one cannot define a wave vector in that direction. Physically, when k is small the B coefficient is larger than A in Equation (82) giving rise to imaginary ω. Note that the anisotropy I is contained in A so that increasing I for small k will result in $A > B$ making ω real.

We show now in Figure 22a the spectrum along the axis $k_x = k_y$ at $T = 0$ for $I = 0$. Again here the frequency is imaginary for small k, as in the previous figure. The spin length $< S^z >$ along the local quantization axis is shown in Figure 22b. Several remarks are in order: (i) At $T = 0$, the spin length is not equal to $1/2$ as in ferromagnets because of the zero-point spin contraction due to antiferromagnetic interactions (see Ref. [31]), its length is $\simeq 0.40$, quite small; (ii) the magnetic ordering is destroyed at $T \simeq 1.2$.

To close the present section, we note that in the case of perpendicular **D** considered above, we did not observe skyrmion textures when applying a perpendicular magnetic field: all spin configurations are no more planar, making the calculation of the SW spectrum more difficult. This problem is left for a future investigation.

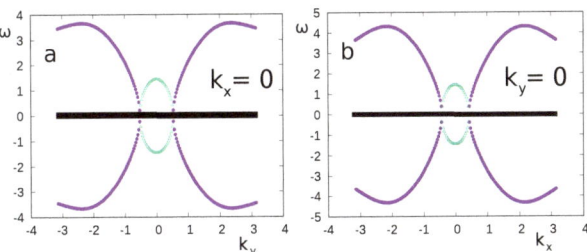

Figure 21. (a) Spin-wave spectrum versus k_y with $k_x = 0$ at $T = 0$ for $I = 0$, (b) Spin-wave spectrum versus k_x with $k_y = 0$ at $T = 0$ for $I = 0$. The magenta curves show the real frequency, while the green ones show the imaginary frequency. See text for comments. Parameters: $D = 0.5$, $J = -1$, $H = 0$ where $\theta = 102$ degrees and $\beta = 156$ degrees (see the spin configuration shown in Figure 20), $\hbar = 1$.

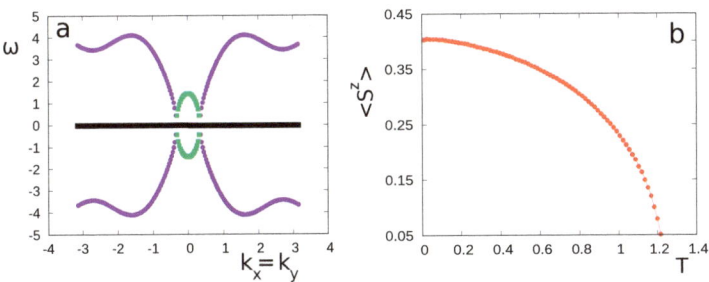

Figure 22. (a) Spin-wave spectrum versus $k_x = k_y$ at $T = 0$ for $I = 0$. The magenta curves show the real frequency, while the green ones show the imaginary frequency. See text for comments, (b) The spin length S^z versus temperature T ($k_B = 1$). Parameters: $D = 0.5$, $J = -1$, $H = 0$ where $\theta = 102$ degrees and $\beta = 156$ degrees (see the spin configuration shown in Figure 20).

6. Other Systems of Non-Collinear Ground-State Spin Configurations: Frustrated Surface in Stacked Triangular Thin Films

In this section, we study by the GF technique the effect of a frustrated surface on the magnetic properties of a film composed triangular layers stacked in the z direction. Each lattice site is occupied by a quantum Heisenberg spin of magnitude 1/2. Let the in-plane surface interaction be J_s which can be antiferromagnetic or ferromagnetic. The other interactions in the film are ferromagnetic. We show in the following that the GS spin configuration is non-collinear when J_s is lower than a critical value J_s^c. The film surfaces are then frustrated. In the frustrated case, there are two phase transitions, one corresponds to the disordering of the two surfaces and the other to the disordering of the interior layers. The GF results agree qualitatively with Monte Carlo simulation using the classical spins (see the original paper in Ref. [39]).

In this section we review some of the results given in the original paper Ref. [39], emphasizing the SW calculation and the important results. The Hamiltonian is written as

$$\mathcal{H} = -\sum_{\langle i,j \rangle} J_{i,j} \mathbf{S}_i \cdot \mathbf{S}_j - \sum_{<i,j>} I_{i,j} S_i^z S_j^z \tag{84}$$

where the first sum is performed over the NN spin pairs $\mathbf{S}_i$ and $\mathbf{S}_j$, the second sum over their z components. $J_{i,j}$ and $I_{i,j}$ are respectively their exchange interaction and their anisotropic one. The latter is small, taken to ensure the ordering at finite T when the film thickness goes down to a few layers, without this we know that a monolayer with vector spin models does not have a long-range ordering at finite T [40].

Let J_s be the exchange between two NN surface spins. We suppose that all other interactions are ferromagnetic and equal to J. We shall use $J = 1$ as the unit of energy in the following.

6.1. Ground State

In the case where J_s is ferromagnetic, the GS of the film is ferromagnetic. When J_s is antiferromagnetic, the situation becomes complicated. We recall that for a single triangular lattice with antiferromagnetic interaction, the spins are frustrated and arranged in a 120-degree configuration [8]. This structure is modified when we turn on the ferromagnetic interaction J with the beneath layer. The competition between the non collinear surface ordering and the ferromagnetic ordering of the bulk leads to an intermediate structure which is determined in the following.

The GS configuration can be determined by using the steepest descent method described below Equation (31). Let us describe qualitatively the GS configuration: when J_s is negative and $J_s < J_s^c$ where $J_s^c(<0)$ is a critical value, the GS is formed by pulling out the planar 120° spin structure along the z axis by an angle β. This is shown in Figure 23.

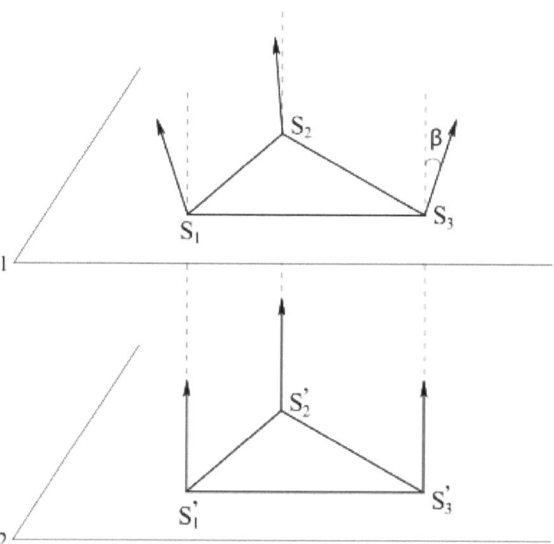

Figure 23. Ground state of the film when J_s is smaller than the critical value J_s^c. See text for description.

Figure 24 shows $\cos \alpha$ and $\cos \beta$ versus J_s obtained by the steepest descent method. As seen for $J_s > J_s^c$, the angles are zero, namely the GS is ferromagnetic. The critical value J_s^c is numerically found between -0.18 and -0.19.

We show in the following that this value can be analytically calculated by assuming the structure shown in Figure 23). We number the spins as in that figure: S_1, S_2 and S_3 are the spins in the surface layer, S_1', S_2' and S_3' are the spins in the second layer. The energy of the cell is

$$\begin{aligned} H_p = &-6[J_s(\mathbf{S}_1 \cdot \mathbf{S}_2 + \mathbf{S}_2 \cdot \mathbf{S}_3 + \mathbf{S}_3 \cdot \mathbf{S}_1) \\ &+ I_s(S_1^z S_2^z + S_2^z S_3^z + S_3^z S_1^z) \\ &+ J(\mathbf{S}_1' \cdot \mathbf{S}_2' + \mathbf{S}_2' \cdot \mathbf{S}_3' + \mathbf{S}_3' \cdot \mathbf{S}_1') \\ &+ I\left(S_1'^z S_2'^z + S_2'^z S_3'^z + S_3'^z S_1'^z\right)] \\ &- 2J(\mathbf{S}_1 \cdot \mathbf{S}_1' + \mathbf{S}_2 \cdot \mathbf{S}_2' + \mathbf{S}_3 \cdot \mathbf{S}_3') \\ &- 2I\left(S_1^z S_1'^z + S_2^z S_2'^z + S_3^z S_3'^z\right), \end{aligned} \qquad (85)$$

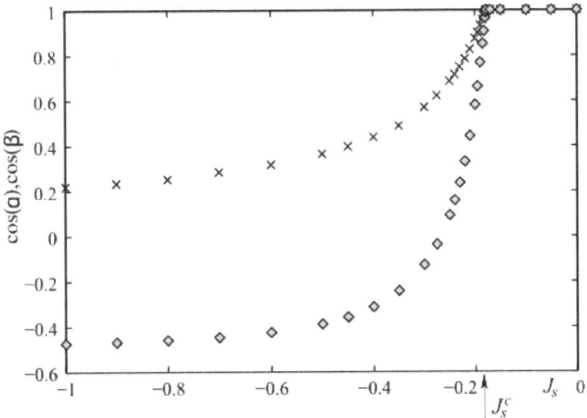

Figure 24. Ground state determined by cos(α) (diamonds) and cos(β) (crosses) as functions of J_s. Critical value of J_s^c is shown by the arrow.

We project the spins on the xy plane and on the z axis. One writes $\mathbf{S}_i = (\mathbf{S}_i^{\parallel}, S_i^z)$. One observes that only surface spins have non-zero xy vector components. Let the angle between these xy components of NN surface spins be $\gamma_{i,j}$ which is in fact the projection of the angle α on the xy plane. By symmetry, we have

$$\gamma_{1,2} = 0, \quad \gamma_{2,3} = \frac{2\pi}{3}, \quad \gamma_{3,1} = \frac{4\pi}{3}. \tag{86}$$

The angles β_i and β_i' of $\mathbf{S}_i$ and $\mathbf{S}_i'$ formed with the z axis are by symmetry

$$\begin{cases} \beta_1 = \beta_2 = \beta_3 = \beta, \\ \beta_1' = \beta_2' = \beta_3' = 0, \end{cases}$$

The total energy of the cell (86), with $S_i = S_i' = \frac{1}{2}$, is thus

$$\begin{aligned} H_p &= -\frac{9(J+I)}{2} - \frac{3(J+I)}{2} \cos \beta - \frac{9(J_s + I_s)}{2} \cos^2 \beta \\ &\quad + \frac{9 J_s}{4} \sin^2 \beta. \end{aligned} \tag{87}$$

The minimum of the cell energy verifies this condition:

$$\frac{\partial H_p}{\partial \beta} = \left(\frac{27}{2} J_s + 9 I_s\right) \cos \beta \sin \beta + \frac{3}{2}(J+I) \sin \beta = 0 \tag{88}$$

One deduces

$$\cos \beta = -\frac{J+I}{9 J_s + 6 I_s}. \tag{89}$$

This solution exists under the condition $-1 \leq \cos \beta \leq 1$. The critical values are determined from this condition. For $I = -I_s = 0.1$, $J_s^c \approx -0.1889 J$ which is in excellent agreement with the results obtained from the steepest descent method.

Now, using the GF method for such a film in the way described in the previous sections, we obtain the full Hamiltonian (84) in the local framework:

$$\begin{aligned}
\mathcal{H} &= -\sum_{<i,j>} J_{i,j} \Big\{ \frac{1}{4}(\cos\theta_{ij}-1)\left(S_i^+ S_j^+ + S_i^- S_j^-\right) \\
&+ \frac{1}{4}(\cos\theta_{ij}+1)\left(S_i^+ S_j^- + S_i^- S_j^+\right) \\
&+ \frac{1}{2}\sin\theta_{ij}(S_i^+ + S_i^-)S_j^z - \frac{1}{2}\sin\theta_{ij} S_i^z \left(S_j^+ + S_j^-\right) \\
&+ \cos\theta_{ij} S_i^z S_j^z \Big\} - \sum_{<i,j>} I_{i,j} S_i^z S_j^z
\end{aligned} \quad (90)$$

where $\cos(\theta_{ij})$ is the angle between two NN spins. We define the two coupled GF, and we write their equations of motions in the real space. Taking Tyablikov's decoupling scheme to reduce higher-order GFs, and then using the Fourier transform in the xy plane we arrive at a matrix equation as in the previous section with the matrix $\mathbf{M}$ is defined as

$$\mathbf{M}(\omega) = \begin{pmatrix} A_1^+ & B_1 & D_1^+ & D_1^- & \cdots \\ -B_1 & A_1^- & -D_1^- & -D_1^+ & \vdots \\ \vdots & \cdots & \cdots & \cdots & \vdots \\ \vdots & C_{N_z}^+ & C_{N_z}^- & A_{N_z}^+ & B_{N_z} \\ \cdots & -C_{N_z}^- & -C_{N_z}^+ & -B_{N_z} & A_{N_z}^- \end{pmatrix}, \quad (91)$$

where

$$\begin{aligned}
A_n^\pm &= \omega \pm \Big[\frac{1}{2}J_n\langle S_n^z\rangle(Z\gamma)(\cos\theta_n+1) \\
&- J_n\langle S_n^z\rangle Z\cos\theta_n - J_{n,n+1}\langle S_{n+1}^z\rangle \cos\theta_{n,n+1} \\
&- J_{n,n-1}\langle S_{n-1}^z\rangle \cos\theta_{n,n-1} - ZI_n\langle S_n^z\rangle \\
&- I_{n,n+1}\langle S_{n+1}^z\rangle - I_{n,n-1}\langle S_{n-1}^z\rangle\Big],
\end{aligned} \quad (92)$$

$$B_n = \frac{1}{2}J_n\langle S_n^z\rangle(\cos\theta_n-1)(Z\gamma), \quad (93)$$

$$C_n^\pm = \frac{1}{2}J_{n,n-1}\langle S_n^z\rangle(\cos\theta_{n,n-1}\pm 1), \quad (94)$$

$$D_n^\pm = \frac{1}{2}J_{n,n+1}\langle S_n^z\rangle(\cos\theta_{n,n+1}\pm 1), \quad (95)$$

where $Z=6$ is the in-plane coordination number, $\theta_{n,n\pm 1}$ denotes the angle between two NN spins belonging to the adjacent layers n and $n \pm 1$, while θ_n is the angle between two NN spins of the layer n, and

$$\gamma = \left[2\cos(k_x a) + 4\cos(k_y a/2)\cos\left(k_y a\sqrt{3}/2\right)\right]/Z.$$

Note that in the above coefficients, we have used the following notations:

(i) J_n and I_n are the in-plane interactions. J_n is equal to J_s for the two surface layers and equal to J for the interior layers. All I_n are taken equal to I.
(ii) The interlayer interactions are denoted by $J_{n,n\pm 1}$ and $I_{n,n\pm 1}$. Note that $J_{n,n-1} = I_{n,n-1} = 0$ if $n=1$ and $J_{n,n+1} = I_{n,n+1} = 0$ if $n=N_z$.

As described in the previous sections, the SW spectrum ω is obtained by solving $\det|\mathbf{M}|=0$. Using ω we calculate the magnetizations layer by layer for typical values of parameters. The results are shown in the following.

6.2. Quantum Surface Phase Transition

Let us show a typical case in the region of frustrated surface where $J_s = -0.5$ in Figure 25. Several comments are in order:

(i) The surface magnetization is very small with respect to the magnetization of the second layer,
(ii) At $T = 0$, the length of the surface spin is about 0.425, much shorter than the spin magnitude 1/2. This is due to the antiferromagnetic interaction at the surface which causes a strong spin contraction. For the second layer, the spins are aligned ferromagnetically, their length is fully 0.5,
(iii) The surface undergoes a phase transition at $T_1 \simeq 0.2557$ while the second layer remains ordered up to $T_2 \simeq 1.522$. The system is thus disordered at the surface and ordered in the bulk, for temperatures between T_1 and T_2. This partial disorder is very interesting. It gives another example of the partial disorder observed earlier in bulk frustrated quantum spin systems.
(iv) One observes that between T_1 and T_2, the first layer has a small magnetization. This is understood by the fact that the strong magnetization of the second layer acts as an external field on the first layer, inducing therefore a small value of its magnetization.

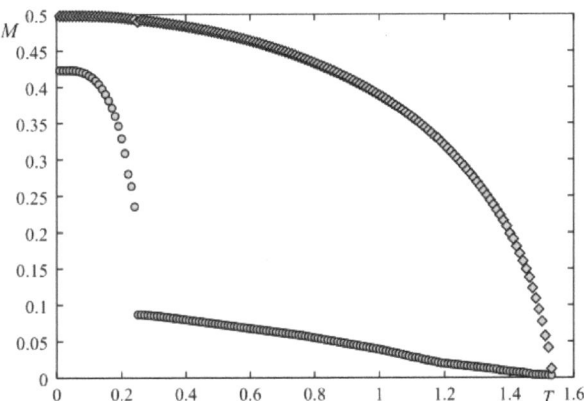

Figure 25. First two layer-magnetizations obtained by the Green's function technique vs. T for $J_s = -0.5$ with $I = -I_s = 0.1$. The surface-layer magnetization (lower curve) is much smaller than the second-layer one. See text for comments.

We plot the phase diagram in the space (J_s, T) in Figure 26. Phase I denotes the surface canted-spin state, phase IIA denotes the partially ordered phase: the surface is disordered while the bulk is ordered. Phase IIB separated from phase IIA by a vertical line issued from $J_s^c \simeq -0.19$ indicates the ferromagnetic state, and phase III is the paramagnetic phase.

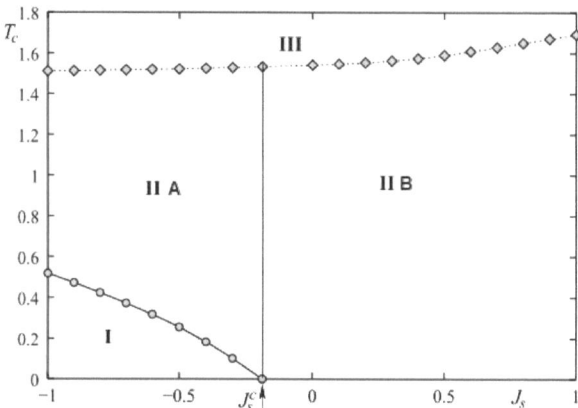

Figure 26. Phase diagram in the space (J_s, T) for the quantum Heisenberg model with $N_z = 4$, $I = |I_s| = 0.1$. See text for the description of phases I to III.

6.3. Classical Phase Transition: Monte Carlo Results

In order to compare with the quantum model shown in the previous subsection, we consider here the classical counterpart model, namely we use the same Hamiltonian (28) but with the classical Heisenberg spin of magnitude $S = 1$. The aim is to compare their qualitative features, in particular the question of the partial disordering at finite T.

We use Monte Carlo simulations for the classical model where the film dimensions are $N \times N \times N_z$, N_z being the film thickness which is taken to be $N_z = 4$ as in the quantum case shown above. We use here $N = 24, 36, 48, 60$ to see the lateral finite-size effect. Periodic boundary conditions are used in the xy planes. We discard 10^6 MC steps per spin to equilibrate the system and average physical quantities over the next 2×10^6 MC steps per spin.

We show in Figure 27 the result obtained in the same frustrated case as in the quantum case shown above, namely $J_s = -0.5$. we see that the surface magnetization falls at $T_1 \simeq 0.25$ while the second-layer magnetization stays ordered up to $T_2 \simeq 1.8$. This surface disordering at low T is similar to the quantum case. Between T_1 and T_2 the system is partially disordered.

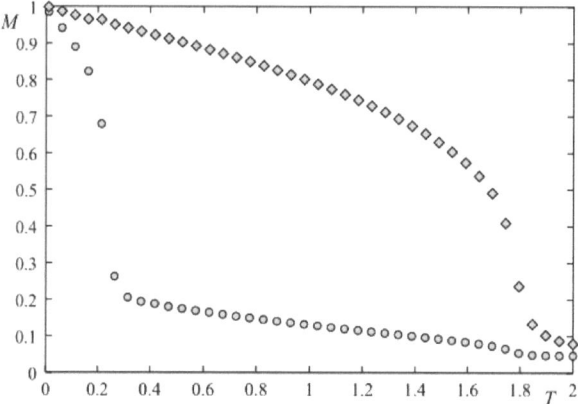

Figure 27. Magnetizations of layer 1 (circles) and layer 2 (diamonds) versus temperature T in unit of J/k_B for $J_s = -0.5$ with $I = -I_s = 0.1$.

Figure 28 shows the phase diagram obtained in the space (J_s, T). It is interesting to note that the classical phase diagram shown here has the same feature as the quantum phase diagram displayed in Figure 26. The difference in the values of the transition temperatures is due to the difference of quantum and classical spins.

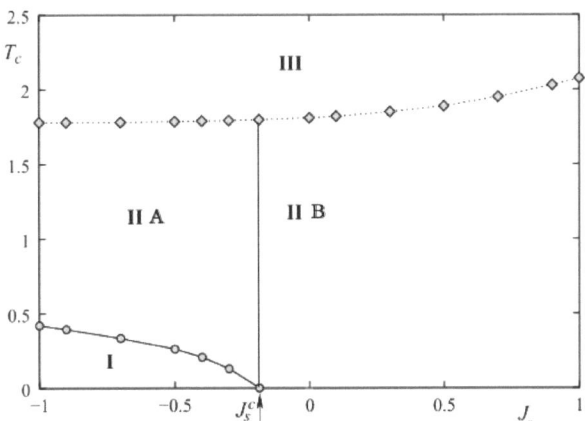

Figure 28. Phase diagram for the classical Heisenberg spin using the same parameters as in the quantum case, i.e., $N_z = 4$, $I = |I_s| = 0.1$. The definitions of phases I to III have been given in the caption of Figure 26.

To close this review, we should mention a few works works where SW in the regime of non-collinear spin configurations have been studied: the frustration effects in antiferromagnetic face-centered cubic Heisenberg films have been studied in Ref. [49], a frustrated ferrimagnet in Ref. [50] and a quantum frustrated spin system in Ref. [51]. These results are not reviewed here to limit the paper's length. The reader is referred to those works for details.

7. Concluding Remarks

As said in the Introduction, the self-consistent Green's function theory is the only one which allows to calculate the SW dispersion relation in the case of non-collinear spin configurations, in two and three dimensions, as well as in thin films. The non-collinear spin configurations are due to the existence of competing interactions in the system, to the geometry frustration such as in the antiferromagnetic triangular lattice, or to the competition between ferromagnetic and/or antiferromagnetic interactions with the Dzyaloshinskii–Moriya interaction. We have shown that without an applied magnetic field, the GS spin configuration is non-collinear but periodic in space. We have, in most cases, analytically calculated them. We have checked them by using the iterative numerical minimization of the local energy (the so-called steepest-descent method). The agreement between the analytical method and the numerical energy minimization is excellent. The determination of the GS is necessary because we need them to calculate the SW spectrum: SW are elementary excitations of the GS when T increases.

The double-fold purpose of this review is to show the method and the interest of its results. We have reviewed a selected number of works according to their interest of the community: helimagnets, materials with the Dzyaloshinskii–Moriya interaction, and the surface effects in thin magnetic films. The Dzyaloshinskii–Moriya interaction gives rise not only a chiral order but also the formation of skyrmions in an applied magnetic field. The surface effects in helimagnets and in films with a frustrated surface give rise to the reconstruction of surface spin structure and many striking features due to quantum fluctuations at low T such as the zero-point spin contraction and the magnetization crossover). We have also seen above the surface becomes disordered at a low T while the bulk remains ordered up to a high T. This coexistence of bulk order and surface disorder in a temperature region is also found in several frustrated systems [8].

To conclude, we say that the Green's function theory for non-collinear spin systems is laborious, but it is worthwhile to use it to get results with clear physical mechanisms lying behind observed phenomena in frustrated spin systems.

Funding: This research received no external funding.

Institutional Review Board Statement: Not applicable.

Informed Consent Statement: Not applicable.

Acknowledgments: The author thanks his former students R. Quartu, C. Santamaria, V. T. Ngo, S. El Hog, A. Bailly-Reyre and I. F. Sharafullin for the collaborative works presented in this review.

Conflicts of Interest: The author declares no conflict of interest.

References

1. Zubarev, D.N. Double-time Green Functions in Statistical Physics. *Sov. Phys. Uspekhi* **1960**, *3*, 320–345. [CrossRef]
2. Diep-The-Hung, J.C.S.; Nagai, L.O. Effects of Surface Spin Waves and Surface Anisotropy in Magnetic Thin Films at Finite Temperatures. *Phys. Stat. Sol.* **1979**, *93*, 351–361. [CrossRef]
3. Yoshimori, A. A New Type of Antiferromagnetic Structure in the Rutile Type Crystal. *J. Phys. Soc. Jpn.* **1959**, *14*, 807. [CrossRef]
4. Villain, J. La structure des substances magnetiques. *Phys. Chem. Solids* **1959**, *11*, 303. [CrossRef]
5. Rastelli, E.; Reatto, L.; Tassi, A. Quantum fluctuations in helimagnets. *J. Phys. C* **1985**, *18*, 353. [CrossRef]
6. Diep, H.T. Low-temperature properties of quantum Heisenberg helimagnets. *Phys. Rev. B* **1989**, *40*, 741. [CrossRef]
7. Quartu R.; Diep, H.T. Partial order in frustrated quantum spin systems. *Phys. Rev. B* **1997**, *55*, 2975. [CrossRef]
8. Diep, H.T.; Giacomini, H. Frustration—Exactly Solved Models. In *Frustrated Spin Systems*, 3rd ed.; Diep, H.T., Ed.; World Scientific: Singapore, 2020; pp. 1–60.
9. Dzyaloshinskii, I.E. Thermodynamical Theory of "Weak" Ferromagnetism in Antiferromagnetic Substances. *Sov. Phys. JETP* **1957**, *5*, 1259.
10. Moriya, T. Anisotropic superexchange interaction and weak ferromagnetism. *Phys. Rev.* **1960**, *120*, 91. [CrossRef]
11. Sergienko, A.I.; Dagotto, E. Role of the Dzyaloshinskii-Moriya interaction in multiferroic perovskites. *Phys. Rev. B* **2006**, *73*, 094434. [CrossRef]
12. Stashkevich, A.A.; Belmeguenai, M.; Roussigné, Y.; Cherif, S.M.; Kostylev, M.; Gabor, M.; Lacour, D.; Tiusan, C.; Hehn, M. Experimental study of spin-wave dispersion in Py/Pt film structures in the presence of an interface Dzyaloshinskii-Moriya interaction. *Phys. Rev. B* **2015**, *91*, 214409. [CrossRef]
13. Heide, M.; Bihlmayer, G.; Blügel, S. Dzyaloshinskii-Moriya interaction accounting for the orientation of magnetic domains in ultrathin films: Fe/W(110). *Phys. Rev. B* **2008**, *78*, 140403(R). [CrossRef]
14. Ederer, C.; Spaldin, N.A. Weak ferromagnetism and magnetoelectric coupling in bismuth ferrite. *Phys. Rev. B* **2005**, *71*, 060401(R). [CrossRef]
15. Cépas, O.; Fong, C.M.; Leung, P.W.; Lhuillier, C. Quantum phase transition induced by Dzyaloshinskii-Moriya interactions in the kagome antiferromagnet. *Phys. Rev. B* **2008**, *78*, 140405(R). [CrossRef]
16. Rohart, S.; Thiaville, A. Skyrmion confinement in ultrathin film nanostructures in the presence of Dzyaloshinskii-Moriya interaction. *Phys. Rev. B* **2013**, *88*, 184422. [CrossRef]
17. Bogdanov, A.N.; Yablonskii, D.A. Thermodynamically stable "vortices" in magnetically ordered crystals: The mixed state of magnets. *Sov. Phys. JETP* **1989**, *68*, 101.
18. Mühlbauer, S.; Binz, B.; Jonietz, F.; Pfleiderer, C.; Rosch, A.; Neubauer, A.; Georgii, R.; Böni, B. Skyrmion Lattice in a Chiral Magnet. *Science* **2009**, *323*, 915. [CrossRef]
19. Yu, X.Z.; Kanazawa, N.; Onose, Y.; Kimoto, K.; Zhang, W.Z.; [CrossRef] Ishiwata, S.; Matsui, Y.; Tokura, Y. Near room-temperature formation of a skyrmion crystal in thin-films of the helimagnet FeGe. *Nat. Mater.* **2011**, *10*, 106. [CrossRef]
20. Seki, S.; Yu, X.Z.; Ishiwata, S.; Tokura, Y. Observation of skyrmions in a multiferroic material. *Science* **2012**, *336*, 198. [CrossRef]
21. Leonov, A.O.; Mostovoy, M. Multiply periodic states and isolated skyrmions in an anisotropic frustrated magnet. *Nat. Commun.* **2015**, *6*, 8275. [CrossRef]
22. Fert, A.; Cros, V.; Sampaio, J. Skyrmions on the track. *Nat. Nanotechnol.* **2013**, *8*, 152. [CrossRef] [PubMed]
23. Xia, J.; Zhang, X.; Ezawa, M.; Tretiakov, O.A.; Hou, Z.; Wang, W.; Zhao, G.; Liu, X.; Diep, H.T.; Zhou, Y. Current-driven skyrmionium in a frustrated magnetic system. *Appl. Phys. Lett.* **2020**, *117*, 012403. [CrossRef]
24. Zhang, X.; Xia, J.; Ezawa, M.; Tretiakov, O.A.; Diep, H.T.; Zhao, G.; Liu, X.; Zhou, Y. A Frustrated Bimeronium: Static Structure and Dynamics. *Appl. Phys. Lett.* **2021**, *118*, 052411. [CrossRef]
25. Mello, V.D.; Chianca, C.V.; Danta, A.L.; Carriç, A.S. Magnetic surface phase of thin helimagnetic films. *Phys. Rev. B* **2003**, *67*, 012401. [CrossRef]

26. Cinti, F.; Cuccoli, A.; Rettori, A. Exotic magnetic structures in ultrathin helimagnetic holmium films. *Phys. Rev. B* **2008**, *78*, 020402(R). [CrossRef]
27. Karhu, E.A.; Kahwaji, S.; Robertson, M.D.; Fritzsche, H.; Kirby, B.J.; Majkrzak, C.F.; Monchesky, T.L. Helical magnetic order in MnSi thin films. *Phys. Rev. B* **2011**, *84*, 060404(R). [CrossRef]
28. Karhu, E.A.; Rößler, U.K.; Bogdanov, A.N.; Kahwaji, S.; Kirby, B.J.; Fritzsche, H.; Robertson, M.D.; Majkrzak, C.F.; Monchesky, T.L. Chiral modulation and reorientation effects in MnSi thin films. *Phys. Rev. B* **2012**, *85*, 094429. [CrossRef]
29. Diep, H.T. Quantum Theory of Helimagnetic Thin Films. *Phys. Rev. B* **2015**, *91*, 014436. [CrossRef]
30. Tyablikov, S.V.V. *Methods in the Quantum Theory of Magnetism*; Plenum Press: New York, NY, USA, 1967.
31. Diep, H.T. *Theory of Magnetism—Application to Surface Physics*; World Scientific: Singapore, 2013.
32. Bland, J.A.C.; Heinrich, B. (Eds.) *Ultrathin Magnetic Structures*; Springer: Berlin, Germany, 1994; Volume I and II.
33. Zangwill, A. *Physics at Surfaces*; Cambridge University Press: London, UK, 1988.
34. Diep, H.T. Quantum effects in antiferromagnetic thin films. *Phys. Rev. B* **1991**, *43*, 8509. [CrossRef]
35. Diep, H.T. Theory of antiferromagnetic superlattices at finite temperatures. *Phys. Rev. B* **1989**, *40*, 4818. [CrossRef]
36. El Hog, S.; Diep, H.T. Helimagnetic Thin Films: Surface Reconstruction, Surface Spin-Waves, Magnetization. *J. Magn. Magn. Mater.* **2016**, *400*, 276–281. [CrossRef]
37. Quartu, R.; Diep, H.T. Phase diagram of body-centered tetragonal helimagnets. *J. Magn. Magn. Mater.* **1998**, *182*, 38–48. [CrossRef]
38. El Hog, S.; Diep, H.T.; Puszkarski, H. Theory of magnons in spin systems with Dzyaloshinskii-Moriya interaction. *J. Phys. Condens. Matter* **2017**, *29*, 305001. [CrossRef]
39. Ngo, V.T.; Diep, H.T. Effects of frustrated surface in Heisenberg thin films. *Phys. Rev. B* **2007**, *75*, 035412. [CrossRef]
40. Mermin, N.D.; Wagner, H. Absence of Ferromagnetism or Antiferromagnetism in One- or Two-Dimensional Isotropic Heisenberg Models. *Phys. Rev. Lett.* **1966**, *17*, 1133; Erratum in *Phys. Rev. Lett.* **1966**, *17*, 1307. [CrossRef]
41. Sharafullin, I.F.; Kharrasov, M.K.; Diep, H.T. Dzyaloshinskii-Moriya interaction in magnetoferroelectric superlattices: Spin waves and skyrmions. *Phys. Rev. B.* **2019**, *99*, 214420. [CrossRef]
42. El Hog, S.; Bailly-Reyre, A.; Diep, H.T. Stability and phase transition of skyrmion crystals generated by Dzyaloshinskii-Moriya interaction. *J. Magn. Magn. Mater.* **2018**, *455*, 32–38. [CrossRef]
43. Sharafullin, I.F.; Diep, H.T. Skyrmion Crystals and Phase Transitions in Magneto-Ferroelectric Superlattices: Dzyaloshinskii-Moriya Interaction in a Frustrated $J_1 - J_2$ Model. *Symmetry* **2020**, *12*, 26. [CrossRef]
44. El Hog, S.; Sharafullin, I.F.; Diep, H.T.; Garbouj, H.; Debbichi, M.; Said, M. Frustrated Antiferromagnetic Triangular Lattice with Dzyaloshinskii-Moriya Interaction: Ground States, Spin Waves, Skyrmion Crystal, Phase Transition. *arXiv* **2022**, arXiv:2204.12248.
45. Keffer, F. Moriya Interaction and the Problem of the Spin Arrangements in β MnS. *Phys. Rev.* **1962**, *126*, 896. [CrossRef]
46. Cheong, S.-W.; Mostovoy, M. Multiferroics: A magnetic twist for ferroelectricity. *Nat. Mater.* **2007**, *6*, 13. [CrossRef] [PubMed]
47. Rosales, H.D.; Cabra D.C.; Pujol, P. Three-sublattice Skyrmions crystal in the antiferromagnetic triangular lattice. *Phys. Rev. B* **2015**, *92*, 214439. [CrossRef]
48. Mohylna, M.; Žukovič, M. Stability of skyrmion crystal phase in antiferromagnetic triangular lattice with DMI and single-ion anisotropy. *J. Magn. Magn. Mater.* **2022**, *546*, 168840. [CrossRef]
49. Ngo, V.T.; Diep, H.T. Frustration effects in antiferrormagnetic face-centered cubic Heisenberg films. *J. Phys. Condens. Matter* **2007**, *19*, 386202.
50. Quartu R.; Diep, H.T. Magnetic properties of ferrimagnets. *J. Magn. Magn. Mater.* **1997**, *168*, 94–104. [CrossRef]
51. Santamaria, C.; Quartu, R.; Diep, H.T. Frustration effect in a quantum Heisenberg spin system. *J. Appl. Phys.* **1998**, *84*, 1953. [CrossRef]

MDPI AG
Grosspeteranlage 5
4052 Basel
Switzerland
Tel.: +41 61 683 77 34

Symmetry Editorial Office
E-mail: symmetry@mdpi.com
www.mdpi.com/journal/symmetry

Disclaimer/Publisher's Note: The title and front matter of this reprint are at the discretion of the Guest Editor. The publisher is not responsible for their content or any associated concerns. The statements, opinions and data contained in all individual articles are solely those of the individual Editor and contributors and not of MDPI. MDPI disclaims responsibility for any injury to people or property resulting from any ideas, methods, instructions or products referred to in the content.

www.ingramcontent.com/pod-product-compliance
Lightning Source LLC
LaVergne TN
LVHW072345090526
838202LV00019B/2483